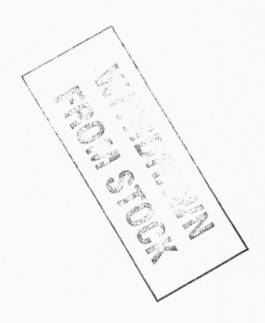

A

METHODS FOR
ECOLOGICAL BIOENERGETICS

IBP HANDBOOK No. 24

Methods for
Ecological Bioenergetics

Edited by

W. GRODZINSKI

R. Z. KLEKOWSKI

A. DUNCAN

BLACKWELL SCIENTIFIC PUBLICATIONS

OXFORD LONDON EDINBURGH MELBOURNE

ISBN 0 632 05690 8

First published 1975

Distributed in the United States of America by
J.B. Lippincott Company, Philadelphia
and in Canada by
J.B. Lippincott Company of Canada Ltd, Toronto

Printed and bound in Great Britain by
A. Brown & Sons Ltd, Hull

Contents

Contents

Contributors

H. DOMINAS *Department of Biochemistry, Nencki Institute of Experimental Biology, Polish Academy of Sciences, Pasteura 3, Warsaw, Poland*

A. DOWGIALLO *Department of Bioenergetics and Bioproductivity, Nencki Institute of Experimental Biology, Polish Academy of Sciences, Pasteura 3, Warsaw, Poland*

A. DROŻDŻ *Department of Animal Ecology, Jagiellonian University, Krupnicza 50, 30-060 Kraków, Poland*

A. DUNCAN *Department of Zoology, Royal Holloway College, London, Englefield Green, Surrey TW20 STY, England*

A. GÓRECKI *Department of Animal Ecology, Jagiellonian University, Krupnicza 50, 30-060 Kraków, Poland*

W. GRODZINSKI *Department of Animal Ecology, Jagiellonian University, Krupnicza 50, 30-060 Kraków, Poland*

S. C. KENDEIGH *Department of Zoology, Vivarium Building, University of Illinois, Healey and Wright Streets, Champaign, Illinois 61820, U.S.A.*

R. Z. KLEKOWSKI *Institute of Ecology, Polish Academy of Sciences, P.O. Lomianki 05-50, Dziekanow Lesny, Nr. Warsaw, Poland*

P. MORRISON *Institute of Arctic Biology, University of Alaska, College, Alaska 99701, U.S.A.*

J. PHILLIPSON *Animal Ecology Research Group, Department of Zoology, University of Oxford, Oxford, England*

J. PINOWSKI *Institute of Ecology, Polish Academy of Sciences, P.O. Lomianki 05-50, Dziekanow Lesny, Nr. Warsaw, Poland*

P. POCZOPKO *Department of Environmental Physiology, Institute of Animal Physiology and Nutrition, Polish Academy of Sciences, 05-110 Jablonna, Nr. Warsaw, Poland*

T. P RUS *Department of Bioenergetics and Bioproductivity, Nencki Institute of Experimental Biology, Polish Academy of Sciences, Pasteura 3, Warsaw, Poland*

K. S AWICKA - K APUSTA *Department of Animal Ecology, Jagiellonian University, Krupnicza 50, 30-060 Kraków, Poland*

G. C. W EST *Institute of Arctic Biology, University of Alaska, College, Alaska 99701, U.S.A.*

Foreword

This volume forms part of the history of **IBP** as well as providing a starting point for many investigations which will be made in the programmes which are to follow **IBP**.

The organizers of the Bioenergetics Workshop held in April 1968, Professor W. Grodzinski and Professor R.Z. Klekowski, brought together a group of visiting lecturers and collaborators who ensured, by their quality and enthusiasm, that the students attending the course would have the capacity to make important contributions to the development of **IBP**. The late publication of this Handbook is in part due to the heavy involvement of the authors in the Programme and to the difficult task of providing precise and clear descriptions of the complex techniques which had been demonstrated in practical sessions.

Although based on the 1968 Workshop, this volume has been carefully revised and brought up to date, as a glance at the references will make clear. The final version owes much to the efforts of Dr Annie Duncan, one of the contributors, who was appointed a co-editor in 1973, in recognition of these efforts.

It is difficult in a Foreword to recapture the atmosphere that made the course a major experience for many of the participants. Professor Grodzinski, in his words of welcome, reminded his audience that they were meeting in the Jagiellonian University of Kraków which had numbered Nicolaus Copernicus (Mikołaj Kopernik) among its students and professors and that the University was over 600 years old. On the other hand, the ecological laboratory (which became the Department of Animal Ecology in 1972) was one of the youngest components of the University. Professor Klekowski was equally warm in welcoming the group for their sessions at the well-established Department of Experimental Hydrobiology of the Nencki Institute of the Polish Academy of Sciences. Certainly, between them the two leaders of the course created an enthusiastic, friendly, scientific atmosphere which has left its mark within **IBP**.

This volume on bioenergetics represents only one of the many important contributions which the **IBP** National Committee of Poland, under the Chairmanship of Professor Petrusewicz, has made to the whole Programme.

It is a pleasure to take this opportunity to record our appreciation of Poland's outstanding role in the PT Section of IBP.

J.B. CRAGG,
Convenor IBP/PT

Introduction

This handbook arose out of an IBP Training Course in Methods of Ecological Bioenergetics organised under the auspices of the PT-IBP and held in Warsaw and Kraków, Poland, from 6-21 April 1968. The course took place very early during IBP and was largely the result of the far-sightedness of Professors Grodzinski and Klekowski in understanding the needs of ecologists embarking on IBP projects. The aim of the course was to provide an opportunity for such ecologists to learn how to measure the bioenergetic parameters relevant to a study on animal productivity. These parameters are the same for invertebrate and vertebrate animals but the techniques and apparatus differed according to the species and size of animal involved. And so the course was held in two centres in Poland well known for their research on ecological energetics, namely the Department of Bioenergetics and Bioproductivity, Nencki Institute of Experimental Biology, Polish Academy of Sciences, Warsaw and the Department of Animal Ecology, Jagiellonian University, Kraków.

Twenty-three ecologists from fifteen countries attended the course selected from 35 candidates proposed by the PT section of IBP National Committees; the countries were Austria, Bulgaria, Canada, Czechoslovakia, Denmark, Finland, France, Japan, Netherlands, Norway, Rumania, Sweden, United Kingdom and United States of America. The course was designed for ecologists from those parts of the world where scientific instruments were not readily available but where IBP was anxious to encourage work on biological productivity; the aim was to demonstrate simple, easily constructed apparatus, given a metal, wood and/or glass workshop. However, most participants in fact came from countries with relatively rich scientific worlds.

The course lasted for twelve working days during which there were 16 hours of lectures, 9 hours of seminars and 59 hours of practical work. The lectures ranged in subject from descriptions of techniques, reviews of general aspects of ecological energetics and detailed discussion of problems of computation of energy budgets, production as well as whole balances of energy flow; these are included in the present volume. The seminars provided a useful opportunity to discuss participants' own investigations. It was generally agreed that by far the most valuable part of the course were the practical sessions in which a variety of calorimeters, respirometers and techniques for feeding studies were demonstrated and actively used by the participants themselves; these also form part of the present volume. Part of the success of the practicals was due to the sixteen Polish ecologists/physiologists who very effectively and in the English language demonstrated the techniques which they themselves were using in their own research. Those who contributed to this volume are listed in the list of contributors, besides that instructors at the course were Drs B. Bolser, J. Gruder, E. Kamler, T. Stachurska and Z. Stępień. There was an invaluable contribution by three invited lecturers, Dr A. Duncan, Dr J. Phillipson (U.K.) and Dr P.R. Morrison (U.S.A.). They discussed some general problems in lectures and also actively participated in seminars and practical sessions (labs.). Also greatly valued was the printed manual which was the forerunner of the present volume provided by the Polish Academy of Sciences for each participant together with the two volumes of *Secondary Productivity of Terrestrial Ecosystems*, edited by Professor K. Petrusewicz.

The delay in publishing the present volume is the price paid when busy scientists are involved in a voluntary organisation like IBP over and above their own job. However, it is also true that the necessity for training ecologists in techniques for the measurement of individual size, growth, feeding and respiration, all basic parameters for IBP, was not understood and there existed no effective central policy decision for encouraging such training.

The real authors of the handbook are the participants of the course as well as instructors and all contributors to the volume. They together initiated an 'international bioenergetic club' which is still working in many countries and is still growing.

General Problems

1

Introduction to Ecological Energetics

J. PHILLIPSON

In most disciplines the growth of vocabulary is attended by special hazards. There is a tendency for users to generate terms meaningful only to themselves, or to attach to established terms a meaning inconsistent with that originally intended. Ecology is no exception, and as this contribution is intended as an introduction to *ecological energetics* it is important to establish what the present writer understands by this term. The chapter entitled 'Physiological approach to ecological energetics' indicates that *bioenergetics* is the study of energy transformations in living organisms, and that such a study may be according to any one of three approaches: the molecular-biochemical, the physiological, and the ecological. It would appear that *ecological energetics* implies the last named approach, which differs from the first in that it is concerned with ecological rather than molecular units. However, it is difficult to distinguish between the physiological and the ecological approach; for the former as well as the latter encompasses whole organisms. Wiegert (1968) suggests that ' . . . the distinction is real and important . . . ' and states that *ecological energetics* ' . . . encompasses the energy costs of the individual in a growing, reproducing population of organisms; . . . ' whereas *physiological energetics* ' . . . deals with the utilization of energy by the resting, post-absorptive individual.' On the other hand Engelmann (1966) quotes several purely physiological studies on non-resting, non-post-absorptive individuals where the results have a potential for use in, rather than make a direct contribution to, ecological studies. Clearly there is a problem of definition, but rather than become embroiled in semantics it would be more profitable to accept under the heading *ecological energetics* any bioenergetic approach which contributes to a fuller understanding of energy transformations by ecological units (be they individuals or biocoenoses) in the environmental conditions normally experienced by them.

The ultimate goal of ecology is a full description and understanding of the biological complexity of the world in which we live, and of which we are an integral part. It should be noted therefore that ecological energetics, being concerned solely with energy transformations by ecological units, is but one aspect of ecology; although one which can make a significant contribution to our understanding of how nature functions.

Ecological energeticists differ in their views as to the ways in which energy transformations in nature should be studied. At one extreme are synecologically minded ecologists, and at the other the autecologically minded ones. Nevertheless, in view of the general acceptance that the current interest in ecological energetics (despite earlier publications) stems from Lindeman's (1942) paper it would seem appropriate to consider the ecosystem approach first.

Sukachev and Dylis (1966, p. 16) feel it inexpedient to recommend 'ecosystem' as a useful term because of its being used in widely differing senses. It is only right and proper therefore that we should consider 'ecosystem' in some detail. As is well known the term was proposed by Tansley (1935) to designate ' . . . the whole *system* (in the sense of physics), including not only the organism complex, but also the whole complex of physical factors forming what we call the environment of the biome.' Later Tansley (1946, p. 207) rephrased his definition without altering its meaning—'A wider conception still is to include with the biome all the physical and chemical factors of the biome's environment or habitat—those factors we have considered under the headings of climate and soil—as parts of one physical *system*, which we may call an *ecosystem*, because it is based on the oikos or home of a particular biome. All the parts of such an ecosystem—organic and inorganic, biome and habitat—may be regarded as interacting factors which, in a mature ecosystem, are in approximate equilibrium: it is through their interactions that the whole system is maintained.' Clearly, the recognition of an ecosystem revolves around the definition of biome.

Tansley (1935, p. 306) understood biome to mean 'The whole complex of organisms present in an ecological unit . . . ' and Tansley (1946, p. 206) ' . . . the whole complex of organisms—both animals and plants—naturally living together as a sociological unit . . .' He thus used 'biome' in the original sense of Clements (1916) where 'The biotic community is regarded as a biotic unit comprising all the species of plants and animals at home in a particular habitat . . . ' 'The biotic community, or *biome*, is fundamentally controlled by the habitat, and exhibits a corresponding development and structure. In its development the biotic formation reacts upon the habitat, and thus produces a succession of biomes . . . ' and Clements (1918) 'Thus it becomes the biome, or mass of plants and animals of a particular area or habitat, on which attention must first be fixed.' Indeed Tansley (1935, p. 299) wrote 'Clements earlier term 'biome' for the whole complex of organisms inhabiting a given region is unobjectionable, and for some purposes convenient . . . ' Given the early definitions of biome it is not surprising that Tansley (1935, p. 299) could state 'These ecosystems, as we may call them, are of the most various kinds and sizes.'

Use and abuse of the original concept of 'ecosystem' has given rise to considerable confusion in the theory and practice of ecological energetics in relation to ecosystems. This has not been aided in any way by the change in meaning of the term biome, a change which seems to have originated with Shelford (1931, p. 456), ' . . . because the larger and more influent animals tend to range throughout units of largest (formational) size including their seral stages the biome or biotic formation is the natural ecological unit . . . ', and Shelford (1932, p. 110), 'The largest natural unit is the biome or biotic formation. These are roughly represented by the large areas of deciduous forest, grassland, tundra, etc., which characterize the North American continent.' One can only assume that the change in meaning arose because Shelford understood Clements (1916) biotic formation, and hence 'biome', to be equivalent in size to the Clementsian *plant formation*. As shown this does not appear to have been Clements' original intention but was later accepted by him (Clements and Shelford, 1939, p. 20). 'Ecosystem' in its original sense then denotes a generalized concept rather than a discrete entity limited by the bounds of a *plant formation*.

Tansley (1935, p. 301) recognized that 'Relatively to the more stable systems the ecosystems are extremely vulnerable, both on account of their own unstable components and because they are very liable to invasion by the components of other systems. Nevertheless some of the fully developed systems—the 'climaxes'—have actually maintained themselves for thousands of years. In others there are elements whose slow change will ultimately bring about the disintegration of the system. This relative instability of the ecosystem, due to the imperfections of its equilibrium, is of all degrees of magnitude, and our means of appreciating and measuring it are still very rudimentary.'

A full appreciation of 'ecosystem' (sensu stricto) makes clear that Lindeman (1942) did not deviate from the original meaning when he wrote 'The *ecosystem* may be formally defined as the system composed of physical-chemical-biological processes active within a space time unit of any magnitude, i.e., the biotic community *plus* its abiotic environment', and 'Natural ecosystems may tend to approach a state of trophic equilibrium under certain conditions, but it is doubtful if any are sufficiently autochthonous to attain, or maintain, true trophic equilibrium for any length of time. The biosphere, as a whole, however as Vernadsky (1929, 1939) so vigorously asserts, may exhibit a high degree of true trophic equilibrium.' Trophic equilibrium, as used, was '. . . roughly defined as the dynamic state of continuous, complete utilization and regeneration of nutrients in an ecosystem without loss or gain from the outside, under a periodically constant energy source—such as might be found in a perfectly balanced aquarium or terrarium.'

Accepting ecosystem in the sense intended by Tansley (1935, 1946), and used by Lindeman (1942) then the present writer is unable to distinguish between *ecosystem* and the *biogeocoenose* of Sukachev (1944, 1945) which is defined in Sukachev and Dylis (1966, p. 26) as 'A biogeocoenose is a combination on a specific area of the earths surface of homogeneous natural phenomena (atmosphere, mineral strata, vegetable, animal, and microbic life, soil, and water conditions), possessing its own specific type of interaction of these components and a definite type of interchange of their matter and energy among themselves and with other natural phenomena, and representing an internally contradicting dialectical unity, being in constant movement and development.' In this contribution, therefore, the two terms are considered to be synonymous.

The important point to note is that neither Tansley nor Sukachev, in their formal definitions, delineate the structure or boundaries of their ecological units in precise terms.

A brief consideration of the different types of thermodynamic systems and their relationship to eco-energetic studies is now necessary. Spanner (1964) and Wiegert (1968), amongst others, have pointed out that thermodynamic systems are of three types; the figure illustrates them in simple diagrammatic

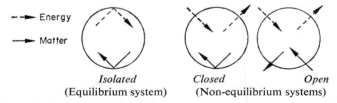

- - ► Energy

——► Matter

| *Isolated* | *Closed* | *Open* |
| (Equilibrium system) | (Non-equilibrium systems) | |

form. The important point to note is that non-equilibrium systems differ from equilibrium systems in that there is a continuous net flow of either matter and/or energy through them. Equilibrium systems require either isolation (adiabatic conditions) or contact with a single fixed reservoir (isothermal conditions).

Early theoretical considerations of ecosystems implied that true trophic equilibrium is the rule, but given such a state it follows that (a) energy enters and leaves the ecosystem in such a manner that energy income equals energy expenditure (b) matter neither enters nor leaves the ecosystem, and (c) within the ecosystem trophic levels can be distinguished readily. From (a) this theoretical ecosystem is in contact with both an energy source (the sun) and an energy sink (outer space); it is not therefore an equilibrium system in thermodynamic terminology but a steady state non-equilibrium system. From (b) this steady state system is of the 'closed' type. An example of the above can be seen in the general formulation of Hutchinson quoted in Lindeman (1942):

$$\frac{\delta \Lambda_n}{\delta_t} = \lambda_n + \lambda_{n\mathrm{l}}$$

where $\Lambda_n = \lambda_n$, and a portion of $\lambda_{n\mathrm{l}}$ (viz. λ_{n+1}) represent matter retained within the system, while the remainder of $\lambda_{n\mathrm{l}}$ represents energy dissipated and hence lost from Λ in particular and the system in general. The word dissipated is Hutchinson's and implies randomization of energy, i.e., heat loss. This change in the ecosystem concept, from a system whose boundaries are open to both matter and energy transfer to one where the boundaries are closed to the transfer of matter, has had a profound effect on the course of eco-energetic thinking.

The fact that most ecosystems; apart from the virtually 'closed' ecosphere (Cole, 1958), the 'closed' experimental microecosystems of Beyers (1962, 1963), and the 'isolated' universe; are 'open' (e.g. Odum, 1957; Teal, 1957, 1962; and Tilly, 1968) is of importance with regard to the applicability of the Second Law of Thermodynamics to ecosystems. Patten (1959) and Wiegert (1964), reported in Phillipson (1966), developed analogies between energy flow through ecological units and the free energy equation; Patten (1959), for example, stated ' . . . a generalized energy flow through the ecosystem may be formulated:

$$\Lambda_{\mathrm{HI}} = \Lambda_{\mathrm{FI}} + \sum_{i=1}^{n} \mathrm{T}\Lambda \mathrm{S_i},$$

where i refers to the trophic level.' Acceptance of a strict analogy is misleading in that the concept of 'ecosystem' changes yet again and is applicable only to 'isolated' *i.e.*, equilibrium systems. One must agree with Wiegert (1968) that this is erroneous, for as Morowitz (1968) points out 'The Second Law asserts that the universe, or any *isolated* section of it, is tending towards maximum entropy; it follows that non-equilibrium systems are not characterized by entropy maxima and need not be considered within the narrow confines of the Second Law'.

Up to the present, attempts to combine the ecosystem concept with classical thermodynamic principles, and to express the results by precise mathematical notation, have lead to a narrowing of, and therefore a change in, the original concept; this does not of course invalidate 'ecosystem' (sensu stricto) as a concept. What we must now consider is the value of the ecosystem concept to ecologists in general, and ecological energeticists in particular.

In general the term 'ecosystem' is of value in that a single word denotes a complex system of inter-related factors, both biotic and abiotic, which can only be understood by viewing the system as a whole. More specifically the fact that Tansley (1935) wrote of an ecosystem as a whole *system* in the sense of physics has lead to model building e.g., energy flow diagrams and, more recently, systems analysis and the employment of Black Box theory (see Watt, 1966). Conceptual model building is not an end in itself but a useful exercise embracing as it does the whole *system* with its component parts. Initial information about the components is obtained from the literature, general field observations, and formal pilot studies; *a priori* considerations allow the insertion of unknowns or inacessibles into the model as 'Black Boxes' (Ashby, 1965). However incomplete the initial model, it simulates the system as a whole and suggests those components and pathways which might profitably be investigated. Working within such a framework ensures that the various sub-model studies can be fitted together in a meaningful manner. Moreover, insertion of the sub-model information modifies the conceptual model, which in turn suggests further possible lines of research. The theory is that, with increased knowledge of the components and variables, the 'Black Boxes' representing components inaccessible to direct observation will become progressively more realistic; such models could have useful predictive value, indicating how the ecosystem can be manipulated in nature so as to produce results optimal to man. As yet no detailed predictive models of any ecosystems are available, although the possibility of their future construction exists.

To be of practical, as well as conceptual, value the limits of a system must be clearly defined; this must be so when it is realized that there is no such thing as *the* ecosystem but only *an* ecosystem. In practice the delineation of specific system limits accounts for ecosystems of varying magnitude, differing degrees of 'open-ness', and unmatched approximations to equilibrium. This, of course, raises the question as to how boundaries should be distinguished and whether different ecosystems are strictly analogous. In theory the concept of trophic levels and the assumption of trophic equilibrium implies definite boundaries. In practice the identification of trophic levels is not easy, Odum (1968) emphasized that the concept of trophic level is not primarily intended for categorizing species, and yet not one study is known to the present author where trophic levels are not delineated by reference to species composition. Detailed quantitative studies of food type would be necessary in order to apportion each 'species' between the different trophic levels, and it is highly likely that the proportional contribution of each species will vary with time. The fact that numbers and biomass do vary with time presents further difficulties in the delineation and quantification of trophic levels.

In some ecosystem studies (e.g., Odum (1957) and Teal (1957)) the above difficulties are circumscribed by (a) the choice of a technically convenient unit of habitat (b) categorization of trophic levels within the unit of habitat according to species composition and (c) the assumption of a steady state system. Clearly, up to the present, the main factor used to delimit a specific ecosystem has been one of habitat. As we have seen, the use of trophic levels according to species composition could be misleading; and steady state systems being such that as the system ages the flows from source to sink become constant (Morowitz, 1968) means that although the ecosphere is in an approximate steady state it is questionable whether this is true for many smaller units. Examples come to mind where income may be greater than expenditure (e.g. seral stages of succession), or where the converse might be true (e.g. senescent systems). In systems where approximation to a steady

state does not hold it is implicit that the efficiency ratio $\dfrac{\lambda_n}{\lambda_{n-1}}$ (Lindeman, 1942)

will not remain constant for any one trophic level (see also Kozlovsky, 1968).

Some ecological energeticists, and I am one of them, have adopted the maxim that complex systems (albeit distinguished by habitat boundaries) should be dissected into a large number of relatively simple unit components (e.g. species populations in the ecological rather than genetic sense) and not into a small number of relatively complex units (e.g. trophic levels). They believe that the evaluation of biological activity, and hence the role, of different species populations, is important for a fuller understanding of ecosystem dynamics; not only can the importance of a given species within a specific ecosystem be elucidated but a comparison can be made of the relative importance of that species in different ecosystems. It is held that the parameters of population sub-models are easier to define than those of trophic level sub-models, and that detailed studies (if carried out within an ecosystems analysis scheme) of food, food consumption, assimilation, respiration and growth of all life stages comprising a population will allow the incorporation of results into a complex but meaningful model.

It could be argued that the labour and time involved in the consideration of all species populations within an ecosystem is prohibitive and that the only feasible course is to concentrate on the 'simplest' system as exemplified by the trophic level concept. However, as indicated, the trophic level approach has severe limitations whereas the results from properly made population studies can be used more easily to model complex systems. Some workers (e.g., Skellam, 1967) are of the opinion that the modelling approach offers little prospect in the immediate future of obtaining reliable quantitative pictures of more than a few animal populations; however short-cut methods are being explored and the one termed 'best estimate' offers certain possibilities (see Phillipson, 1970).

The population approach requires in the first instance that a unit of habitat, which with its biotic/abiotic complex fits the generalized concept of ecosystem, be chosen. Thereafter a list of common species and some idea of their food habits and movement across the system boundaries will allow the construction, however crude, of a generalized flow diagram. *A priori* considerations, general observations, and pilot studies will give some idea as to which species and species interactions might be important in promoting energy flow, and it is on these that initial work should be concentrated.

It is not the function of this handbook to deal with population density estimation but with the measurement of energetic parameters which can be used in conjunction with population density data. As pointed out earlier ecological energetic parameters can be examined within the confines of the First Law of Thermodynamics without recourse to the Second Law. This facet is dealt with in detail by Wiegert (1968) but may be summarized according to the First Law as:

$$\Delta H_s = (H_1 - H_2) + (Q_1 - Q_2) + (W_1 - W_2)$$

where ΔH_s = overall change in enthalpy of the system, or the change in heat content at constant pressure and temperature

H_1 = enthalpy content of matter entering the system.

H_2 = enthalpy content of matter leaving the system.

Q_1 = heat energy entering the system.

Q_2 = heat energy leaving the system

W_1 = work done on system by environment

W_2 = work output of system

The net work exchange ($W_1 - W_2$) is generally negligible and is disregarded in energy budgets thus:

$$\Delta H_s = (H_1 - H_2) + (Q_1 - Q_2)$$

This formula can be equated with the equation normally used to describe ecological energy transformations (IBP News 10, 1968)

$$C = P + R + F + U$$

where in energy terms:

C. Consumption	=	Total intake of food energy during a specified time interval.
P. Production	=	Energy content of the biomass of materials digested during a specified time interval (regardless of whether they all survive to the end of that interval) less that respired or rejected.
R. Respiration	=	The energy equivalent of that part of assimilation (the sum of production and respiration) which is converted to heat and loss in life processes.
F. Egesta	=	The energy content of that part of consumption which is not digested.
U. Excreta	=	The energy content of that part of the digested material which is passed from the body. (This does not include reproductive and secretory products).

The components of the two equations may be equated when it is recognized that in the long term any biological system must be in heat balance thereby rendering Q_1 mathematically equivalent to zero hence:

$$\Delta H_s = \{H_1 - H_2\} + \{Q_1 - Q_2\}$$
$$P = \{C - (F + U)\} + \{-R\}$$

As Wiegert (1968) states 'The energy budget is thus put on a sound thermo-dynamic basis, and, on the assumption that net work-exchanges between the living system and the surroundings are negligible, it quantitatively describes ecological transfers of energy.'

This handbook is primarily concerned with the methodology and techniques of measuring and estimating the various components of the energy budget equation $C = P+R+F+U$, so that the results can be used in an ecologically meaningful manner within the concept of ecosystem. That is it must never be forgotten, however small our sub-models, that ecosystems in their complexity are such that everything affects everything else.

References

ASHBY W.R. (1965) *An introduction to cybernetics*. University Paperbacks, Methuen, London. pp. 295.

BEYERS R.J. (1962) Relationship between temperature and the metabolism of experimental ecosystems. *Science* **136**, (3520): 980–982.

BEYERS R.J. (1963) The metabolism of twelve laboratory microecosystems. *Ecol. Monogr.*, **33**: 281–306.

CLEMENTS F.E. (1916) The development and structure of biotic communities. *Printed programme of Ecol. Soc. Amer. New York meeting* Dec. 27–29, 1916, p. 5. Reprinted in full in *J. Ecol.*, **5**: 120–121, 1917.

CLEMENTS F.E. (1918) Scope and significance of palaeo-ecology. *Bull. Geol. Soc. Amer.*, **29**: 369–374.

CLEMENTS F.E. & SHELFORD V.E. (1939) *Bioecology*. Wiley, New York. pp. 425.

COLE L.C. (1958) The ecosphere. *Sci. Amer.* **198** (4): 83–96.

ENGELMANN M.D. (1966) Energetics, terrestrial field studies, and animal productivity. *Adv. Ecol. Res.* **3**: 73–115.

KOZLOVSKY D.G. (1968) A critical evaluation of the trophic level concept. *Ecology*, **49**: 48–60.

LINDEMAN R.L. (1942) The trophic-dynamic aspect of ecology. *Ecology*, **23**: 399–418.

MOROWITZ H.J. (1968) *Energy flow in biology*. Academic Press, New York and London, pp. 179.

ODUM E.P. (1968) Energy flow ecosystems: a historical review. *Am. Zoologist*, **8**: 11–18.

ODUM H.T. (1957) Trophic structure and productivity of Silver Springs, Florida. *Ecol. Monogr.*, **27**: 55–112.

PATTEN B.C. (1959) An introduction to the cybernetics of the ecosystem: the trophic-dynamic aspect. *Ecology*, **40**: 221–231.

PHILLIPSON J. (1966) *Ecological energetics*. Edw. Arnold, London. pp. 57.

PHILLIPSON J. (1970) The 'best estimate' of respiratory metabolism: its applicability to field situations. *Pol. Arch. Hydrobiol.* **17**: 31–41.

SHELFORD V.E. (1931) Some concepts of bioecology. *Ecology*, **12**: 455–467.

SHELFORD V.E. (1932) Basic principles of the classification of communities and habitats, and the use of terms. *Ecology*, **13**: 105–120.

SKELLAM J.G. (1967) Productive processes in animal populations considered from the biometrical standpoint. In *Petrusewicz, K.* (ed.). *Secondary productivity of terrestrial ecosystems (principles and methods)*. Pánstwowe Wydawnictwo Naukowe, Warsaw and Cracow. pp. 59–82.

SPANNER D.C. (1964) *Introduction to thermodynamics*. Academic Press, London and New York, pp. 278.

SUKACHEV V.N. (1944) Principles of genetic classification in biogeocoenology *Zh. obshch Biol.*, 6. (quoted in Sukachev and Dylis, 1966).

SUKACHEV V.N. (1945) Biogeocoenology and phytocoenology. *Dok Akad. Nauk SSSR*, 30. (quoted in *Sukachev* and *Dylis*, 1966).

SUKACHEV V.N. & DYLIS N.V. (1966) *Fundamentals of Forest Biogeocoenology* (English transl.). Oliver and Boyd, Edinburgh and London, pp. 672.

TANSLEY A.G. (1935) The use and abuse of vegetational concepts and terms. *Ecology*, 16: 284–307.

TANSLEY A.G. (1946) *Introduction to plant ecology*. Allen and Unwin, London. pp. 260.

TEAL J.M. (1957) Community metabolism in a temperate cold spring. *Ecol. Monogr.* 27: 283–302.

TEAL J.M. (1962) Energy flow in the salt marsh ecosystem of Georgia. *Ecology*, 43: 614–624.

TILLY, L.J. (1968) The structure and dynamics of Cone Spring. *Ecol. Monogr.* 38: 169–197.

WATT K.E.F. (1966) *The nature of systems analysis.* in WATT K.E.F. (ed.) Systems analysis in ecology. Academic Press. New York and London. pp. 1–14.

WIEGERT R.G. (1968) Thermodynamic considerations in animal nutrition. *Am. Zoologist*, 8: 71–81.

2
Physiological Approach to Ecological Energetics

R. Z. KLEKOWSKI AND A. DUNCAN

2.1 Introduction

Both authors of this section of the Handbook have been concerned mainly with aquatic species of animals and here find themselves contributing to a book mainly designed for terrestrial ecologists! Moreover both authors in recent work have approached ecological problems from a physiological point of view whereas the problems with which the I.B.P. is mainly concerned are based on population or community units. However, we both believe that in order to understand the productive processes in nature we must also understand the underlying physiological pattern of response of an animal species throughout its life cycle to varying environmental conditions and this applies both to aquatic and terrestrial animal species.

Bioenergetics is the study of energy transformations in living organisms; ecological energetics is considering this subject in terms of species-populations in nature and the subsequent effects in the communities with which these are associated within the ecosystems.

There are several approaches to bioenergetics, approaches which deal with the same events but at different levels. The first is the molecular-biochemical level, dealing with cells and sub-cellular structures, and at present the most fundamental level since the site of energy transformation is inside the cell. The second level of approach can be called physiological because it deals with whole organisms. The present section has much in common with this point of view, except that the main interest is in animals not divorced from their normal habitat or with experiments usually designed to be meaningful in terms of the animal's normal conditions of life. The third level of approach is ecological in which the interest is in energy transformation under field conditions; the basic unit studied in this type of work may vary—it may be species throughout its life cycle (Richman, 1958), the species population (Smalley, 1960), the biocenosis (MacFadyen, 1948), the ecosystem (Odum,

15

1957), a single trophic level (if you can identify it!) (Lindeman, 1942), or a single food chain (Golley, 1960).

What we have called the physiological approach to ecological energetics may be defined as physiological autecology, mostly based on laboratory studies but with some preliminary attempts to transfer laboratory results to field situations, or to interpret field events with the aid of some physiological constants.

2.2 Metabolic pathway resulting in the liberation and accumulation of energy from the main food components.

The source of energy for animals are the carbohydrates, fats and proteins forming the food substance. By means of an oxidative process (except in anaerobic animals), the breakdown of these complex compounds is achieved and results in the reduction of molecular oxygen to water, the release of carbon dioxide and the production of ATP chemical energy available for work; these end-products of the oxidative process do not all appear at the same point in the series of very complex reactions but represent the visible end result. Figure 2.1 modified from one given in Lehninger (1965) gives a highly simplified but still nevertheless complicated diagram of this series of events and is a diagram designed for ecologists and not for biochemists.

Once food is ingested, that is taken into the alimentary canal of the animal, a process of enzymatic digestion begins. During this process, the various components of food are hydrolysed into their basic units or 'building blocks' that is carbohydrates into monosaccharides, fats to glycerol and fatty acids, and protein to amino acids. This process takes place inside the alimentary canal and the digestive products pass through the gut wall into either the blood or the lymph for distribution to the cells of the body; this is called absorption. About 1% of the total energy liberated from the food is released during this process of hydrolysis in the gut and goes to waste. All subsequent processes of energy conversion takes place within cells during 'cellular respiration' the fuel for which are these basic building blocks.

The next problem is to convert by oxidation these simple food components into a form acceptable to the Krebs cycle, that is, into simple two carbon compounds such as acetic acid which, however, is more usually combined in the cell with Co-enzyme A to form the especially reactive form known as Acetyl Coenzyme A or Active Acetate (Fig. 2.1). The process whereby this

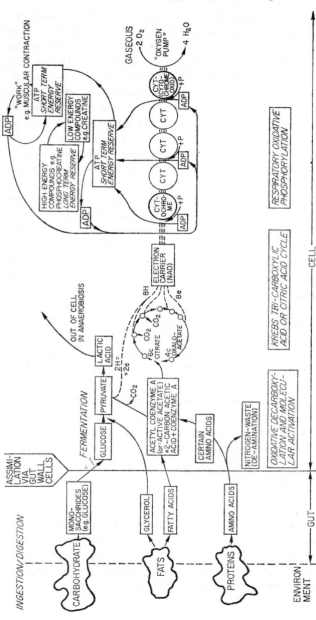

Fig. 2-1. 'Master plan' of the oxidation of the major foodstuffs: carbohydrate, fats and proteins in heterotrophic animal cells (after Lehninger, 1965, somewhat modified).

In addition

Phase II: *Krebs cycle* 2-carbon acetyl Co-A combines with 4-carbon oxalo-acetic acid to produce 6-carbon citric acid§ during 8 major transformations 2 molecules of CO^2 appear, 4-carbon oxaloacetic acid is regenerated and 8 hydrogens (8 electrons) are liberated.

Phase III: *Phosphorylation* electron transport by successive reduction (hydrogenation) and oxidation (dehydrogenation) along respiratory chain of electron carriers (cytochrome). There is a decrease in energy along the chain and formation of ATP from ADP and inorganic phosphate.

C

takes place is called 'oxidative decarboxylation and molecular activation'. Dealing with carbohydrates first, as animals depend on carbohydrates for much of their energy, let us take glucose as our substrate as it is one of the monosaccharides often resulting from carbohydrate hydrolysis. During the metabolism of glucose, a chain of fermentative reactions occur which result in the formation of pyruvic acid and lactic acid. In the process of oxidative decarboxylation and molecular activation, the pyruvic acid is converted to the simple two carbon acetic acid which, however, combines in the cell to form Acetyl CoA as described above; carbon dioxide is released as a waste product of the metabolism and two hydrogens are supplied to the electron carriers associated with the Krebs cycle, as shown in the diagram. Some pyruvic acid is further metabolised to form lactic acid and some may be re-formed from lactic acid but we will not go into the details of this. In the case of fats, which are hydrolysed into fatty acids and glycerol, the conversion to Acetyl CoA occurs along two different pathways; the glycerol via the fermentative cycle and the fatty acid directly. With proteins the picture is much more complex and as these are more rarely used as respiratory fuels it is not necessary to go into details; however, some amino acids are converted directly into Acetyl CoA.

The next important stage in the process of oxidation of food substances and liberation of energy is the Krebs tricarboxylic (or citric) acid cycle which acts as the cell's 'metabolic or energy dynamo' (Lipmann, 1941—from Hoar, 1966). This cycle occurs almost universally in aerobic and, somewhat modified, in anaerobic animals. During the cycle, the bulk of the energy contained in the Acetyl CoA is released and immediately passed on to electron acceptors associated with the Krebs cycle. How is this achieved so that the liberated energy is not dissipated to entropy? It is accomplished by the combination of the two carboacetyl CoA with the four carbon oxalo-acetic acid to form the six carbon citric acid, a process involving a series of eight major transformations during which two molecules of carbon dioxide are released, four-carbon oxalo-acetic acid is reformed and eight hydrogen (or eight electrons) are liberated.

It is important, of course, that this newly liberated energy is not allowed to dissipate to entropy and this is achieved by their immediate capture by the electron carriers associated with the Krebs cycle, for example NAD (or nicotinamide adenine dinucleotide). NAD occurs in many cells where it acts as an electron carrier rather while ATP acts as a phosphate carrier. It is the nicotinamide portion of the molecule which can accept electrons and, in its reduced form, donate electrons to electron acceptors; nicotinamide is also a vitamin of the B class and, as such, must be present in the diet of animals which otherwise would suffer from dietary deficiency disease, pellagra, in which there is a defect in electron transfer to and from NAD (Lehninger, 1965). Once captured by NAD, the electrons are passed along a series of respiratory electron carriers, consisting of flavoprotein and cytochromes, until they finally reduce gaseous oxygen to water, without which final step, the whole complex sequence of events would come to a halt.

Along this respiratory chain of electron carriers, a series of successive reductive and oxidative reactions resulting in the transfer of electrons and also in a decrease in energy as ATP energy is formed. This process whereby ADP together with inorganic phosphate is recharged to form ATP in which energy is conserved for immediate use is called respiratory oxidative phosphorylation.

The main function of the process of respiratory oxidative phosphorylation is to conserve the energy liberated from the Krebs cycle and captured by NAD in a form readily available for immediate work. At three points along the respiratory chain of carriers there occurs a release of energy through the coupling of inorganic phosphate with ADP, as shown in Fig. 2.2. At the final cytochrome link, the enzyme cytochrome oxidase catalyzes the reduction of molecular oxygen acting as the final acceptor is essential for the action of the whole system.

How is it possible for energy to be conserved in ATP and still be readily available for work? In order to understand this, we must examine a group of phosphorylated compounds called energy-rich compounds. Most biochemical reactions result in the release of some energy, usually in very small amounts, some reactions require energy from outside in order to take place. However, a few reactions are associated with the release of large amounts of energy and these always involve one of these energy-rich substances. One of these compounds is ATP or adenosine triphosphate whose molecular structure is given in Figure 2 in a simplified form (Lehninger, 1965).

As can be seen, ATP consists of three kinds of building blocks, an adenine base linked with a sugar D-ribose to which a series of phosphate groups are added. Without the phosphate groups, the compound is adenosine; with one, two or three attached phosphate groups, we have adenosine monophosphate (AMP), adenosine diphosphate (ADP) and adenosine triphosphate (ATP). The free energy in these mononucleotide compounds is stored or conserved in their phosphate bonds which is indicated as P, the usual symbol denoting

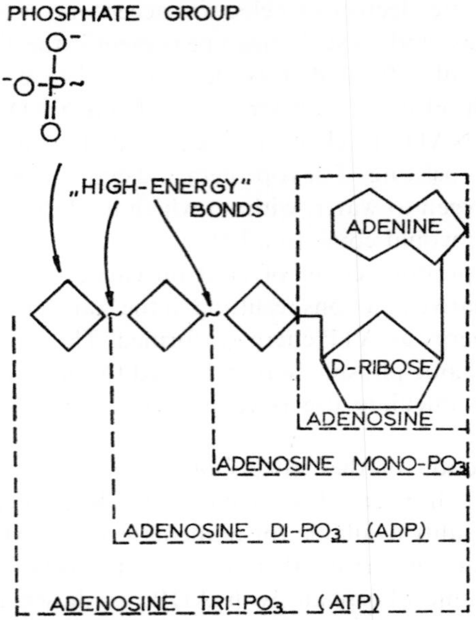

PHOSPHATE GROUP

```
        O⁻
        |
  ⁻O-P~
        ||
        O
```

Fig. 2.2. The structure of ATP, ADP, and AMP.

an energy-rich bond. In ATP, most of the free energy is conserved in the terminal phosphatic bond, and whereas the others are relatively stable and resistant to hydrolysis, the terminal bond of ATP is relatively unstable and will give up its energy readily upon hydrolysis. Note that the energy rich bond is a hydrolytic bond and not an atomic one. Moreover, the amount of energy released by hydrolysis of the terminal phosphate bond is much greater in ATP (7000 cal/mole) than in ADP or AMP (between 1000 and 2000 cal/mole; Lehninger, 1965). Thus ATP-energy forms a short term conservation of energy, ready for immediate demands as for example in muscular action. The main function of ATP is the capture and transfer of high energy bonds.

ATP is only one of a series of phosphorylated compounds having relatively high free energies of hydrolysis. It is in fact not one of those with the highest free energies of hydrolysis but has rather an intermediate value (7000 cal/mole). Some such as phosphoenolpyruvate (12,800 cal/mole) or phospho-creatine (10,500 cal/mole) have much higher values and others, such as glucose-1-phosphate (5000 cal/mole) or glucose-6-phosphate (3300 cal/mole) have lower ones. This intermediate position in the scale of free energies of hydrolysis is an important feature of ATP, since the ADP-ATP system can act as an intermediate bridge or linking system for transfer of phosphate groups from the high energy compounds which readily give up their phosphate groups to the low energy ones which are 'energised' or 'charged' for chemical work inside the cell; this linking system is shown in Fig. 2.3. Thus ATP is

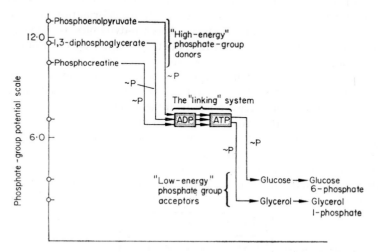

Fig. 2.3. The transfer of phosphate groups from high-energy donors to low-energy acceptors via the ATP-ADP system (after Lehninger, 1965).

important not only as a short term store of energy available for immediate work but also as the means whereby the cell's longer term energy reserves can be called upon when needed during emergencies, as for example during prolonged active locomotion. These long term cellular reserves of energy were formed earlier from oxidation of food substances inside the cell and the transfers of their phosphate groups via ATP to low energy compounds such as glucose-1-phosphate permits the continuation of chemical work, e.g., formation of sucrose from glucose and fructose. Another example of such utilization of energy reserves is 'degradation' of phosphocreatine to creatine with utilization of released energy via ADP-ATP system for muscle contraction (Fig. 2.1).

What has been written about the molecular mechanism of energy transfer in living organisms is only a very simplified version giving a general orientation in what is a series of very complex reactions.

Chapter 2

2.3 Energy balances in general terms

Although the cellular release of energy from food for growth and for work consists of a complicated series of biochemical reactions, it is possible to produce some equations balancing the energy transformations involved on the assumption that the first Law of Thermo-dynamics is applicable to biological systems (Wiegert, 1968; introductory chapter by Dr. Phillipson in this handbook), systems such as organisms, populations, single trophic levels (according to Lindeman, 1942) or to whole ecosystems (according to Odum and Odum, 1955). The theoretical basis for these equations are as follows: 1. All forms of energy, apart from heat, are inter-convertible under the conditions of isothermy and constant pressure in living organisms; 2. During transformation of energy, energy is neither created nor destroyed; 3. Reactions occur spontaneously only when there is also some degradation of energy to heat. All the constituents must be measured in units of energy such as calories or joules or in units convertible to energy units (see p. 275). The quantities of energy in the equation should balance. Figure 2.4

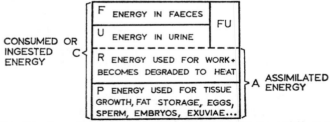

Fig. 2.4. The fate of energy in an organism (after Kleiber, 1961, somewhat modified). All values in calories per unit time. (see below for additional notes).

presents a generalized diagram of what happens to the energy of ingested food in an organism, taken from Kleiber (1961) and somewhat modified. The figure is fully labelled and needs very little additional description; the symbols used are those defined earlier in the handbook (p. 11).

F for faeces or egesta and U for nitrogenous waste or excreta are given separately in Fig. 2.4, although in terrestrial molluscs, insects and birds, the faeces are passed out together with solid nitrogenous waste and are difficult to separate (see section 5C, pp. 185). The same applies to aquatic situations where the nitrogenous waste from both a predator and its prey is passed into the water, together with any soluble substances from faecal material, a

FOOD ENERGY (gross energy, ingestion or food intake C Consumption	→ DIGESTED ENERGY i.e. food hydrolyzed in gut to sugars, fatty acids and glycerol and amino acids and then absorbed through the gut wall into the blood or lymph	→ METABOLIS- ABLE ENERGY	→ ENERGY IN PRODUCTION i.e. for body growth repair of tissues, fat reserves, for eggs, sperm and embryos. P Production
	↳ EGESTED ENERGY faecal waste egestion. F Faeces	↳ ENERGY IN URINARY WASTE i.e. urea, uric acid, ammonia, creatine and other nitrogenous (toxic) metabolic by-products excreted in urine or via the skin and resulting from deamination of amino acids before utilization as respiratory fuels. U urinary and other nitrogenous wastes	↳ ENERGY REQUIRED FOR MAINT- TENANCE OF LIFE i.e. to carry out work of various kinds: mechanical work, work in chemical synthesis, work in active transport, work in conversion of food into active metabolites (S.D.A.*) etc. R (Respiration)

$$P + R = A \text{ (Assimilation) I.B.P.— News } \mathbf{10}; 1969)$$

*S.D.A. Rubner's 'Specific Dynamic Action', also named the 'Calorigenic Effect of Food' or the 'Heat Increments' has been transferred in this diagram to the energy required for maintenance work. It has been studied mostly in homoiothermic animals, birds and mammals, in which the basal energy expenditure may be raised by 15–40% after the ingestion of protein and by lesser proportions after the ingestion of fats and carbohydrates. This heat may be utilized to regulate body temperature in homoiotherms but in poikilotherms is waste heat. It is not evolved during the process of digestion where the energy loss is only about 1% of the total energy liberated but may represent the extra work involved in metabolic inter-conversion and storage of food molecules, as for example, deamination and formation of nitrogenous wastes with protein molecules or cost of resynthesis of macromolecules and active transport. If this explanation is correct; then this heat energy is already accounted for in energy measured by respiration —R. SDA has been studied mainly in homoiotherms (Brody, 1945; Kleiber, 1961; Hoar, 1966); only scanty data are available for poikilotherms, e.g. fish (Davies, 1964, 1966, 1967); copepods (Conover, 1966).

situation solvable only be collecting the urine directly from the kidney ducts, a technically difficult matter. For productivity studies, these difficulties can be avoided by treating F and U together as FU. However, it may be useful to quantify U separately from F in order to assess to what extent protein is being utilized as a respiratory substrate (Blažka, 1966) which may be a much more frequent event in nature than quoted in textbooks.

A for assimilation, as defined by I.B.P. News 10, is the sum of production and respiration $(A = P + R)$ or, as defined by IBP Handbook number 3 (ed. Ricker, 1968) is the food absorbed less the excreta $(A = C - F - U)$ and represents the physiologically useful energy. Digestion and absorption (refered to as D whenever necessary), on the other hand, refers to that part of the food that is digested in the alimentary canal and absorbed into the body through its wall; $D = A + U = P + R + U = C - F$.

2.3.1 'Instantaneous' energy budget

From Fig. 2.4, we may construct some equations, using the symbols presented there, namely,

$$C = P + R + U + F, \tag{1}$$

or where F and U are impossible to separate,

$$C = P + R + FU; \tag{2}$$

where P may consist of P_g, that is the growth of body tissues, and P_r, the growth of all the organic material which is passed out into the environment during the life of the animal, for example, reproductive products, milk, exuvia, moulted hair, feathers, reptile skin, spiders webs, silk, beeswax etc. Often no measure of U is attempted or some approximate value obtained either from the literature or by means of an informed guess is applied.

If we take an organism at one moment in its life cycle, it is theoretically possible to measure simultaneously all the parameters indicated in the above equations, either as an instantaneous rate, i.e., dx/dt, or as a quantity per experimental period such as twenty-four hours, i.e., $\Delta x/\Delta t$, from which a calculated rate can be obtained. If all these values are converted into common units of energy or energy equivalents the equation should balance (Fig. 2.5).

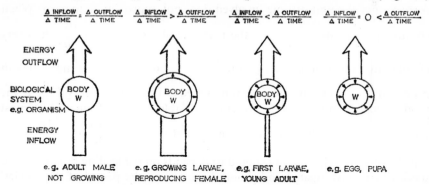

Fig. 2.5A—Energy flow diagrams in organisms with different forms of energy balances, e.g. *Tribolium*.

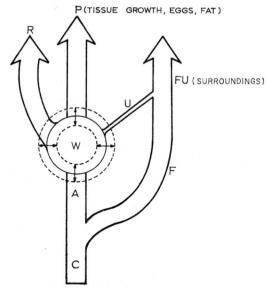

Fig. 2.5B—Energy flow diagram, identifying energy parameters (after Engelmann, 1966).

Such an energy budget for a moment in time during the animal's life cycle, given in terms of measured or calculated rates and energy units, provides what may be termed an 'instantaneous budget', although this term is not correct since the unit of time involved does not approximate to zero. When an attempt is made to measure the true rate of consumption, $\dfrac{dC}{dt}$, it is very possible that this will *not* be equal to $\dfrac{dA}{dt} + \dfrac{dF}{dt}$ because of the time delay in processing as for example when animals feed at infrequent intervals or if the food is retained in the digestive tract for variable periods of time. Thus, most often what is actually and usefully measured is $\dfrac{\Delta C}{\Delta t}$, $\dfrac{\Delta A}{\Delta t}$ and $\dfrac{\Delta F}{\Delta t}$.

The decision as to how long t should be depends on the feeding biology of the species being investigated and should be either t_C, the period of the species' feeding cycle, or t_F, the period of the defaecation cycle, in which case an energy balance is likely to be achieved, or t_R, the period of the respiration cycle. This question of how long t should be is discussed in more detail one pp. 106-108.

2.3.2 'Dynamic' energy budgets.

In fact, we are dealing with organisms in which various reactions take place continuously and at rates which change under different external conditions or with increase in age or change in state of development. A more realistic presentation of this dynamic state may be given in terms of changes in rates of inflow and outflow of energy in an organism, which has been attempted diagramatically in Fig. 2.5 A, B, using as examples different stages in the life cycle of *Tribolium* sp. whose energy budget is given later in the handbook (Chapter 8). Thus, where these two rates are equal, there exists a condition of dynamic steady state such as may occur in certain stages, as for example in the non-growing adult male *Tribolium*. However, unbalanced situations exist when the rate of inflow is greater than the rate of outflow, as in periods of rapid growth in *Tribolium* larvae or reproducing females, or when the rate of outflow is greater than inflow of energy, as in periods when the food supply is inadequate, e.g., for very young *Tribolium* larvae or very young adults, or when no food is taken in at all (as in *Tribolium* eggs or pupal stages), in which case there is a decrease in the body's energy reserves to cover maintenance costs. These examples of balanced and unbalanced situations have been taken from a *Tribolium* budget but are also part of the normal life cycle in many other species which need to be included in any complete life cycle budget.

Such simple diagrams of energy 'flowing' through an organisms cannot cope with the idea of energy 'outflowing' at a different stage in the life cycle compared with when it entered the organism; for example, maintenance energy is dissipated as heat soon after work is performed but may have been derived from food assimilated a short time before or from fat stored much earlier; or, egg, sperm or embryo production 'flows out' of the organism somewhat earlier than the body-growth production which is 'liberated' only upon death. This difficulty can be seen in Fig. 2.5B which shows a simple 'energy flow' diagram along the lines used by (Engelmann, 1966) where he subdivided and named the various processes involved in energy flow in an organism; the inflow of energy consists of consumed food energy and the outflow of energy consists of energy in faeces, urine, respiratory heat and production.

Such a schematic presentation of energy flow may also be attempted for a species-population but again only with considerable difficulties. In Fig. 2.6 the biological system consists of the standing crop of the population denoted by *B* (its biomass) or *N* (its numerical density). Assuming that the population exists in a steady state (with constant age composition, natality, growth rate and mortality) and the equation of energy balance is expressed for a defined period, then, all the parameters of equation (1) must be estimated both for the survivors present during the period of study and for those that died or were born during that period (commonly assumed that these lived for half of the period in question). Thus, in our steady-state population, a balanced equation for a defined period can be obtained from the measurement of not only the population's inflowing (*C*) and outflowing (*R*, *F* and *U*) energy but also must include measurement of the new production stored in the population as reproductive and body growth-energy as well as energy eliminated from the population, in predation and decomposition of naturally dying individuals, during the defined period.

Thus, even with steady-state populations we are dealing with a very complex situation and in nature such populations are probably rather rare and we can expect to be dealing with populations in an unbalanced state where the population biomass or numerical density is either increasing or decreasing due not only to increased recruitment of young or predation but also to immigration and emigration of individuals to or from other populations. Still further complications are shown in Fig. 2.7 which demonstrates

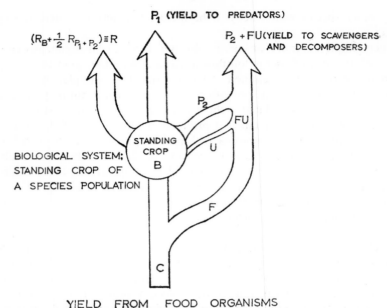

YIELD FROM FOOD ORGANISMS

Fig. 2.6. Energy balance in a species population in a steady state for a defined period time.

the situation in certain species which re-ingested their own faeces, a process called 'refaecation', as in hares (Eden, 1940; Piekarz, 1961), in *Galleria mellonella*, the wax moth (Reamur, 1737, in Wlodawer 1945), in *Palaeomonetes pugic* (Johannes and Satomi, 1966) or in *Asellus aquaticus* (Prus 1971, 1972); or in other species which are cannibalistic and eat their own species, as in pike (Munro, 1957), *Tribolium* (Park, 1934; Park *et al.*, 1965; Prus, 1968) or *Lestes* (Fischer, 1961).

2.3.3 'Cumulative' energy budgets

Some of these difficulties in expressing the dynamic state of energy balance in organisms with infinite capacity for variation is somewhat overcome in the 'cumulative energy budget' first proposed by Klekowski *et al.* (1967); more recently some fragments of this concept have been examined in mathematical terms by Shushkina *et al.* (1968) and discussed on the ecological background (Klekowski, 1970).

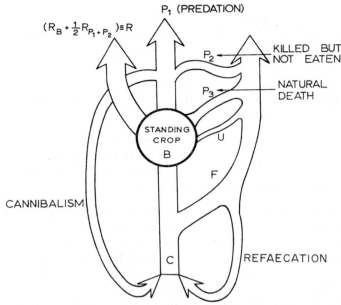

Fig. 2.7. Energy balance in a species population (as in Fig. 2.6) with cannibalism and refaecation.

A cumulative energy budget for an individual or average individual can be produced when all the parameters in equation (1) are cumulated from the beginning of the life cycle (time τ_o) to a determined period in the life cycle or to the end of life (time τ_n), where τ_o is the moment when the newly laid egg or newly born young appears and τ_n the moment of death or the end of an identifiable developmental stage.

The cumulated or integrated elements of a cumulative energy budget may be defined as follows:

$$C_c = \int_{\tau_o}^{\tau_n} C\,(\tau)\ dt \doteqdot \sum_{\tau_o}^{\tau_n} C \qquad \text{cumulated food consumption from } \tau_o \text{ to } \tau_n \qquad (3)$$

$$R_c = \int_{\tau_o}^{\tau_n} R\,(\tau)\ dt \doteqdot \sum_{\tau_o}^{\tau_n} R \qquad \text{cumulated respiration from } \tau_o \text{ to } \tau_n \qquad (4)$$

$$F_c = \int_{\tau_o}^{\tau_n} F(\tau)\, dt \simeq \sum_{\tau_o}^{\tau_n} F \qquad \text{cumulated unassimilated food from } \tau_o \text{ to } \tau_n \qquad (5)$$

$$P_c = \int_{\tau}^{\tau_n} P(\tau)\, dt \simeq \sum_{\tau_o}^{\tau_n} P = B_g + \sum_{\tau_o}^{\tau_n} P_r = B_g + B_r \qquad (6)$$

cumulated production of body growth (B_g) reproductive and other products (B_r) from τ_o to τ_n.

$A_c = P_c + R_c$, cumulated assimilation from τ_o to τ_n.

An important difficulty arises in connection with the calculation of a cumulative budget for one generation. The fertilized eggs belong to one generation but the cost of their production 'was paid' for by the parents belonging to the previous generation. In cumulative budgets, the inclusion or exclusion of this egg-production-costs is reflected in the cumulated parameters not only throughout the life cycle of one generation but also of all subsequent ones and this may have important ecological implications for those species-populations with more than one generation per year. Moreover, the actual assessment of the 'cost' of egg (and theoretically sperm production) is a relatively complex matter. Most of the energy budgets given as examples in the following section have been calculated excluding the cost of egg or young production; such budgets can be usefully distinguished as 'intra-generation budget'. Where the budget includes the cost of egg or young production paid for by the adults of the previous generation, such a budget is called an 'inter-generation budget'.

2.3.4 A discussion on the efficiencies U^{-1}, K_1 and K_2

The basic parameters of energy budgets, C, P and R, can be related to each other in the form of non-dimensional ratios or percentages usually termed coefficients of efficiency or simply efficiencies. As a consequence of the existence of the two types of energy budget just described, namely instantaneous and cumulative, each particular efficiency may be calculated using either instantaneous (C_i, P_i and R_i) or cumulative (C_c, P_c and R_c) parameters. In this section is considered the nature of the information given by these two types of efficiencies and how they change throughout the life cycle of a species, since this has ecological implications.

The concept of biological efficiency was first introduced by Terroine (1922) for microorganisms. Later, the efficiencies U^{-1}, K_1 and K_2, were developed theoretically by Ivlev (1939a, b, 1945, 1966) and later extensively applied to hydrobiological research by Winberg (1962, 1964, 1965, 1967, 1968). Other efficiencies are considered in a later section (p. 131).

In an instantaneous energy budget, what is measured in practice is $\dfrac{\Delta C}{\Delta t}$, $\dfrac{\Delta P}{\Delta t}$, $\dfrac{\Delta R}{\Delta t}$ so that instantaneous efficiencies can be defined as:

$$U_i^{-1} = \frac{P_i + R_i}{C_i} = \frac{A_i}{C} \quad \text{— instantaneous coefficient of assimilation} \tag{8}$$
efficiency;

$$K_{1i} = \frac{P_i}{C_i} \quad \text{— *instantaneous coefficient of utilization of} \tag{9}$$
consumed energy for growth (Winberg, 1962);

$$K_{2i} = \frac{P_i}{P_i + R_i} = \frac{P_i}{A_i} \quad \text{— instantaneous coefficient of utilization energy for}$$
growth (Winberg, 1962).

*other verbal definitions are given on p. 264.

At an individual level, such instantaneous efficiencies are useful in characterizing the physiological-biochemical properties of a given species particularly as these change with development, physiological state or environmental condition (e.g. Ivlev, 1966; Klekowski and Shushkina, 1966). However, where the energy budget is unbalanced, for example during periods of 'negative production' (Petrusewicz, 1967) when an animal is either underfed or starving and loses body weight, instantaneous efficiencies cannot be calculated. In such cases, however, cumulative coefficients of efficiences can be obtained, calculated from the cumulated energy parameters of a cumulative energy budget; these are defined as follows:

$$(11) \quad U_c^{-1} = \frac{P_c + R_c}{C_c} = \frac{A_c}{C_c} \quad \text{— cumulative coefficient of assimilation efficiency;}$$

$$(12) \quad K_{1c} = \frac{P_c}{C_c} = \frac{*W_g + W_r}{C_c} \quad \text{— cumulative coefficient of utilization of an individual energy for growth; *(where W_g is body weight of an individual and W_r cumulated weight of its sexual products);}$$

$$(13) \quad K_{2c} = \frac{P_c}{P_c + R_c} = \frac{W_g + W_r}{A_c} \quad - \quad \text{cumulative coefficient of utilization of assimilated energy for growth.}$$

This method of measuring an energy budget consecutively throughout the species' developmental cycle and of expressing efficiencies provide two kinds of useful information. Firstly, it sums the bioenergetic characteristic of the species, namely, shows how well or badly it converts the biomass of food eaten throughout its life cycle into biomass of body growth or biomass of reproductive products. It measures the total 'cost' (R) of accumulating energy as growth or reproductive products. Cumulative efficiencies are therefore better indices than instantaneous efficiencies of the species capacity to be an 'energy carrier' between its food and its predator, particularly as the predator may consume it at any point in its developmental cycle and its efficiency as an energy carrier varies throughout its life span. The second kind of information provided by cumulative energy budgets is the possibility of reading off how much energy is required for maintenance and production and how much is unassimilated in order that a species can attain a certain stage or size of development. Later, it will be demonstrated how such information can be applied to a field population of known density, age/size composition and mortality.

2.3.5 Examples of laboratory cumulative energy budgets.

The following examples of both terrestrial and aquatic species have been chosen to demonstrate different feeding types.

(a) A filter feeder:

Simocephalus vetulus (Cladocera) (Klekowski and Ivanova, in lit.) (Fig. 2.8). Food, Chlorella; temperature, 22°C; the parameters measured, C, P_g, R together with an approximate evaluation of P_r. Body growth occurred during the first eight days and practically stopped during the period when the young were produced.

The cumulative assimilation efficiency, U_c^{-1}, varied considerably during the twenty days investigated. During the period of exponential body growth, it increased rapidly to a maximal value of 75% just before the onset of reproduction and then decreased during the reproductive period down to a value of about 30%. This decline indicates that the instantaneous assimilation

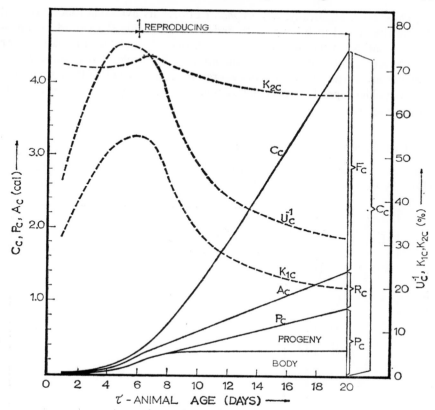

Fig. 2.8. Cumulative energy budget for an average animal *Simocephalus vetulus* (after Klekowski and Ivanova in lit.).

efficiency also decreased during this period, possibly due to the greater necessity to supply the developing oocytes in the ovary with specific 'building materials' such as the essential amino acids or fatty acids (see p. 80) needed for embryonic growth later.

K_{2c} had a high value and was approximately constant at 65-70% throughout the twenty days, values being somewhat higher during the period of body growth. Since $K_{1c} = K_{2c}$. U_c^{-1} (see p. 143) and in *Simocephalus* K_{2c} can be taken to be approximately constant, any changes in U_c^{-1} will be reflected in the value of K_{1c}, so that these two efficiencies (K_{1c} and U_c^{-1}) vary in a similar way throughout the life cycle of this species.

D

Thus, bioenergetically, *Simocephalus* utilizes its available food very well during the period of body growth and less well during the period of reproduction, although the proportion of food, once assimilated, that is used for production is constantly high throughout its life cycle. Thus young *Simocephalus* are better at transferring energy from their food to their predators than the older reproducing females, a point of some significance in the habitats where they occur.

(b) A stored-product deposit feeder:

Tribolium castaneum (Coleoptera) (Klekowski *et al.*, 1967, Chapter 8) (Fig. 2.9A). Food, flour mixed with yeast; temperature, 29°C; parameters

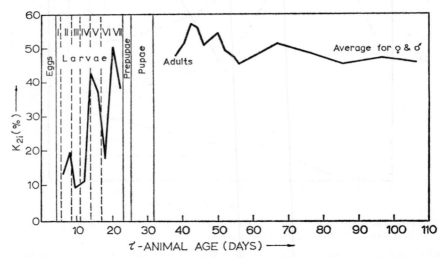

Fig. 2.9A—Changes of instantaneous coefficient of net production efficiency (K_2) during the development of *Tribolium castaneum* (after Klekowski *et al.*, 1967).

measured, R, P_g, P_r, about which more detail is given on Chapter 8.

This species does not feed during pupation and illustrates the difficulty of calculating instantaneous efficiencies for periods of physiological starvation discussed earlier (p. 26). Thus, in Fig. 2.9A, the instantaneous K_{2i} reaches a level of up to 50% during the pre-imaginal growth, cannot be calculated for the pupa and attains value up to 60% during egg production in the adults whereas the cumulative K_{2c} (Fig. 2.9B) is never greater than 30-40% at all periods throughout the life cycle.

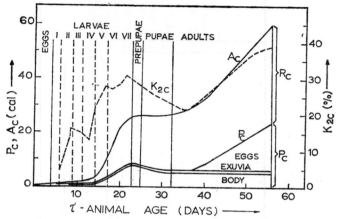

Fig. 2.9B—Cumulative energy budget for an average animal *Tribolium castaneum* (replotted after Klekowski *et al.*, 1967).

(c) A stored-product deposit feeder:

Rhisoglyphus echinopus (Acarina) (Klekowski, Stępień in lit.; Klekowski, 1970) (Fig. 2.10). Food, rye germ; temperature, 25°C; parameters measured, C, R, Pg, P$_r$. NB!—Only 'inter-generation budget' is presented.

During the pre-imaginal development of this species Fig. 2.10A, there occurs within each instar a period of feeding and active growth (the active larva, active protonympha, active tritonympha) which alternates with a non-feeding stage with loss in body weight (the egg, resting larva, resting protonympha, resting tritonympha). This species thus illustrates a complex life cycle with alternating periods of 'positive' and 'negative' production or active feeding and physiological starvation. Instantaneous efficiencies can therefore be calculated only for the periods of active feeding and growth and are valid for that period only whereas cumulative efficiencies provide a realistic picture for every moment of the life cycle.

In Fig. 2.10B, the instantaneous U_i^{-1} is highest in the active larval stage whereas it is not until just before the end of the active protonymphal stage does the cumulative assimilation efficiency (U_c^{-1}) attain its maximal value of 73% because the cumulative assimilation efficiency for the eggs is low.

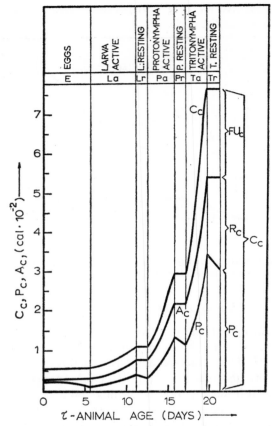

Fig. 2.10. Inter-generation energy budget for an average animal (♀ and ♂) *Rhisopglyphus echinopus* (after Klekowski and Stępień in lit.).
Fig. 2.10A Cumulative budget for preimaginal development.

Figure 2.10C shows that the instantaneous K_{1i} varies greatly during the pre-imaginal development but is always high because these stages must store energy to cover the subsequent non-feeding period. The minimal value of K_{1c} (20%) is found in the first larva hatching from the egg and the maximal (40-50%) in the active protonympya and tritonympha. Thus these two life stages are the most efficient 'energy carriers' to their predators, in this case, predatory mites. Another period in the life cycle with high K_{1c} values is

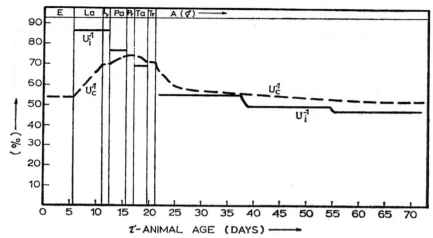

Fig. 2.10B Changes of assimilation efficiency.

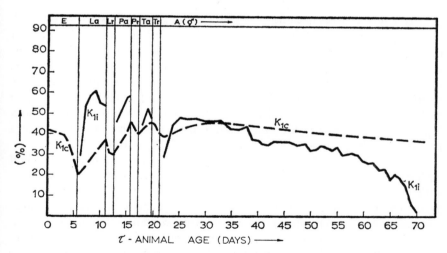

Fig. 2.10C Changes of efficiency of utilization for growth of consumed energy.

during egg production. This implies that egg production is bioenergetically very economical since the consumption per unit production is very low, which is expressed by the coefficient $1/K_{1c}$. Similarly, broiler fowl are killed for human consumption during the period when their $1/K_{1c}$ coefficient is lowest.

Figure 2.10D shows that K_{2i} is also very high (80%) from the 25th to

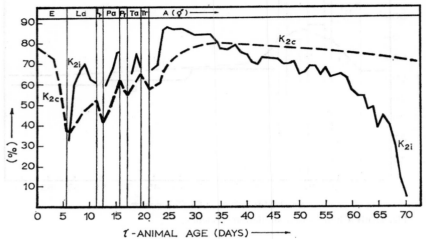

t -ANIMAL AGE (DAYS) ———▶

Fig. 2.10D Changes of efficiency of utilization for growth of assimilated energy.

30th day of life when both growth-production and egg-production is most intense. Such a high level is probably near to the absolute upper limit of this coefficient according to Winberg (1968). The cumulative K_{2c} for the whole life cycle is also high probably because the various non-feeding periods are relatively short and the loss of energy during them relatively small compared with the body's stored reserves. The very marked decline in the values of both K_{1i} and K_{2i} just before the end of adult life is the result of decreased egg production, although a few eggs are produced right up to the end.

Another feature of interest is illustrated in this species; during the imaginal period (Fig. 2.10E) the egg production per mean parent is very high compared with the energy accumulated in the body of the parent, being six times as great. The mean weight or calorific content of an egg is thus relatively high compared with the mean weight or calorific content of a mature adult. Therefore, the cost of production of one egg will be appreciable and is paid for by the parent in its consumption and assimilation. In most species, it is possible to exclude this cost to the parent of egg production and to cumulative the parameters from zero, that is within one generation only; the efficiencies derived from these parameters can be usefully distinguished, where necessary, as, intra-generational efficiencies and are those normally employed. However, in species such as *Rhisoglyphus echinopus*, egg costs to the parent-generation

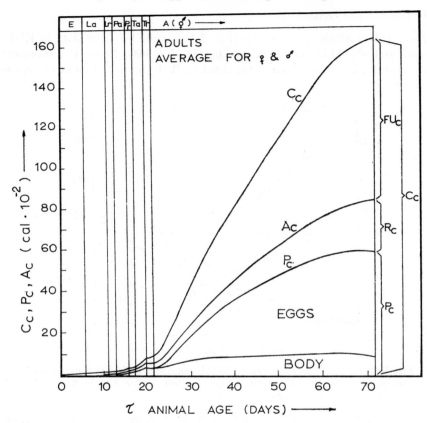

Fig. 2.10E Cumulative budget for whole individual life cycle.

are too high to be excluded and the cumulation of parameters for one generation must start with costs to the previous generation; efficiencies derived
from such parameters can be distinguished as 'inter-generational' efficiencies.
This difficulty, which arises from a real situation in nature, is not a problem
in the calculation of instantaneous efficiencies. The cost was calculated from
the adult consumption and assimilation minus respiration cumulated for a
period of half the duration of imaginal life, this period of half adult life was
adopted as the mortality rate, which in cultures was relatively constant.

(*d*) *A crustacean predatar:*

Macrocyclops albidus (Copepoda) (Fig. 2.11), compiled from data in Klekowski and Shushkina (1966) Shushkina *et al.* (1968). Food, Paramecium;
temperature, 21°C; parameters measured C, R, P_g up to the onset of reproduction.

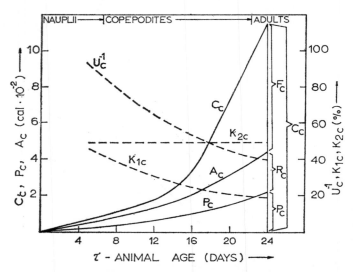

Fig. 2.11. Cumulative energy budget for *Macrocyclops albidus* growing in food
concentration lg/m³; before reproduction (replotted after data in Klekowski
and Shushkina, 1966, Shushkina *et al.* 1968).

The cumulative assimilation efficiency U_c^{-1} decreased from more than
90% down to about 40% during the larval development of the animal whereas
K_{2c} remained constant at about 50%. The utilization of food is therefore,
considerably better in the younger nauplii and copepodites than in the older
copepodites and adults, suggesting that the younger stages act as better
'energy carriers' to the predatory zooplankton and fish feeding upon them
than the older stages; a similar conclusion to that reached for *Simocephalus
vetulus.*

(*e*) *An insect predator:*

Lestes sponsa (Odonata) (Fischer, 1972a) (Fig. 2.12a, b). Food, cladocera and

Tubifex worms; temperature, 22°C; parameters measured; C,R,P$_g$ for larval stages only.

Figure 2.12A shows that the daily energy parameters (i.e. 24 hourly) are very variable which would result in very variable instantaneous efficiencies.

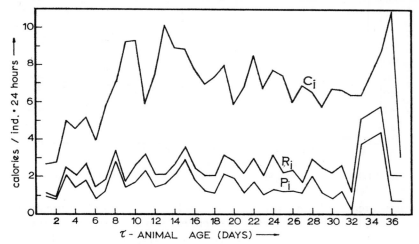

Fig. 2.12A—Daily consumption, respiration and production of *Lestes sponsa* (after Fischer, 1972a).

The cumulation of these daily parameters (Fig. 2.12B) produce a much more uniform and stable picture which is also reflected in the relative constancy of the efficiences U_c^{-1}, K_{1c}, K_{2c}. The cumulative energy budget therefore illustrates well the hemi-metabolic character of the larval development of *Lestes sponsa*, during which no dramatic changes occur in its physiology or way of life such as take place in Holometabola or in mites. The level of U_c^{-1} as well as K_{1c} are rather low for predatory animals which may be related to their culture conditions of excess food. Compared with the active predator, *Macrocyclops albidus*, *Lestes* showed a much higher K_{2c} (65-80%). This is probably associated with its habit of 'sedentary hunting' in which it sits hidden and waits for prey to appear, so that in excess food, it carried out the minimum of movement. Maintenance cost were therefore rather low.

2.3.6 Application of laboratory cumulative energy budgets to field populations.

Application of energy parameters of laboratory budgets to field populations

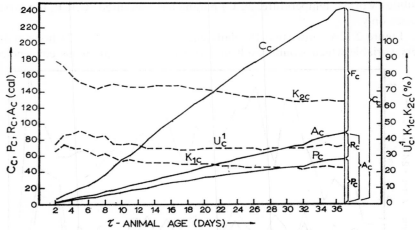

Fig. 2.12B—Cumulative energy budget for whole larval development of *Lestes sponsa* (after Fischer, 1972a).

per unit area or volume provide the possibility of not only estimating the population production but also of obtaining some assessment of the other population parameters such as consumption, respiration etc. These latter population parameters are difficult to measure in the field unlike population production which can be calculated from changes in field numbers, age structure, mean body weights and field developmental times. Without some knowledge of the levels of these other population parameters, it is difficult to assess the effects on the ecosystem by the population's removal of food or return of faeces or excreta or its minimal energy demands in the form of assimilated energy.

In order to understand how it may be possible to apply cumulative energy budgets to the calculation of energy balances in field populations, it is necessary first to discuss two common approaches to the estimation of population production (cf. Klekowski, 1970).

(a) By 'sum of biomass increments'

Most frequently, population production is understood as the 'sum of increments' of all the individuals together with their sexual and other organic products present in the population (Phillipson, 1967; Petrusewicz, 1967; Winberg, 1962, 1967, 1968; MacFadyen, 1967). This is calculated for a series of sampling intervals during the population's active existence or for a whole year where some individuals die during the sampling interval, their production up to the moment of death, or some arbitrary proportion of the sampling interval, is included.

Thus, the production of one age/instar group of the population for the sampling interval L_{n-1} to $L_n = \Delta t$ is expressed by the formula (cf. Phillipson, 1967):

$$P'_{t,n-1} - t_n = \left(N'_{t,n} \cdot \frac{\Delta \overline{W'}}{\Delta t}\right) + \left(\frac{\Delta N'}{\Delta t} \cdot \frac{1}{2} \cdot \frac{\Delta \overline{W'}}{\Delta t}\right) \tag{14}$$

where:

$P'_{t,n-1} - t_n$	is	the production for the age/instar group for the time interval Δt;
$N'_{t,n}$	is	the numbers of individuals of the age/instar group surviving at the moment t_n;
$\dfrac{\Delta N'}{\Delta t}$	is	the numbers of individuals of the age/instar group that died during the interval Δt;
$\dfrac{\Delta \overline{W'}}{\Delta t}$	is	the mean weight (or calorific) increment of the survivors of the age/instar group during the interval Δt.

Where more than one age/instar group is present during the sampling interval Δt, their production can be calculated in a similar manner and summed to give the population production for the interval concerned. Instead of weight increments or number decrements, it may be possible to apply determined growth and mortality rates in the above formula. Where the weight increment and number decrement is not known for each of the age/instar group present, it may be possible to obtain a weight increment for a mean individual ($\Delta \overline{W}/\Delta t$) and a decrease in the numbers of the total population during the sampling interval Δt. The following formula will then apply:

$$P_{t,n-1}-t_n = \left(N_{t,n} \cdot \frac{\Delta \overline{W}}{\Delta t}\right) + \left(\frac{\Delta N}{\Delta t} \cdot \frac{1}{2} \cdot \frac{\Delta \overline{W}}{\Delta t}\right) \tag{15}$$

where*:

$P_{n-1} - t_n$ is the total population production for the time interval Δt;

*Interrelations between symbols/units in formulas 14 and 15 are as follows:

$$N' + N'' + \ldots = N; \quad \frac{(\overline{W'} \cdot N') + (\overline{W''} \cdot N'') + \ldots}{N' + N'' + \ldots} = \frac{B}{N} = \overline{W}$$

B is biomass of the total population.

$N_{t,n}$ is the numbers of individuals in the total population surviving
            at the moment t_n;

$\dfrac{\Delta N}{\Delta t}$ is the numbers of individuals in the total population that died
            during the interval Δt;

$\dfrac{\Delta \overline{W}}{\Delta t}$ is the mean weight (or calorific) increment per mean individual
            of survivors in the total population during the interval Δt.

This formula will provide only rough estimates of population production
and will be difficult to apply to populations with over-lapping generations.

  Where the total population production is to be calculated for a time
interval $(t_o - t_n)$ which consists of more than one sampling interval (Δt), it
will be necessary to integrate the periods of growth of the population biomass
for the whole time $(t_o - t_n)$; in practice, the successive periods of growth
may be summed.

$$\int_{to}^{tn} P\,|t|\,dt \simeq \sum_{to}^{tn} P = \sum_{to}^{tn} \left[\left(N_{t,n} \cdot \frac{\Delta \overline{W}}{\Delta t} \right) + \left(\frac{\Delta N}{\Delta t} \cdot \frac{1}{2} \cdot \frac{\Delta \overline{W}}{\Delta t} \right) \right] \tag{16}$$

(b) By the 'sum of eliminated biomass'

Another approach is to consider production as the 'sum of eliminated
biomass', that is, the total energy contained in dead individuals together with
their dead reproductive and other non-living organic matter products. This
represents the population energy which is transferred to other trophic levels,
by predation or decomposition (unless cannibalism or lactation is involved);
such an understanding of production has been rather rarely adopted in
ecology (e.g. Cooper, 1965; Bocock *et al.*, 1967).

  Production as a 'sum of eliminated biomass' can be expressed by the
following equation:

$$P_{t,n-1-t,n} = \left(\frac{\Delta N}{\Delta t} \cdot \frac{\overline{W}_{t,n-1} + \overline{W}_{t,n}}{2} \right) + \Delta B_r \tag{17}$$

where:

$\dfrac{\Delta N}{\Delta t}$ — the numbers of individuals in the population that died
            during the interval Δt;

$W_{t,n-1}$, $W_{t,n}$ — mean weights (or calorific values) of individuals at the time t_{n-1} and t_n.

B_r — biomass (or calorific value) of reproductive products dead during Δt and of other non-living organic matter products 'lost' by all members of the population during the interval Δt.

The population biomass increment per mean individual and the population number decrement is known. This method of calculation provides an estimate of the eliminated population production yielded to other trophic levels during the interval of time Δt. The total population production yielded to other trophic levels during a longer period ($t_o - t_n$) will be given by:

$$\int_{to}^{tn} P\,|t|\,dt \simeq \sum_{to}^{tn} P = \sum_{tn}^{to} \left[\left(\frac{\Delta N}{\Delta t} \cdot \frac{\overline{W}_{t,n-1} + \overline{W}_{t,n}}{2} \right) + \Delta B_r \right] \qquad (18)$$

Another approach can be adopted for calculating the annual production of a species from the 'sum of eliminated biomass'. In species where one or more cohorts are distinguishable within the period of active existence of a species, and where the overlap in time of successive stages is minimal or unravellable, it is possible to construct a numerical budget or life table for one cohort, from egg to adult. This involves determining, from a series of successive samples, the number in the cohort entering a developmental stage, starting with the egg, and the number that survive each stage. The sum of these gives the total numbers of each developmental stage in the cohort and, by subtraction, the numbers that died in each stage plus the numbers of the final survivors. An excellent description on the construction such age-specific life tables is given in Southwood (1966) to which the reader is referred. Once such a numerical budget for a cohort has been obtained, it is relatively easy to read off from a cumulative laboratory budget for an individual of the species, the appropriate cumulated parameters and mutiply them by the numbers of individuals of each stage from the life table.

Both instantaneous parameters for different stages and cumulated parameters for the same individuals studied throughout their life cycle can be applied to a field population per unit area or volume in order to obtain some idea of its energy balance. These parameters for C,R,F, or FU as well as P_g and P_r can be applied to equations 16 or 18: the instantaneous parameters applied to equation 16 are usually measured by hour or by day and must be converted to the increment values $\left(\dfrac{\Delta C}{\Delta t}, \dfrac{\Delta R}{\Delta t}, \text{and } \dfrac{\Delta FU}{\Delta t} \right)$, whereas these

increments may be read off directly from the cumulative budget diagram, provided that the instar duration is known for the time interval concerned and an appropriate adjustment to the time scale is made.

Now it is possible to assess the usefulness of these two approaches to the calculation of production. The incremental formulae (14) and (16) provide an assessment of the 'instantaneous' relationship of a population to its food and environment, including such factors as CO_2 exchange in soil animals or secreted metabolities. The elimination formulae (17) and (18) on the other hand, characterizes the cumulated energy available to the predator, and decomposer, having taken into account those periods in the life cycle when the animal is respiring its own body tissue, for example, during dormancy, under-feeding or starvation. Because the rates of life processes vary throughout a life cycle, instantaneous efficiencies such as U_i^{-1}, K_{1i} and K_{2i} apply only to one moment in time whereas these efficiencies in their cumulated form $(U_c^{-1}, K_{1c}, K_{2c})$ are *not* associated with any one particular interval but reflect the relation of various processes during the period from birth to the extant age. Thus, the elimination formulae provide an estimate not only of the population's 'final product' (P_c) together with some assessment of the 'final cost' (R_c) of maintenance and 'final waste' (FU_c) passed to other trophic levels but also of the 'final' relationship between these processes. For example, the cumulative K_{1c} quantifies the ratio between the cumulated food consumed by a population and the biomass it produces, that is, it characterizes that species-population as an 'energy carrier' between its consumed food and its predator (or decomposer).

2.3.7. An example of a field population cumulative budget.

An example is presented here of the application of individual cumulative energy budgets to field census data in order to calculate the population cumulative energy balance, both throughout the year and finally at the end of its active existence. The species is a phalangid, *Oligolophus tridens*, investigated as a field population living in the neighbourhood of Durham, England, by Dr. J. Phillipson, who had already calculated the annual population production and metabolism from a formula similar to equations (14) and (16). During the Training Course in Warsaw, Dr. Phillipson very kindly allowed us to use his original data in order to attempt the calculation of the annual population energy parameters from individual cumulative budgets and with the aid of equation (18). Dr. Phillipson is, of course, in no way responsible for this use of his data, so generously supplied.

In fact, *O. tridens* is an ideal species for such an exercise since it has one generation per year and successive monthly samples revealed predominantly one developmental stage. It was therefore possible to plot the individual as well as the population energy budget against sampling data rather than against individual age, because these were directly related. In species with several generations per year, particularly if these over-lapped in time, the budgets would have to be calculated for an individual life cycle or one cohort. In this particular case, the individual energy budget for a mean female and a mean male (Fig. 2.13 C, D) were derived from the field data rather than from a laboratory study following closely the development of several individuals throughout their life span, which is the specific feature of cumulative energy budgets.

The steps in the computation were as follows:

(1) From the numbers of different instars collected at monthly intervals, it was possible to determine the average duration of each instar in the field (Fig. 2.13 A), the start being taken when 50% of the population consisted

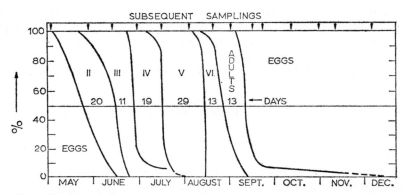

Fig. 2.13. Cumulative energy budget for a field population of *Oligolophus tridens* (plotted after data of Dr. J. Phillipson).
Fig. 2.13A—Calculation of average duration of subsequent instars.

of that instar, as well as their mortality (Fig. 2.13 B). The date of appearance of various instars was also noted.

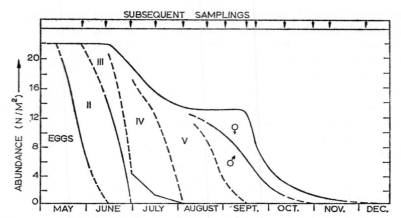

Fig. 2.13B—Estimation of mortality and age- instar composition from population figures.

(2) From laboratory measurements of mean weight, calorific value and respiratory rates of different instars, a cumulative energy budget for a mean female and a mean male from this particular population was calculated, cumulated for the various sampling dates when they occurred (Fig. 2.13 C, D).

(3) From these two individual budgets and the population census data for various sampling dates, the cumulative population energy balance was computed on the basis of equation (18), the number of individuals that died during each sampling interval $\left(\dfrac{\Delta N}{\Delta t} \right)$ was read off from survivorship curves (Fig. 2.13 B).

(4) The values available for consumption covered only the earlier part of the life cycle.

The cumulative energy budgets (Fig. 2.13 C, D) illustrate in a convenient way several interesting features of this species, which have already been pointed out by Dr. J. Phillipson, The most striking feature is the greater proportion of energy that goes to cover maintenance metabolism compared with that stored in growth; this applies to both sexes and is quantified in the relatively low values of the individual K_{2c} as well as its decrease throughout the life cycles.

Another marked characteristic of the individual budgests is the much greater total cumulated assimilation (A_c) of females compared with the males,

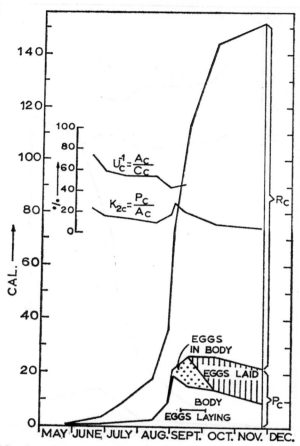

Fig. 2.13C—Cumulative energy budget and efficiencies for an average female.

being almost twice higher. Individual females are making much greater demands on the available food supply than the males. This difference is not shown in the earlier instars but is associated with greater body growth of the female, the onset of ovarian growth and laying of eggs. The greater respiratory energy demands of females may be related to increased hunting activity for food to cover their higher production ($P_g + P_r$) as well as any activity involved in laying eggs. The fact that females lived longer than males (the last female was collected during December and the last male during

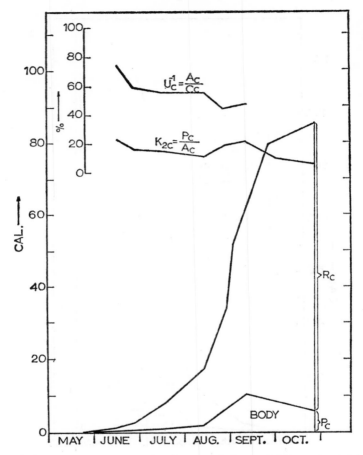

Fig. 2.13D—Cumulative energy budget and efficiencies for an average male.

November) also contributes to the higher cumulated respiration (R_c) and so also assimilation (A_c) of females.

One would expect that the cumulated consumption (C_c) of a female would also be much greater in the female but consumption data was available only for the earlier part of the life cycle. During this early period, the cumulative assimilation efficiency U_c^{-1} reached a similar level in the two sexes and decreased from a maximum of more than 70% to about 40%.

Fig. 2.13 B, C, D show that the male production consists mostly of body growth whereas that of the female is subdivided into somewhat greater body growth and egg production. The egg production starts off inside the body of the female and later after spawning, in the environment. In both sexes, there is some loss of body production towards the end of their life; Dr. Phillipson suggests that senescent individuals were inefficient hunters and showed higher respiratory rates and so were forced to metabolise their own body tissue.

Fig. 2.13 E shows the population cumulative budget obtained by

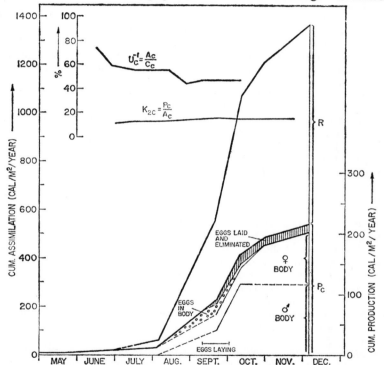

Fig. 2.13E—Cumulative energy budget for a population from 'sum of eliminated biomass'.

applying the above individual budgets to the population census data; notice that the scale for cumulated assimilation is different from the scale for cumulated production. The production scale was expanded in order to be able to demonstrate more clearly the various forms of production. The population budget in this form is ready for considering the role of this particular species in its ecosystem.

References to Chapters 2, 4 and 7a.

ALLEE W.C. *et al.* (1949) *Principles of animal ecology.* B. Saunders, Philadelphia. 837p.
ANDRZEJEWSKA L. (1967) Estimation of the effects of feeding of the sucking insect *Cicadella viridis* L. (Homoptera-Auchenorrhyncha) on plants. In Petrusewicz K. (ed.) *Secondary productivity of terrestrial ecosystems.* p. 791–799. PWN, Warszawa-Kraków.
BEADLE L.C. (1961) Adaptations of some aquatic animals to low oxygen levels and to anaerobic conditions. *Symp. Soc. exp. Biol.,* **15,** 120–131.
BERTALANFFY L.VON & MULLER I. (1941) Untersuchungen über die Gesetzlichkeit des Wachstums. VIII—Die Abhängigkeit des Stoffwechsels von der Körpergrösse und der Zusammenhang von Stoffwechseltypen und Wachstumstypen. *Riv. Biol.,* **35,** 48–95.
BERTALANFFY L.VON (1964) Basic concepts in quantitative biology of metabolism. *Heloländer wiss. Meeresunters.,* **9,** 5–37.
BERTHET P. (1964) L'activite des oribatides (Acari: Oribatei) d'une chenaie. *Mem. Inst. r. Sci. nat. Belg.,* No. 152, 152p.
BETTS M.M. (1954) Experiments with an artificial nestling. *Br. Birds,* **47,** 229–231.
BETTS M.M. (1955) The food of titmice in oak woodlands. *J. anim. Ecol.,* **24,** 282–323.
BLAXTER K.L. (ed.) (1965) *Energy metabolism.* Academic Press, London-New York. 450p.
BLAXTER K.L. (1967). *The energy metabolism of ruminants.* (2 ed.) Hutchinson, London. 329p.
BLAŽKA P. (1960) Maximalni a aktivni metabolismus ryb. Studies on comparative Physiology of Metabolism. *Babak's collection,* **16** (6), 27–29.
BLAŽKA P. (1966) Metabolism of natural and cultured population of *Daphnia* related to secondary production. *Verh. int. Ver. Limnol.,* **16,** 380–385.
BOCOCK K.L. *et al.* (1967) Estimation of annual production of millipede population. In: Petrusewicz K. (ed.) *Secondary productivity of terrestrial ecosystems.* p. 727–739. PWN, Warszawa-Kraków.
BRAY J.R. (1961) Measurement of leaf utilization as an index of minimum level of primary consumption., *Oikos* **12,** 71–74.
BRODY S. (1945) *Bioenergetics and growth, with special reference to the efficiency comples in domestic animals.* Reinhold Publishing Corp., New York. 1023p.
BROOKE M.M. & PROSKE H.O. (1946). Precipitin test for determining natural insect predators of immature mosquitoes. *J. Nat. Malar. Soc.,* **5,** 45–56.
BUCK J. & KEISTER M. (1955). Cyclic CO_2 release in diapausing *Agapema* pupae. *Biol. Bull.,* **109,** 144–163.
CHADWICK L.E. & GILMOUR D. (1940). Respiration during flight in *Drosophila repleta* Wollston; the oxygen consumption considered in relation to the wing-rate. *Physiol. Zool.,* **13,** 398–410.

CHAUVIN R. (1956) *Physiologie de l'insecte. Le comportement les grandes fonctions ecophysiologie.* (Ed. 2). Dr W. Junk Publ. Hague. 916p.

CLARKE G.L. (1946) Dynamics of production in a marine area. *Ecol. Monogr.*, **16**, 323–335.

CLARKE K.U. (1957) The relationship of oxygen consumption to age and weight during the post-embryonic growth of *Locusta migratoria* L. *J. exp. Biol.*, **34**, 29–41.

COLLARDEAU C. (1961) Influence de la temperature sur la consommation d'oxygene de quelques larves Trichopteres. *Hydrobiologia*, **18**, 252–264.

COLLARDEAU-ROUX C. (1966) Influence de la temperature sur la consommation d'oxygene de *Micropterna testacea* (Gmel.) (Trichoptera, Limnophilidae). *Hydrobiologia*, **27**, 385–394.

CONOVER R.J. (1966) Assimilation of organic matter by zooplankton. *Limnol. Oceanogr.*, **11**, 338–345.

COOPER W.E. (1965) Dynamics and production of a natural population of a freshwater amphipod, *Hyalella azteca*. *Ecol. Monogr.*, **35**, 377–394.

CORBETT J.L. *et al.* (1960) Excretion of chromium sesquioxide administered as a component of paper to sheep. *Br. J. Nutr.*, **14**, 289–299.

CRAMPTON E.W. (1965) *Applied animal nutrition.* W. H. Freeman, San Francisco. 458p.

CRAMPTON E.W. & LLOYD L.E. (1960) *Fundamentals of nutrition.* W. H. Freeman, San Francisco. 494p.

CROSSLEY D.A. (1963) Consumption of vegetation by insects. In Schultz V. and Klement A.W. Jr (eds.) *Radioecology*, p. 431–440. Chapman and Hall, London.

CROWELL H.H. (1941) Alimentation and utilization of food body by the armyworm *Prodenia eridania*. Ohio St. Univ. *Doct. Diss.*, **34**, 131–136.

CUMMINS K.W. & WUYCHECK T.C. (1971) Caloric equivalents for investigations in ecological energetics. *Mitt. int. Ver. Limnol.* 18.

DARWIN CH. (1898) *The formation of vegetable mould through the action of worms with observations on their habits.* D. Appleton and Co. New York. 326p.

DAVIES P.M.C. (1964) The energy relations of *Carassius auratus* L. I. Food input and energy extraction efficiency at two experimental temperatures. *Comp. Biochem. Physiol.*, **12/1/**, 67–79.

DAVIES P.M.C. (1966) The energy relations of *Carassius auratur* L. II. The effect of food, crowding and darkness on heat production. *Comp.Biochem.Physiol.* **17/3/**, 983–995.

DAVIES P.M.C. (1967) The energy relations of *Carassius aurutus* L. III. Growth and the overall balance of energy. *Comp. Biochem. Physiol.*, **23**(1), 59–63.

DAVIES P.S. (1966) Physiological ecology of *Patella*. I—The effect of body size and temperature on metabolic rate. *J. mar. biol. Ass. U.K.*, **46**, 647–658.

DAVIES P.S. (1967) Physiological ecology of *Patella*. II—Effect of environmental acclimation on the metabolic rate. *J. mar. biol. Ass. U.K.*, **47**, 61–74.

DAVIES R.W. (1969) The production of antisfera for detecting specific triclad antigens in the gut contents of predators. *Oikos*, **20** (2), 248–260.

DAVIS R.A. & FRAENKEL G. (1940) The oxygen consumption of flies during flight. *J. exp. Biol.*, **17**, 402–407.

DEMPSTER J.P. (1960) A quantitative study of the predators on the eggs and larvae of the broom beetle, *Phytodecta divacea* Forster, using the precipitin test. *J. anim. Ecol.*, **29**, 149–167.

DEMPSTER J.P., RICHARDS O.W. & WALOFF N. (1959) Carabidae as predators on the pupal stage of the Chrysomelid beetle *Phytodecta divacea* (Forster). *Oikos*, **10**, 65–70.

DINEEN C.F. (1953) An ecological study of a Minnesota pond. *Am. Midl. Nat.*, **50**, 349–376.

DIXON M. (1952) *Manometric methods as applied to the measurement of cell respiration and other processes.* (3 ed.) Univ. Press, Cambridge. 165p.

DOWNE A.E.R. & WEST A.S. (1954) Progress in the use of the precipitin test in entomological studies. *Canad. Ent.* **86**, 181-4.

DRÓŻDŻ A. (1967) Food preference, food digestibility and the natural food supply of small rodents. In Petrusewicz K. (ed.) *Secondary productivity of terrestrial ecosystems.* p. 323–330. PWN, Warszawa-Kraków.

DRÓŻDŻ A. (1968) Digestibility and assimilation of natural foods in small rodents. *Acta theriol.*, **13**, 367–389.

DUSPIVA F. (1936) Beiträge zur enzymatischen Histochemie. XXI—Die proteolytischen Enzyme der Kleider und Wachsmottenraupen. Hoppe-Seyler's *Z. physiol. Chem.*, **241**, 177–200.

EDEN A. (1940) Coprophagy in rabbits. *Nature*

ENGELMANN M.D. (1961) The role of soil Arthropods in the energetics of an old field community. *Ecol. Monogr.*, **31**, 221–238.

ENGELMANN M.D. (1966) Energetics, terrestrial field studies, and animal productivity. *Adv. ecol. Res.*, **3**, 73–115.

EVANS A.C. & GOODLIFFE E.R. (1939) The utilization of food by the larva of the mealworm *Tenebrio molitor* L. *Proc. R. ent. Soc. Lond. Ser. A.*, **14**, 57–62

FISCHER Z. (1961) Cannibalism among the larvae of the dragonfly *Lestes* nympha Selys. *Ekol. pol. Ser. B*, **7**, 33–39.

FISCHER Z. (1966) Food selection and energy transformation in larvae of *Lestes sponsa* (Odonata) in astatic waters. *Verh. int. Ver. Limnol.*, **16**, 600–603.

FISCHER Z. (1967) Food composition and food preference in larvae of *Lestes sponsa* (L.) in astatic water environments. *Pol. Arch. Hydrobiol.*, **14**, 59–71.

FISCHER Z. (1968) Food selection in grass carp (*Ctenopharyngodon idella* Val.) under experimental conditions. *Pol. Arch. Hydrobiol.*, **15**, 1–8.

FISCHER Z. (1970) The elements of energy balance in grass carp (*Ctenopharyngodon idella* Val.) Part I, *Pol. Arch. Hydrobiol.* **17**, 421-434.

FISCHER Z. (1972a) The energy budget of *Lestes sponsa* (Hans. during its larval development *Pol. Arch. Hydrobiol.* **19**, 215-222.

FISCHER Z. (1972b) The elements of energy balance in grass carp (*Ctenopharyngodon idella* Val.) Part II. Fish fed with animal food. *Pol. Arch. Hydrobiol.* **19**, 65-82.

FOX C.J.S. & MACLELLAN C.R. (1956) Some Carabidae and Staphylinidae shown to feed on a wire worm, *Agnotes sputator* (L.) by the precipitin test. *Can. Ent.*, **88**, 228–231.

FRAENKEL C. & BLEWETT M. (1944) The utilization of metabolic water in insects. *Bull. ent. Res.*, **35**, 127–139.

FRAENKEL G. & BLEWETT M. (1947) Linoleic acid and arachidonic acid in the metabolism of two insects, *Ephestia kuehniella* (Lep.) and *Tenebrio molitor* (Col.). *Biochem. J.*, **41**, 475–478.

FRY F.E.J. (1947) Effects of the enviornment on animal activity. *Univ. Toronto Stud. biol. Ser.* No. 55. 62p.

FRY F.E.J. (1957) The aquatic respiration of fish. In: Brown M.E. (ed.) *The physiology of fishes.* Vol. 1—*Metabolism*, Part 1, 1–63. Academic Press. New York.

FRY F.E.J. & HART J.S. (1948a) Cruising speed of goldfish in relation to water temperature. *J. Fis. Res. Bd Canada*, **7**, 169–175

FRY F.E.J. & HART J.S. (1948b) The realtion of temperature to oxygen consumption in the goldfish. *Biol. Bull.*, **94**, 66–77.

GERKING S.D. (1962) Production and food utilization in a population of bluegill sunfish. *Ecol. Monogr.*, **32**, 31–78.

GILLIES M.T. (1958) A simple autoradiographic method for distinguishing insects labelled with phosphorus—32 and sulphur—35. *Nature, Lond.*, **182**, 1683–1684.

GLICK D. (1961) *Quantitative chemical techniques of histo–and cytochemistry.* Vol. 1, p. 123–289. J. Wiley and Sons, New York–London.

GOLLEY F.B. (1960) Energy dynamics of a food chain of an old-field community. *Ecol. Monogr.*, **30**, 187–206.

GOLLEY F.B. (1967) Methods of measuring secondary productivity in terrestrial vertebrate populations. In Petrusewicz K. (ed.) *Secondary productivity of terrestrial ecosystems* p. 99–119. PWN, Warszawa–Kraków.

GOLLEY F.B. & Gentry J.B. (1964) Bioenergetics of the southern harvester ant, *Pogonomyrmex badius. Ecology*, **45**, 217–225.

GYLLENBERG G. (1969) The energy flow through a *Chorthippus parallelus* (Zett.) (Orthoptera) population on a meadow in Tvärminne, Finland. *Acta zool. fenn.*, **123**, 1–74.

HALCROW K. & BOYD C.M. (1967) The oxygen consumption and swimming activity of the Amphipod *Gammarus oceanicus* at different temperatures. *Comp. Biochem. Physiol.*, **23**, 233–242.

HALL R.R. *et al.* (1953) Evaluation of insect predator-prey relationship by precipitin test studies. *Mosquito News*, **13**, 199–204.

HANNA H.M. (1957) A study of the growth and feeding habits of the larvae of four species of caddis flies. *Proc. R. ent. Soc. Lond. Ser. A*, **32**, 139–146.

HARKER J.E. (1958) Diurnal rhythms in the animal kingdom. *Biol. Rev.*, **33**, 1–52.

HARTLEY P.H.T. (1948) The assessment of the food of birds. *Ibis*, **90**, 361–381.

HEALEY I.N. (1965) Studies on the production biology of soil Collembola, with special reference to a species of *Onychiurus*. Thesis Ph.D. University College of Swansea.

HEALEY I.N. (1967) The energy flow through a population of soil Collembola. In: Petrusewicz K. (ed.) *Secondary productivity of terrestrial ecosystems.* p. 695–708. PWN, Warszawa–Kraków.

HEMMINGSEN A.M. (1960) A energy metabolism as related to body size and respiratory surfaces, and its evolution. *Rep. Steno meml Hosp.*, 9 part **2**, 1–110.

HIRATSUKA E. (1920) Recherches on the nutrition of the silk worm. *Bull. imp. serie. Exp. Stn.*, Tokyo, **1**, 257–315

HOAR W.S. (1966) *General and comparative physiology.* Prentice-Hall, Inc., Englewood Cliffs. 815p.

HOLIŠOVA V., PELIKAN J. & ZEJDA J. (1962) Ecology and population dynamics in *Apodemus microps* Kret and Ros. (Mamm.: Muridae). *Acta Acad. Sci. Cechosl.*, **34**, 493–540.

HOUSE H.L. (1958) Nutritional requirements of insects associated with animal parasitism. *Expl. Parasit.*, **7**, 555–609.

HUBBELL S.P., SIKORA A. & PARIS O.H. (1965) Radiotracer, gravimetric and calorimetric studies of ingestion and assimilation rates of an Isopod. *Hlth. Phys*, **11**, 1485–1501.

ITO T. & FRAENKEL G. (1966) The effect of nitrogen starvation of *Tenebrio molitor* L. *J. ins. Physiol.*, **12**, 803–817.

IVLEV V.S. (1938) O prevrasceni energii pri roste bespozvonocnyh Energy transformation in growing invertebrates. *Bjul. MOIP, Otd. Biol.*, **47**, 267–277.

IVLEV V.S. (1939a) Energeticeskij balans karpov. Energy balance in the carp. *Zool. Zh., Mosk.*, **18**, 303–318.

IVLEV V.S. (1939b) Transformation of energy by aquatic animals. *Int. Rev. ges Hydrobiol. Hydrogr.*, **38**, 449–458.

IVLEV V.S. (1945) Biologiceskaja produktivnost' vodoemov. The biological productivity of waters. *Usp. Sovr. Biol.*, **19** (1), 98–120. Tranl *Ser. of J. Fish Res. Bd Can.*, **23**, 1727–1759, 1965

IVLEV V.S. (1954) Zavisimost' intensivnosti obmena u ryb ot vesa ih tela. Size dependence of metabolism intensity in fish. *Fiziol. Zh., Mosk.*, **40**, 717–721. Russian.

IVLEV V.S. (1955) Eksperimental' naja ekologija pitanija ryb. Experimental ecology of fish nutrition. *Piscepromiz–dat, Moskva.* 250p. Russian.

IVLEV V.S. (1966) Elementy fizjologiceskoj gidrobiologii. Elements of physiological hydrobiology. In: *Fizjologija morskih zivotnyh.* Izdat. 'Nauka', Moskva.

JOHANNES R.E. & SATOMI M. (1966). Composition and nutritive value of fecal pellets of a marine crustacean. *Limnol. Oceanogr.*, **11**, 191–197

JONGBLOED J. & WIERSMA C.A.G. (1935) Der Stoffwechsel der Honigbiene während des Fliegens. *Z. vergl. Physiol.*, **21**, 519–533.

KAMLER E. (1970) The main parameters regulating the level of energy expenditure in aquatic animals. *Pol. Arch. Hydrobiol.*, **17**, 201–216.

KARPEVIČH A.F. (1957) Potreblenie kisloroda morskimi rybami pri razlicnom ih fizjologiceskom sostojani. Oxygen uptake in marine fish indifferent physiological conditions. *Vopr. Ikhtiol.*, No. **10**, 131–138. Russian.

KAUSCH H. (1969) The influence of spontaneous activity on the metabolic rate of starved and fed young carp *Cyprinus carpio* L. *Verhmint. Ver. Limnol.*, **17**, 669–679.

KENNEDY C.H. (1950) The relation of american dragonfly-eating birds to their prey. *Ecol. Monogr.*, **20**, 103–142.

KHMELEVA N.N. (1967) Transformacija energii u *Artemia salina* (L.). Energy transformation in *Artemia salina* L. *Dokl. Akad. Nauk SSSR*, **175**, 934–937. Russian.

KHMELEVA N.N. & JURKEVICH G.N. (1968) Energeticeskij obmen *Artemia salina* (L.) i ego osobennosti v rjadu rakoobraznyh. Energy metabolism in *Artemia salina* (L.) and its peculiar features as compared to those in the rest Crustacea. *Dokl. Akad. Nauk SSSR*, **183**, 978–981. Russian.

KLEIBER M. (1947) Body size and metabolic rate. *Physiol. Rev.*, **27**, 511–541.

KLEIBER M. (1961) *The fire of life. An introduction to animal energetics.* J. Wiley, New York–London. 454p.

KLEIBER M. et al. (1951) Radiophosphorus (p^{32}) as tracer for measuring endogenous phosphorus in cow's faeces. *J. Nutr.*, **45**, 253–264

KLEINZELLER A., MALEK J. & VRBA R. (1954) Manometricke metody a jeich pouziti v biologii a biochemii. *Stat. Zdravotnicke Naklad.*, Praha. 391p.

KLEKOWSKI R.Z. (1961) Gas compression in the lung of desiccating snails *Coretus corneus* L. and *Limnea stagnalis* (L.). *Pol. Arch. Hydrobiol.*, **9**, 361–381.

KLEKOWSKI R.Z. (1970) Bioenergetic budgets and their application for estimation of production efficiency. *Pol. Arch. Hydrobiol.*, **17**, 55–80.

KLEKOWSKI R.Z. & IVANOVA M.B. in litt. Elements of energy budget of *Simocephalus vetulus*.

KLEKOWSKI R.Z. & STĘPIEŃ Z. in litt. Elements of energy budget of *Rhisoglyphus echinopus* (Acarina).

KLEKOWSKI R.Z. & Shushkina E.A. (1966) Ernährung, Atmung, Wachstum und Energie Umformung in *Macrocyclops albidus* (Jurine). *Verh. int. Ver. Limnol.*, **16**, 399–418.

KLEKOWSKI R.Z., PRUS T. & ZYROMSKA-RUDZKA H. (1967) Elements of energy budget of *Tribolium castaneum* (Hbst) in its developmental cycle. In: Petrusewicz K. (ed.) *Secondary productivity of terrestrial ecosystems.* p. 859–879. PWN, Warszawa–Kraków.

KLEKOWSKI R.Z., WASILEWSKA L. & PAPLINSKA E. (1972) Oxygen consumption by soil-inhabiting nematodes. *Nematologica*.

KLOTZ I.M. (1967) Energy changes in biochemical reactions. Academic Press, New York–London, 108p.

KNIGHT R.H. & SOUTHON H.A.W. (1963) A simple method for marking haematophagous insects during the act of feeding. *Bull. ent. Res.*, **54**, 379–382.

KOSTELECKA-MYRCHA A. & MYRCHA A. (1964) The rate of passage of foodstuffs through the alimentary tracts of certain Microtidae under laboratory conditions. *Acta Theriol.*, **9**, 37–53.

KOZLOVSKY D.G. (1968) A critical evaluation of the trophic level concept. I—Ecological efficiencies. *Ecology*, **49**, 48–59.

KROGH A. (1914) The quantitative relation between temperature and standard metabolism in animals. *Int. Z. phys.–chem. Biol.*, **1**, 491–508.

KROGH A. (1916) *The respiratory exchange of animals and man.* Longmans Green and Co., London–New York.

KROGH A. (1941) *The comparative physiology of respiratory mechanisms.* Univ. of Pennsylvania Press, Philadelphia. 172p.

KROGH A. & WEIS-FOGH T. (1951) The respiratory exchange of the desert locust (*Schistocerca gregaria*) before, during and after flight. *J. exp. Biol.*, **28**, 344–357.

LEHNINGER A.L. (1965) *Bioenergetics. The molecular basis of biological energy transformations*. W. A. Benjamin, New York–Amsterdam. 258p.

LEVANIDOV V.J. (1949) Znacenije allohtonnogo materiala kak piscevogo resursa v vodoeme na primere vodjanogo oslika *(Asellus aquaticus)*. The role of allochtonous material as food resurse in a water body, as examplified by water hog louse *(Asellus aquaticus)*. *Tr. Vses. Gidrobiol. Obsc.*, **1**, 100–117.

LEWIS C.T. & WALOFF N. (1964) The use of radioactive tracers in the study of dispersion of *Orthotylus virescens* (Douglas and Scott) (Miridae, Heteroptera). *Entomologia exp. appl.* **7**, 15–24.

LINDEMAN R.L. (1942) The trophic-dynamic aspect of ecology. *Ecology*, **23**, 399–418.

LIPMANN F. (1941) Metabolic generation and utilization of phosphate band energy. *Adv. Enzymol.*, **1**, 99–162.

LOCKER A. (1961) Das Problem der Abhängigkeit des Stoffwechsels von der Körpergrösse. *Naturwissenschaften*, **48**, 445–449.

LOUGHTON B.G., DERRY C. & WEST A.S. (1963) Spiders and the spruce budworm. *Mem. ent. Soc. Can.*, **31**, 249–268.

LUICK J.R. & LOFGREEN G.P. (1957) An improved method for the determination of metabolic fecal P. *J. anim. Sci.*, **16**, 201–206.

MACFADYEN A. (1948) The meaning of productivity in biological systems. *J. anim. Ecol.*, **17**, 75–80.

MACFADYEN A. (1961) A new system for continuous respirometry of small air-breathing invertebrates under near-natural conditions. *J. exp. Biol.*, **38**, 323–341.

MACFADYEN A. (1963) *Animal ecology: aims and methods*. Pitman and Sons, London. 344p.

MACFADYEN A. (1967) Methods of investigation of productivity of invertebrates in terrestrial ecosystems. In: Petrusewicz K. (ed.) *Secondary productivity of terrestrial ecosystems*. p. 383–412. PWN, Warszawa–Kraków.

MCGINNIS A.J. & KASTING R. (1964a) Digestion in insects, colorimetric analysis of chronic oxide used to study food utilization by phytophagous insects. *J. agric. Fd. Chem.*, **12**, 259–262.

MCGINNIS A.J. & KASTING R. (1964b) Chronic oxide indicator method for measuring food utilization in a plant-feeding insect. *Science*, **144**, 1464–1465.

MCLEESE D.W. (1956) Effects of temperature, salinity and oxygen on the survival of the American lobster. *J. Fish. Res. Bd Can.*, **13**, 247–272.

MCNEILL S. & LAWTON J.H. (1970) Annual production and respiration. *Nature, Lond.*, **225**, 472–474.

MANN K.H. (1964) The pattern of energy flow in the fish and invertebrate fauna of the River Thames. *Verh. int. Ver. Limnol.*, **15**. 485–495.

MANN K.H. (1965) Energy transformation by a population of fish in the River Thames. *J. Anim. Ecol.*, **34**, 253–275.

MANN K.H. (1969) The dynamic of aquatic ecosystems. *Adv. ecol. Res.*, **6**, 1–71.

MCNEILL S. (1969) Ph.D. Thesis. Univ. Lond.

MELLANBY K. (1939) The functions of insect blood. *Biol. Rev.*, **14**, 243–260.

MORRISON P.B. (1956) *Feeds and Feeding. A handbook for the students and stockmen*. 22nd ed. Clinton, Ithaca, N.Y. 1165p.

Moss R. (1967) Probable limiting nutrients in the main food of red grouse (*Lagopus lagopus scoticus*). In: Petrusewicz K. (ed.) *Secondary productivity of terrestrial ecosystems*. p. 369–379. PWN, Warszawa–Kraków.

Movčan V.A. (1953) O vnutridovyh otnosenijah u ryb. Intraspecific relations by fishes. *Agrobiologija*, 1953, (3), 81–90. Russian.

Mulkern G.B. & Anderson J.F. (1959) A technique for studying the food habits and preferences of grasshoppers. *J. econ. Ent.*, **52**, 342.

Munro W.R. (1957) The pike of Loch Choin. *Fresh wat. Salm. Fish Res.* **16**, 16 pp.

Nakamura M. (1965) Bio-economics of some larval populations of pleurostict Scarabaeidae on the flood plain of the River Tamagawa. *Jap. J. Ecol.*, **15**, 1–18.

Newell R.C. (1969) Effect of fluctuations in temperature on the metabolism of intertidial invertebrates. *Am. Zool.*, **9**, 293–307.

Newell R.C. (1970) *Biology of intertidal Animals.*, Logos Press Ltd.

Newell R.C. & Northcroft H.R. (1965) The relationship between cirral activity and oxygen uptake in *Balanus balanoides*. *J. mar. biol. Ass. U.K.*, **45**, 387–403.

Newell R.C. & Northcroft H.R. (1967) A re-interpretation of the effect of temperature on the metabolism of certain marine invertebrates. *J. Zool., Lond.*, **151**, 277–298

Odum E.P. (1959) *Fundamentals of ecology.* 2nd ed. Saunders Comp., Philadelphia. 384p.

Odum E.P. & Odum H.T. (1955) Trophic structure and productivity of a Windward coral reef community on Eniwetok Atoll. *Ecol. Monogr.*, **25**, 291–320.

Odum E.P. & Smalley A.E. (1959) Comparison of population energy flow of a herbivorous and a deposit-feeding invertebrates in a salt march ecosystem. *Proc. natn. Acad. Sci. USA*, **45**, 617–622

Odum H.T. (1957) Trophic structure and productivity of Silver Springs, Florida. *Ecol. Monogr.*, **27**, 55–112.

Park T. (1934) Observations on the general biology of the flour beetle *Tribolium confusum*. *Quart. Rev. Biol.*, **9**, 36–54

Park T. et al. (1965) Cannibalistic predation in populations of flour beetles. *Physiol. Zool.*, **38**, 289–321.

Patten B.C. (1959) An introduction to the cybernetics of the ecosystem: the trophic-dynamic aspect. *Ecology*, **40**, 221–231.

Petal J. (1967) Productivity and the consumption of food in the *Myrmica laevinodis* Nyl. population. In: Petrusewicz K. (ed.) *Secondary productivity of terrestrial ecosystems*. p. 841–857. PWN, Warszawa–Kraków.

Petrusewicz K. (1967) Concepts in studies on the secondary productivity of terrestrial ecosystems. In: Petrusewicz K. (ed.) *Secondary productivity of terrestrial ecosystems* p. 17–49. PWN, Warszawa–Kraków.

Petrusewicz K. & Macfadyen A. (1970) Eds. *Productivity of Terrestrial animals. Principles and Methods.*, IBP Handbook No. 13. Blackwell Sc. Pub.

Phillipson J. (1960a) A contribution to the feeding biology of *Mitopus moria* (F) (Phalangida). *J. Anim. Ecol.*, **29**, 35–43.

Phillipson J. (1960b) The food consumption of different instars of *Mitopus moria* (F) (Phalangida) under natural conditions. *J. Anim. Ecol.*, **29**, 299–307.

PHILLIPSON J. (1966) *Ecological energetics.* E. Arnold Publ., London, 57p.

PHILLIPSON J. (1967) Secondary productivity in invertebrates reproducing more than once in a lifetime. In: Petrusewicz K. (ed.) *Secondary productivity of terrestrial ecosystems.* p. 459–475. PNW, Warszawa–Kraków.

PHILLIPSON J. (1970) The 'best estimate' of respiratory metabolism: its applicability to field situations. *Pol. Arch. Hydrobiol.,* **17**, 31–41

PHILLIPSON J. & WATSON J. (1965) Respiratory metabolism of the terrestrial isopod *Oniscus asellus* L. *Oikos,* **16**, 78–87.

PIEKARZ R. (1961) Wpływ koprofagii na szybkość przechodzenia treści pokarmowej u królika domowego i dzikiego. The influence of coprophagy on the retendices time in the gut of domestic and wild rabbits.

POLIKARPOV G.G. (1965) *Radioekologija morskih organizmov. (Radioecology of aquatic organisms).* Moskwa, Atomizdat. Engl. transl. ed. by V. Schultz and A. W. Kelement, Amsterdam 1966, North Holland Publ. Co. New York 1966, Reinhold Book Division.

PRECHT H., CHRISTOPHERSON J. & HENSEL H. (1955) *Temperatur und Leben.* Springer, Berlin-Göttingen-Heidleberg.

PROSSER C.L. & BROWN F.A. (1961) *Comparative animal physiology.* 2nd ed. W. B. Saunders Co., Philadelphia–London. 688p.

PRUS T. (1968) Some regulatory mechanisms in populations of *Tribolium confusum* Duval and *Tribolium castaneum* Herbst. *Ekol. pol. Ser. A,* **16**, 335–374.

PRUS T. (1970) Calorific value of animals as an element of bioenergetical investigations, *Pol. Arch. Hydrobiol.,* **17** (30) 1/2: 183–199.

PRUS T. (1972) in print. Energy requirement, expenditure and transformation efficiency during development of *Asellus aquaticus* L. (Crustacea, Isopoda). *Pol. Arch. Hydrobiol.,* **19**, 97-112

PRUS T. (1971) The assimilation efficiency of *Asellus aquaticus* L. (Crustacea, Isopoda). *J. Freshwater Biol.,* **1**, 287–305.

REEVES W.C., BROOKMAN B. & HAMMON W.M. (1948) Studies on the flight range of certain *Culex* mosquitoes, using a fluorescent-dye marker, with notes on *Culiseta* and *Anopheles. Mosquito News,* **8**, 61–69.

REICHLE D.E. & CROSSLEY D.A. (1967) Investigation on heterotrophic productivity in forest insect communities. In: Petrusewicz K. (ed.) *Secondary productivity of terrestrial ecosystems.* p. 563–587. PWN. Warszawa–Kraków.

REYNOLDSON T.B. & YOUNG J.O. (1963) The food of four species of lake-dwelling triclads. *J. Anim. Ecol.,* **32**, 175–191.

RICH E.R. (1956) Egg cannibalism and fecundity in Tribolium. *Ecology,* **37**, 109–120.

RICHMAN S. (1958) The transformation of energy by *Daphnia pulex. Ecol. Monogr.,* **28**, 273–291.

RICKER W.E. (1968) (ed.) *Methods for the Assessment of Fish Production in Freshwaters.* IBP Handbook No. 3. Blackwell Scientific Publications, Oxford.

RIGLER F.H. (1971) in I.B.P. Handbook No. 17, 228-256, 264-269.

ROEDER K. (ed.) (1953) *Insect Physiology.* John Willey and Sons, New Tork, 1100p.

Roux C. & Roux A.L. (1967) Temperature et metabolisme respiratoire d'especes sympatiques de gammares diu group pulex (Crustaces, Amphipodes). *Annls. Limnol.*, **3**, 3–16.

Rubner M. (1883) Uber den Einfluss der Kórpergrósse auf Stoff-und Kraft-Wechsel. *Z. Biol.*, **19**, 535–562.

Sarrus & Remeau (1839) Memoire adresse a l'Academie Royale. *Bull. Acad. r. med. Belg.*, **3**, 1094–1100.

Schiemer F. & Duncan A. (1974) The oxygen consumption of a freshwater benthic nematodes *Tobrilus gracilis* (Bastian, 1865). *Oecologia* (Berlin) **15**, 121–126.

Schneiderman H.A. & Williams C.N. (1955) An experimental analysis of the discontinuous respiration of the cecropia silkworm. *Biol. Bull.*, **109**, 123–143.

Scholander P.F. *et al.* (1953) Climatic adaptation in Arctic and tropical poikilotherms. *Physiol. Zool.*, **26**, 67–92.

Schultz Fr.N. (1930) Zur Biologie des Mehlwurms (*Tenebrio molitor*). I—Mitteilung: Der Wasserhaushalt. *Biochem. Z.*, **227**, 340–353.

Shinoda O. (1931) On the starch digestion in the silkworm. *Annales zool. jap.*, **13**, 117–125.

Shushkina E.A., Anisimov S.I. & Klekowski R.Z. (1968) Calculation of production efficiency in planktonic copepods. *Pol. Arch. Hydrobiol.*, **15**, 251–261.

Slobodkin L.B. (1959) Energetics in *Daphnia pulex* population. *Ecology*, **40**, 232–243.

Slobodkin L.B. (1960) Ecological energy relationships at the population level. *Am. Nat.*, **94**, 213–236.

Slobodkin L.B. (1962) Energy in animal ecology. *Adv. ecol. Res.*, **1**, 69–101.

Slobodkin L.B. & Richman S. (1961) Calories/gm. in species of animals. *Nature*, **191**, 299.

Smalley A.E. (1960). Energy flow of a salt marsh grasshopper population. *Ecology*, **41**, 672–677.

Smith D.S. (1959) Utilization of food plants by the migratory grasshopper *Melanopus bilituratus* (Walker) (Orthoptera: Acaridae) with some observations on het nutritional value of the plants. *Ann. ent. Soc. Am.*, **52**, 674–680.

Snedecor G.W. (1956) *Statistical Methods*. Iowa State Univ. Press, Ames, Iowa, U.S.A.

Sonleitner F.J. (1961) Factors affecting egg, cannibalism and fecundity in populations of adult *Tribolium castaneum* Herbst. *Physiol. Zool.*, **34**, 233–255.

Soo-Hoo C.F. & Fraenkel G. (1966a) The selection of food plants in a polyphagous insect, *Prodenia eridania* (Cramer). *J. Insect Physiol.*, **12**, 693–709.

Soo-Hoo C.F. & Fraenkel G. (1966b) The consumption, digestion and utilization of food plants by a polyphagous insect, *Prodenia eridania* (Cramer). *J. Insect Physiol.*, **12**, 711–730.

Sorokin J.I. & Panov D.A. (1966) The use of C^{14} for the quantitative study of the nutrition of fish larvae. *Int. Rev. ges. Hydrobiol.*, **51**, 743–753.

Sorokin J.I. (1968) The use of C^{14} in the study of nutrition of aquatic animals. *Int. Ver. Mitteilungen* No. **16**, 1–41.

SORYGIN A.A. (1939) Pitanie, izbiratel 'naja sposobnost' v piscevyh vzaimootnosenijah Gobiidae Kaspijskogo morja. Feeding behaviour and food preference ability in nutritional relationships of some Gobiidae of the Caspian Sea. *Zool. Zh.*, Mosk., 18 No. **1** Russian.

SOUTH A. (1965) Biology and ecology of *Agriolimax reticulatus* (Mull.) and other slugs: Spatial distribution. *J. Anim. Ecol.*, **34**, 403–417.

SOUTHWOOD T.R.E. (1966) *Ecological methods with particular reference to the study of insect populations.* Methuen and Co., London. 391p.

SPECTOR W.S. (ed.). (1956) *Handbook of biological data.* W. B. Saunders, Philadelphia. 584p.

STACHURSKA T. (in litt.) Metabolism and mortality in *Polydesmus complanatus* heavily infected with Gregarines.

STEWARD D.R.M. (1967) Analysis of plant epidermis in faeces: A technique of studying the food preferences of grazing herbivores. *J. appl. Ecol.*, **4**, 83–111.

STRASKRABA M. (1968) Der Anteil der höheren Pflougen an der Produktion der Gewässer. *Mitt. Int. Verein. Limnol.* **14**, 955–957.

STROGANOV N.S. (1956) Fizjologiceskaja prisposobljaemost' ryb k temperature sredy. Physiological adaptation to environmental temperature in fish. Izd. *Akad. Nauk SSSR*, Moskva, 151p. Russian.

STROGANOV N.S. (1962) Ekologiceskaja fizjologija ryb. Ecological physiology of fish. *T. I. Izd. Moskovskogo Universiteta, Moskva.* 444p. Russian.

SURA-BURA B.I. & GREGEAN V.L. (1956) O primenenii ljuminescentnogo analiza ryb k izucenij migracii nasekomyh. The use of fluorescent analysis in studies of insect migration. *Ent. Obozr.*, **35**, 760–763. Russian.

SUSHCHENIA L.M. (1962) Kolicestvennye dannye o pitanii i balans energii *Artemia salina* (L.). Quantitative data on the nutrition and energy balance in *Artemia salina* (L.) *Dokl. Akad. Nauk SSSR*, **143**, 1205–1207. Russian.

TERROINE E.F. & WURMSER R. (1922) L'energie de croissance. I–Le *développement* de *l'Aspergillus niger*. *Bull. Soc. Chim. Biol.*, **4**, 519–567.

UMBREIT W.W., BURRIS R.H. & STAUFFER J.F. (1964) *Manometric techniques. A mannual describing methods applicable to the study of tissue metabolism.* 4th ed. Burges Publ. Co., Minneapolis. 305p.

VAN DYNE G.M. & MEYER J.H. (1964) A method for measurements of forage intake of grazing livestock using microinjection techniques *J. Range Mgmt.*, **17**, 204–208.

VARLEY G.C. (1967) Estimation of secondary production in species with an annual life-cycle. In: Petrusewicz K. (ed.) *Secondary productivity of terrestrial ecosystems.* p. 447–457. PWN, Warszawa–Kraków.

VERNBERG F.J. & VERNBERG W.B. (1970) *The Animal and the Environment.* Holt Rinehart & Winston Inc. New York. 398 pp.

WALLWORK J.A. (1958) Notes on the feeding behaviour of same forest soil Acarina. *Oikos*, **9**, 260–271.

WEBB H.M. & BROWN F.A. (1959) Timing long-cycle physiological rhythms. *Physiol. Rev.*, **39**, 127–161.

WEIS-FOGH T. (1952) Fat combustion and metabolic rate of flying locusts (*Schistocerca gregaria* Forskal). *Phil. Trans. R. Soc. Ser. B*, **237**, 1–36.

WEST A.S. (1950) The precipitin test as an entomological tool. *Can. Ent.*, **82**, 241–244.

WIEGERT R.G. (1964) Population energetics of meadow spittlebugs (*Philaenus spurmarius* L.) as affected by migration and habitat. *Ecol. Monogr.*, **34**, 225–241.

WIEGERT R.G. (1965) Energy dynamics of the grasshopper populations in old field and alfalfa field ecosystems. *Oikos*, **16**, 161–176.

WIEGERT R.G. (1968) Thermodynamic considerations in animal nutrition. *Am. Zool.*, **8**, 71–81.

WIGGLESWORTH V.B. (1943) The rate of haemoglobin in *Rhodnius prolixus* (Hemiptera) and other blood-sucking Arthropods. *Proc. roy. Soc. London, Ser. B*, **131**, 313–339.

WILLIAMS O. (1955) The food of mice and shrews in a Colorado montane forest. *Univ. Colorado Stud. Biol.*, **3**, 109–114.

WILLIAMS O. (1962) A technique for studying Microtine food habits. *J. Mammal.*, **43**, 365–368.

WINBERG G.G. (1950) Intensivnost' obmena i razmery rakoobraznyh. The intensity of metabolism and body size of Crustacea. *Z. Obshch. Biol.*, **11**, 367–380. Russian.

WINBERG G.G. (1956a) Intensivnost' obmena i piscevye potrebnosti ryb. Rate of metabolism and food requirements of fish. Russian. *Izd. Belgosuniversiteta, Minsk.* (Engl. transl.: *Fish Res. Bd Can. Transl. Ser.*, No. 194).

WINBERG G.G. (1956b) O zavisimosti intensivnosti obmena clenistonogih ot veliciny tela. On dependence of metabolic rate on body dimension in arthropods. *Uc. Zap. Beloruss. gos. Univ., Ser. Biol.*, No. **26**, 243–254. Russian.

WINBERG G.G. (1962) Energeticeskij princip izucenija troficeskih svjazej i produktivnosti ekologiceskih sistem. Energetic principle in investigation of trophic relations and productivity of ecological systems. *Zool. Zh., Mosk.*, **41**, 1618–1630. Engl. summ.

WINBERG G.G. (1964) Puti kolicestvennogo izucenija potreblenija i usvoenija pisci vodnymi zivotnymi. The pathways of quantitative study of food consumption and assimilation by aquatic animals. *Zh. Obshch Biol.*, **25**, 254–266. Engl. summ.

WINBERG G.G. (1965) Bioticeskij balans vescestva i energii i biologiceskaja produktivnost' vodoemov. Biotic balance of matter and energy and the biological productivity of water basins. *Gidrobiol. Zh.*, **1**, 25–32. Engl. summ.

WINBERG G.G. (1966) Skorost' rosta i intensivnost' obmena u zivotnyh. The rate of growth and intensity metabolism in animals. *Usp. sovrem. Biol.*, **61**, 274–293. Russian.

WINBERG G.G. (1967) Osnovnye napravlenija v izucenii bioticeskogo balansa ozer. Basic concepts in the study of biotic balance of lakes. In: Krugovorot vescestva i energii v ozernyh vodoemah p. 132–147. *Izd. "Nauka", Moskva.* Russian.

WINBERG G.G. (1968) Zavisimost' skorosti razvitija ot temperatury. The dependence of the rate of development on temperatures. In: Winberg G.G. (ed.) Metody opredelenija produkcii vodnyh zivotnyh. *Izd 'Vysejsaja Skola', Minsk.* Russian.

WINBERG G.G. & BELIAZKAYA I.S. (1959) Sootnosenie intensivnosti obmena i vesa tela u presnovodnyh brjuhonogih molljuskov. Relation between metabolic rate and body weight in freshwater gastropods. *Zool. Zh., Mosk.*, **38**, 1146–1151. Engl. summ.

WLODAWER P. (1954) O trawieniu i metaboliźmie wosku u mola woskowego (Galleria mellonella) Digestion and metabolism of wax by the waxmoth. *Pr. Lódź. Tow. Nauk.*, *Wydz. III*, No. **29**, 30pp. Engl. summ.

WORYTKIEWICZ K. in print. *Epiderma atlas.*

YONGE C.M. (1928) Feeding mechanisms in the invertebrates. *Biol. Rev.*, **3**, 21–76.

YONGE C.M. (1937) Evolution and adaptation in the digestive system of the Metazoa. *Biol. Rev.*, **12**, 87–115.

YOUNG J.O., MORRIS I.G. & REYNOLDSON T.B. (1964) A serological study of *Asellus* in the diet of lake-dwelling triclads. *Arch. Hydrobiol.*, **60**, 366–373.

YURKIEWICZ W.J. & Smyth R. (1966) Effects of temperature on oxygen consumption and fuel utilization by the sheep blowfly. *J. Insect Physiol.*, **12**, 403–408.

ZAIDENOV A.M. (1960) K izuceniju peremescenij komnatnyh muh (Diptera, Muscidae) pri pomosci ljuminescentnogo metoda markirovki v g. Cite. Study of house fiy (Diptera, Muscidae) migrations in Chita by means of luminescent tagging. *Ent. Obozr.*, **39**, 574–584. Russian. Transl.: *Ent. Rev.*, *Wash.*, **39**, 406–414.

ZEUTHEN E. (1947) Body size and metabolic rate in the animal kingdom with special regard to the marine microfauna. *C. r. Trav. Lab. Carlsberg, Ser. Chim.*, **26**, 17–161.

ZEUTHEN E. (1953) Oxygen uptake as related to body size in organisms. *Quart. Rev. Biol.*, **28**, 1–12.

ZEUTHEN E. (1970) Rate of living as related to body size in organisms. *Pol. Arch. Hydrobiol.*, **17**, 21–30.

ZWICKY K. & WIGGLESWORTH V.B. (1956) Course of oxygen consumption during the moulting cycle of *Rhodnius prolixus* Stal. (Hemiptera). *Proc. R. Ent. Soc., Lond.*, **31**, 153–160.

3
Energy Flow Through a Vertebrate Population

W. GRODZINSKY

3.1 Introduction

Energy flow, or assimilation, can be expressed in two well known bioenergetic equations (Petrusewicz, 1967). On the one hand assimilation is the sum of net production, i.e. energy incorporated in body tissues, and respiration, i.e. energy used for the cost of maintenance ($A = P + R$). On the other hand assimilation is the difference between the energy of food intake and the energy lost with rejects i.e. faeces and urine ($A = C - FU$). These general equations have been further developed in detail for computation of energy flow through mammal populations (Golley, 1962; Grodzinski et al., 1966). Using the terminology adopted in previous IBP Handbooks (Petrusewicz and Macfadyen, 1970), equations which are applicable to a vertebrate population per unit area and for an interval of time (T) can be written as follows:

$$A = K_b (\overline{N} \times \overline{W} \times \theta_b) + K_m (\overline{N} \times \overline{W} \times M \times T) \qquad (1)$$

$$A = K_c (N \times C \times T) - K_e (N \times FU \times T) \qquad (2)$$

where:

$\overline{N}$ —average numbers of individuals per unit area during time T;

$\overline{W}$ —mean body weight (fresh) of an individual during time T, expressed as the average individual weight in the population or for the trappable part of the population;

θ_b —turnover of biomass during time T, that is that fraction of the population biomass produced during the time interval T (where $\theta_b = \dfrac{P}{B}$ and $\overline{B} = \overline{N} \times \overline{W}$)

K_b —caloric value per unit fresh weight of animal body;

M —average daily metabolic rate, measured as daily oxygen consumption per unit body weight;

65

F

K_m —caloric equivalent of oxygen;

C —daily consumption or food intake for an average individual;

K_c —caloric value of food;

FU —daily egestion of faeces plus excretion of urine by an average individual;

K_e —caloric value of rejecta (faeces plus urine);

T —period of time (in days) for which energy flow balance is computed.

Equation 2 is given in terms of individuals; however, it can be easily expressed in terms of population biomass by using weight specific consumption and rejecta applied to biomass (as in equation 1) instead of to numbers of individuals.

While preparing data for such a balance three requirements have to be taken into consideration: (1) All data should be converted into energetic units viz. calories or joules; (2) All data should refer to a definite period of time (usually a year); (3) The balance should represent a population or another ecological unit which lives in a definite area, e.g. one hectare.

In order to compute the full balance of energy flow through a vertebrate population all the above mentioned parameters have to be determined. Though a part of them must be derived from field studies, most will have to be estimated in a bioenergetic laboratory. The field population parameters include numbers $(\overline{N})$, biomass $(\overline{B})$ and turnover rate (θ_b). Population numbers is a basic value used for calculation of both production and respiration. The remaining bioenergetic parameters are usually determined under the laboratory conditions using a minimum of three techniques: respirometry (metabolic rate—M), calorimetry (caloric values of mammalian production, food and excrements— K_b, K_c, K_e) and feeding balance (food consumption, faeces and urine production—C, FU).

Classical studies on energy flow through vertebrate populations were completed by Golley (1960) and Odum, Connell and Davenport (1962). During the International Biological Programme the energetics of many vertebrate populations was investigated. Most studies dealt with small homoiotherms, mammals and birds (Pearson, 1964; Kale, 1965; Grodzinski *et al.*, 1966a, 1970; Chew and Chew, 1970; Pinowski, 1968; Grodzinski, 1971; Hansson, 1971; Holmes and Sturges, 1973). Good balance of energy flow are also available for the populations of large homoiotherms (mammals—Petrides and Swank, 1966; Buechner and Golley, 1967) as well as for poikilothermic vertebrates (e.g. Mann, 1965). A number of data from these studies will be discussed and summarized in the present chapter. The general methods and principles of measuring secondary productivity in terrestrial vertebrate populations are also described in other publications (Golley, 1967; Golley and Buechener, 1969; Petrusewicz and Macfadyen, 1970).

A simple computation of energy flow balance in a vertebrate population is given in this chapter, using rodent populations in a beech forest as an example. The calculations will be simplified as much as possible and perhaps even excessively. The emphasis will be on the bioenergetic parameters which constitute the main subject of the handbook.

3.2 Balance of energy flow

Productivity of small rodents inhabiting the mixed beech forest (*Fagetum carpaticum*) in Ojców National Park (Southern Poland)* was the subject of teamwork during four years. We will use data from several publications of our Department of Animal Ecology and summarized in a joint paper (Grodzinski *et al.*, 1970). The assimilation (energy flow) will be estimated using the first (1) bioenergetic equation, as a sum of production and respiration, whereas consumption will be computed using the second (2) equation.

The beech forest in Ojców has a very simple community of small rodents in which bank vole (*Clethrionomys glareolus* Schr.) and yellow-necked field mouse (*Apodemus flavicollis* Melch.) decidedly predominate in terms of numbers. These two species constitute almost 97% of all rodents; the rest being *Apodemus agrarius* Pall., *Pitymys subterraneus* Sel. Long, and *Muscardinus avellanarius* L. (Grodzinski, Pucek and Ryszkowski, 1966; Bobek, 1969). For the sake of clarity all the computations will be carried out only on the two predominating species of voles and mice.

Populations of small rodents were sampled on a "Standard-Minimum" plot (5·76 ha) by the method of prebaiting and intensive removal (Grodzinski, Pucek and Ryszkowski, 1966) during three annual cycles (1965–1968). The rodent numbers were later calculated by the method of maximum likelihood (Zippin, 1956). This number can be expressed in term of trappable animals (N_{tr}) neglecting young staying in nests, or else in term of all animals born in the population (N_b). In this period voles and mice completed nearly

---

*Participants of the course visited the field study area in Ojców National Park during a special trip.

one full population cycle from the high density in 1965 to a low one in 1967/68 (Fig. 3.1). Average density of trappable rodents during the three years cycle

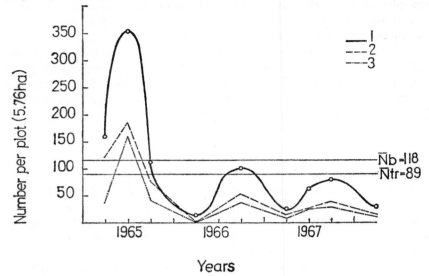

Years

Fig. 3.1. Numbers of small rodents on a "Standard-Minimum" trapping plot (5·76 ha) in beech forest during a three-year cycle (curves represent trappable animals). Horizontal lines show mean density of trappable rodents ($\overline{N}_{tr}$) and of all born rodents ($\overline{N}_b$). 1—all rodents, 2—bank voles (*Clethrionomys glareolus*), 3—field mice (*Apodemus flavicollis*). (After Grodziński *et al.*, 1970).

was approximately 16 animals per hectare, 9 voles and 7 mice. The mean number of all rodents born was consequently higher and on average amounted to some 21/ha. In the first year (1965) the mean numbers of trappable voles reached 15·6/ha and those of mice, 10·9/ha. In subsequent years (1966, 1967) the number of voles decreased to 5·8 and 4·5/ha while the density of trappable mice was reduced to a level of 3·9 and 4·4/ha (Bobek, 1969; Grodzinski *et al.*, 1970).

3.3 Estimation of production

Net production (*P*) of a population consists of both production due to reproduction (*P$_r$*) and that due to individual growth (*P$_g$*) (Petrusewicz and

Macfadyen, 1970). Theoretically, the production can be calculated by multiplying the average standing crop and turnover rate ($P = \overline{N}B \times O_b$). However, in practice the determination of biomass turnover rate is next to impossible, and also there is no satisfactory relationship between the turnover rate of biomass and turnover of individuals (Petrusewicz, 1966) Therefore several other approaches are recommended (Petrusewicz and Macfadyen, 1970—pp. 69–96). For the estimation of production in small mammal population three methods were successfully applied. These are methods using: (1) the growth-survival curve, (2) the turnover (using the individual turnover rate) and (3) the weight-gain (Hansson, 1972). In one example we will use the first method, which is the most exact, but requires several additional population estimates, like age structure, natality (reproduction) and mortality (life tables), as well as the individual growth curves. This method was developed by Allen (1950). A simplified example of computation by the second method will also be given, even though this approach is not entirely correct.

The age of all rodents captured in beech forest was estimated with the accuracy of approximately one month, in mice on the basis of wear of molars, and in voles by measuring the length of roots of the first molar. The number of newborn rodents was calculated by multiplying the number of pregnant or lactating females by the mean litter size and correcting for elimination. This allowed constructing the life tables for populations of both voles and mice in consecutive years (Bobek, 1969) From these tables the survivorship curves for both populations were drawn (Fig. 3.2). The individual growth curves for these animals were also determined under the natural conditions (Fig. 3.3). Starting from the sum of rodents born in a given year, the growth-survivorship curves of their cohorts were plotted and hence the net production was easily computed (Fig. 3.4). On this plot the area between the curve and the coordinate axes represents the net production in term of weight units (grams). The net production in the population of voles and mice computed in this way amounted on average to 616 and 495 g of animal biomass ("mouse meat") per 1ha during a year, or a total of 1,111 g/ha-year. These values represent the mean for whole population cycle. The minimum vole and mouse production in the cycle was only 469 g/ha-year, while the maximum production was as high as 2,083 g/ha-year.

The caloric value of the body of voles and mice from this area was studied in detail by Górecki (1965, 1967). Górecki found seasonal changes in energy values of dry weight and of fresh weight (biomass) of these rodents (Fig. 3.5, see also Chapters 9A, 9B). The seasonal fluctuations of the caloric value of

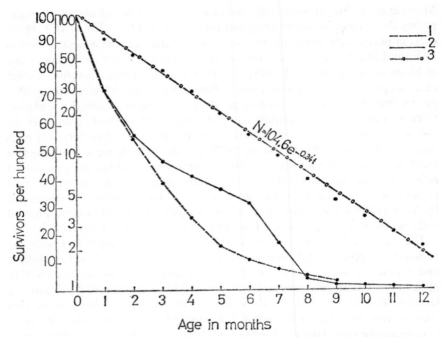

Fig. 3.2. Survivorship curves of field mice (*Apodemus flavicollis*) from beech forest. Curves for the high (1) and low (2) population densities are plotted on a linear scale while the regression line for the whole material (3) on a semi-log scale. (After Bobek, 1969).

the body of voles and mice are related mainly to the annual fat cycle (Fig. 3.6), which was studied by Sawicka-Kapusta (1968—see also Chapter 9C). Mean annual caloric equivalents of the biomass of *Clethrionomys glareolus* and *Apodemus flavicollis* are nearly identical, 1·454 and 1·450 kcal/g fresh weight, respectively (cf. Chapter 9A—Table 1). Using these values the net production could be expressed in terms of energy units. The production of both rodent populations was, on the average 1,613 kcal/ha-year, of which voles accounted for 896 kcal, and mice for 717 kcal.

Production due to reproduction constitutes as much as 56·1% of total production in voles and 48·2% in mice. Hence at least half of the production in these populations arises in the period of pregnancy and nursing, whereas the rest is accounted for by production due to individual growth.

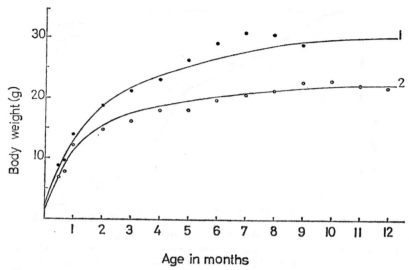

Age in months

Fig. 3.3. Individual growth curves of bank voles (2) and field mice (1) determined under field conditions in the beech forest. (After Bobek, 1969).

Whenever the necessary population parameters are not available, it is possible to estimate the production from the turnover rate of individuals (Grodzinski *et al.*, 1966a; Hansson, 1972). In this case the mean numbers of trappable animals and their biomass are multiplied by turnover of individuals rather than by turnover of biomass ($P = \overline{N}_{tr} \times \overline{W} \times \theta_N$). This method is theoretically incorrect but gives results similar to those obtained with the growth-survival curve method discussed above.

For the populations of voles and mice in beech forest the production can be easily computed in this way. The following four parameters are necessary: average numbers ($\overline{N}$), mean body weight ($\overline{W}$), turnover rate of individuals (θ_N) and caloric value of rodent biomass (K_b). The first and the last value were already discussed. Mean body weight in the populations of voles and mice was somewhat different in the years of normal density (1966–68) and in an outbreaking year (1965). The mean weight of all trappable voles during the whole period was 17·7 g and of mice 23·4 g. Average life span was also changing with the population density and was significantly shorter during a year of high density. In voles it decreased from 3·2 to 2·2 months, and in mice from 3·8 to 2·9 months. Hence the turnover rate of numbers (θ_N) in

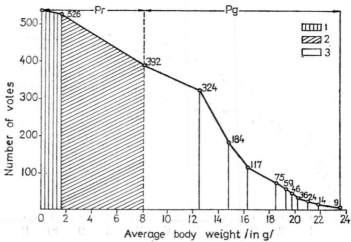

Fig. 3.4. Estimation of net production in the population of bank voles (*Clethrionomys glareolus*) in the beech forest, based on their growth-survivorship curves. The values for all cohorts were summed. Production due to reproduction (P$_r$)—hatched area: 1—during pregnancy, 2—during nursing by the mother; production due to growth (P$_g$)—open area: 3—during independent life. The area of the first trapezium represents the production of new born, the areas of other trapeziums separated by continuous lines—production in consecutive months of life. Numbers refer to individuals surviving until the beginning of next month. (After Bobek, 1969).

the two populations was on the average 3·9 and 3·1 per year, respectively (Bobek, 1969; Grodzinski *et al.*, 1970).

Below are the actual calculations.

Net production of vole and mouse population in kcal/ha-year:

	$\overline{N}$ No/ha		$\overline{W}$ g		θ_N x/year		K_b kcal/g		P kcal/ha-year
Bank vole (*C. glareolus*)	9	×	18	×	3·9	×	1·454	=	918·6
Field mouse (*A. flavicollis*)	7	×	23	×	3·1	×	1·450	=	723·7

The estimations of net production using these two methods yielded comparable results. Production calculated by turnover method is in this case slightly overestimated, even though it does not include the production

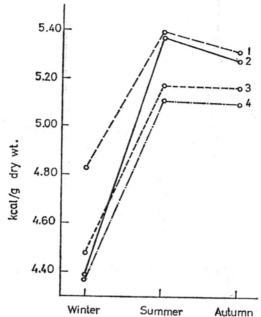

Fig. 3.5. Seasonal changes in energy values per gram dry weight of body for four species of small rodents. Field mice: 1—*Apodemus agrarius*, 2—*Apodemus flavicollis*. Voles: 3—*Clethrionomys glareolus*, 4—*Microtus arvalis*. (Redrawn from Górecki, 1965).

of young animals which died in nests before they became trappable. Turnover of individuals was however in this case much higher than the turnover of biomass (Bobek, 1969) and therefore the production was overestimated. Nonetheless this relatively small value of net production should be kept in mind when it is later compared with the much bigger value of respiration. A very rough method of estimating production from respiration will be discussed later.

3.4 Estimation of respiration

A principal parameter in the computation of respiration (cost of maintenance) is metabolic rate (M) or, more precisely, daily energy budget (DEB) based

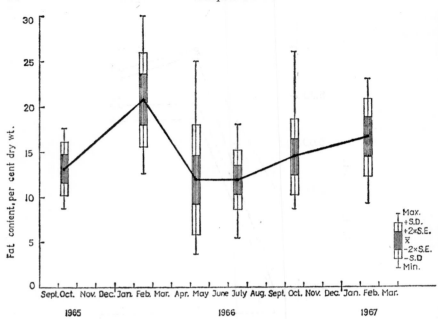

Fig. 3.6. Annual cycle of body fat in field mice (*Apodemus flavicollis*) from the beech forest. The curve (solid line) joins mean values for all seasons. The range of variation is represented by vertical line (Max.—Min.), standard deviation (S.D.) by the open rectangle and two standard errors (S.E.) from the mean by the hatched rectangle. (After Sawicka-Kapusta, 1968).

on the metabolic rate converted into energy units. Population respiration is a simple function of population biomass ($\overline{NW}$) multiplied by the daily energy budget expressed in terms of kcal/g-day.

Three measures of metabolic rate were used for ecological purposes (Grodzinski and Górecki, 1967; Gessaman, 1973). They differ in the conditions of measurements, namely (1) ambient temperature, (2) level of animal activity, (3) absorptive state of the animal, and (4) the duration of measurement. Basal metabolic rate (BMR) or standard metabolic rate (SMR) represent the minimal value of heat production within the region of thermoneutrality when the animal is quiet and is in a post-absorptive state. Resting metabolic rate (RMR) is usually measured at temperatures below the zone of thermoneutrality when the animal is at rest but not post-absorptive. Resting metabolic rate (RMR) exceeds the BMR by the value of specific dynamic action (SDA) and usually by some cost of thermoregulation. Both these metabolic rates are measured over a short period (1 to 2 hours). Average daily metabolic rate (ADMR) is the mean of measurements made over 24 hours at a temperature similar to that of the animal's natural habitat. This measure contains the basal metabolism, metabolic equivalents of thermoregulation and activity, as well as energy of SDA. Daily metabolic rate is not a very accurate determination, but it probably is the most natural and ecological measure of metabolic rate possible under the laboratory conditions. The technique of measuring ADMR is described in Chapter 10B, whereas measuring of RMR is discussed in Chapter 10C (see also Chapter 10A).

The values of ADMR are usually lower than resting metabolism since during the short run of RMR a small wild animal is always stressed and somewhat active (Grodzinski, 1969). During the 24 hr run small mammals exhibit a daily rhythm of both metabolic rate and activity (Fig. 3.7). Conse-

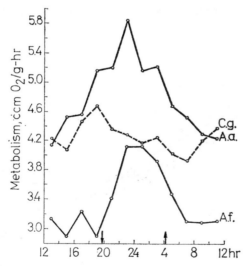

Fig. 3.7. Daily rhythm of metabolic rate (oxygen consumption) in field mice (*A.a.*—*Apodemus agrarius*, A.f.—*Apodemus flavicollis*) and bank vole (C.g.—*Clethrionomys glareolus*) measured at 20°C. Sunset and sunrise are marked by arrows. (Compiled from Górecki, 1968, 1969, and Gebczyński, 1966).

quently, ADMR is an average value which represents periods of natural quietness and rest and of voluntary activity in quite a big chamber.

Chapter 3

ADMR turned out to be consistently related to body size, but the exponent of ADMR regression on weight is close to −0·50 and usually differs significantly from the well known exponent −0·25 (i.e. 0·75−1·00) for basal metabolic rates (Kleiber, 1961). This was determined in the analysis of relationship ADMR/body size in several species of voles, mice and squirrels (Hansson and Grodzinski, 1970; Grodzinski, 1971 a, b; Drożdż *et al.*, 1971; Górecki, 1971).

All these logarithmic functions of allometric forms have a general formula:

$$ADMR = a \, W^{\,b-1} \qquad (3)$$

Actual range of body weight within a rodent species is however limited and therefore interspecific function can have a greater and more general significance. Two general functions of ADMR/body weight were recently computed, separately for small rodents and insectivores (French, Grant and Grodzinski, msc). These computations include 52 available data on daily metabolic rate for 36 rodent species and 8 shrews (Fig. 3.8). The regression

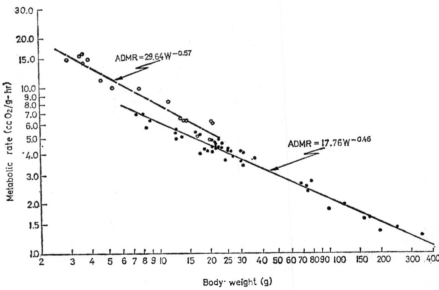

Fig. 3.8. Relationship between average daily metabolic rate (ADMR in ccm O_2/ g-hr) and body weight (W in g) for insectivores (broken line) and rodents (solid line). (From French, Grant and Grodziński, msc.).

line for rodents ranges from 7 g pocket mouse (*Perognathus*) to 370 g hamster (*Cricetus*) and insectivore regression, from 3 g lesser shrew (*Sorex*) to 21 g short-tailed shrew (*Blarina*). Slopes of these two regressions are close to -0.50 (Fig. 3.8) and significantly different from exponent -0.25 (0.75 per individual). Relationship of ADMR and body size can have a different slope than function of BMR, because daily metabolic rate includes also heat production for thermoregulation (thermoconductance), the energy of activity, as well as some behavioural thermoregulation (e.g. nest insulation, group effect etc.).

Several models of daily energy budgets (DEB) have already been constructed for small homoiothermic animals. They are based on BMR (McNab, 1963), RMR (Chew and Chew, 1970), or ADMR (Grodzinski and Górecki, 1967). Those models were critically compared and discussed in an excellent review by Gessaman (1973). Daily energy budget based on ADMR is a strictly empirical model. It contains only two calculated corrections: for thermoregulation when animal is active out-of-the-nest, and for the energetic cost of female reproduction. This budget was originally constructed only for average body weight in given population of animals (Grodzinski, 1966; Grodzinski and Górecki, 1967; Górecki, 1968). It was later extended and expressed as a function of animal body size (Hansson and Grodzinski, 1970; Grodzinski, 1971 a, b; Górecki, 1971; Drożdż et al., 1971).

The method of computation of such a budget (DEB) will be demonstrated below using an example of bank vole in the beech forest at Ojców National Park (Górecki, 1968). Read the following description of these calculations given by Górecki, and also inspect Table 3.1.

Table 3.1. Daily energy budget (*DEB*) in bank vole during winter and summer day (after Górecki, 1968)

	Winter (kcal/g/day)	Summer (kcal/g/day)
ADMR (20°C) in the nest, including group effect (13%)	20 hr × 3.65 ccm O_2/g/hr = 0.351	19.5 hr × 4.29 ccm O_2/g/hr = 0.401
Metabolic rate during periods of out-of-the-nest activity (0°C or 15°C)	4 hr × 7.03 ccm O_2/g/hr = 0.135	4.5 hr × 5.57 ccm O_2/g/hr = 0.121
Cost of reproduction	—	6.5% *ADMR* = 0.034
Corrected *ADMR* × av. b.w. *DEB* in kcal/vole–day	0.486 × 21.0 g 10.2	0.556 × 19.0 g 10.6

ADMR of bank vole determined by oxygen consumption at the temperature of 20°C is 4·29 ccm/g/hour in summer and 3·65 ccm/g/hour in winter. This corresponds to 0·494 and 0·420 kcal/g/day, respectively. Seasonal differences in the intensity of heat production, body insulation, and also in environmental climate are quite remarkable. In winter and in summer these parameters differ most radically. This is why the construction of at least two models of daily bioenergetics (DEB) in voles: summer and winter budgets— is necessary (Table 3.1). During the summer, vole stays outside the nest for 4·5 hours, while in winter—for 4 hours. The level of metabolism in voles at nest temperature of 17–19°C is higher than ADMR measured at 20°C by 10–12%. However, behavioural thermoregulation in nest lowers the metabolic rate in voles on average by 13% and entirely eliminates this correction. Thus it is a fully justified simplification to accept ADMR values for the time the animal stays in the nest. Environmental temperature in beech forest in Ojców is approximately 0°C in winter and 15°C in summer. The increase in heat production due to the difference between the laboratory temperature (20°C) and the prevailing environmental temperature was 6·04% per °C in summer and 5·29% per °C in winter. Thus the oxygen consumption outside the nest is almost doubled (increased by 92%) in winter, while in summer it is increased by 30% only. During hours of outside activity the energy consumption on a summer day reaches 1·204 kcal/g and on a winter day—1·350 kcal/g.

Correction for breeding concerns exclusively the summer DEB. Sex ratio in bank vole population is assumed to be 1:1 and percent of reproducing females is usually 20–25%. When the correction for the period of pregnancy and lactation (on an average 58% increase in metabolism rate; Kaczmarski, 1966) will be distributed among all individuals in population, it will amount to about 6·5% (Table 3.1).

ADMR together with the corrections discussed above is distinctly higher in summer, than in winter (0·556 and 0·486 kcal/g/day). The mean body weight in adult voles in beech forest was about 21 g in winter and 19 g in summer. DEB calculated for such animals amounts in summer to 10·6, and in winter to 10·2 kcal/vole–day. The fact that the winter DEB is less expensive than the summer DEB is probably due to the acclimatization of metabolism to lower temperatures, to the better insulation of body, to the reduction of activity outside the nest in winter, and also to the thermal economics of a nest. The winter budget in voles is also not burdened with the cost of breeding.

The corrections for thermoregulation during activity outside the nest are relatively small but require some general comments. These corrections can be made in absolute values (cc $O_2 \rightarrow$ kcal) or in percent. The latter method is more convenient since both the activity and the reproduction can also be expressed as percent of the total (i.e. fraction of each 24 hour period and reproducing fraction of the population).

Thermoregulation in small rodents was thoroughly discussed by Hart (1971). Heat production for thermoregulation expressed by thermoconductance, is also a function of mammal body size (Herreid and Kessel, 1967). It is computed from temperature-metabolism curves below the thermoneutral zone (Fig. 3.9) and can be expressed both in cal/g/hr–°C and in percent (Morrison and Ryser, 1951; Grodzinski and Górecki, 1967; Herreid and Kessel, 1967).

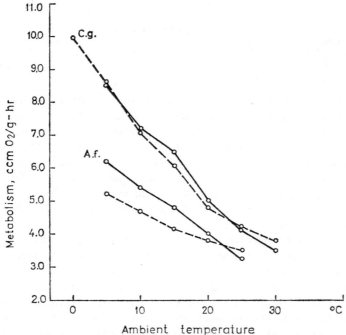

Fig. 3.9. Metabolic rates (oxygen consumption) of bank voles (C.g.— *Clethrionomys glareolus*) and field mice (A.f.—*Apodemus flavicollis*) in relation to ambient temperatures. Solid lines represent summer values, broken lines— winter values. (Redrawn from Górecki, 1968, and Gębczyński, 1966).

Cost of maintenance of a reproducing female rodent increases somewhat during pregnancy and is dramatically elevated during lactation (Fig. 3.10).

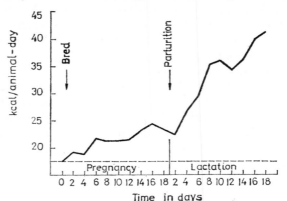

Fig. 3.10. Cost of maintenance of female bank voles (*Clethrionomys glareolus*) during pregnancy and lactation. The level of energy requirements in non-reproducing females of the same size is shown by broken line. (After Kaczmarski, 1966).

Bioenergetics of reproduction was studied only in three species of small rodents (*Clethrionomys glareolus, Microtus arvalis, Mus musculus*) and the average increase of assimilation during gestation and lactation was 58–82% (Kaczmarski, 1966; Trojan and Wojciechowska, 1967; Migula, 1969; Myrcha, Ryszkowski and Walkowa, 1969). By subtracting from assimilation the production of litters, placentae and fetal membranes by a female, it is possible to calculate that in these voles and mice the additional metabolic effort (respiration) of a reproducing female is approximately 61% (Grodzinski and Wunder, in press).

Daily energy budgets for bank vole and field mouse in beech forest can also be constructed as a function of their body weights. Daily metabolism of bank voles from Southern Poland was extensively studied by Górecki (1968), whereas a similar study on field mice from North-eastern Poland was carried out earlier by Gebczynski (1966). Employing all ADMR data from the studies of Gebczynski (1966) and Górecki (1968), two equations of DEB were developed (Grodzinski et al., 1970). They represent mean annual values for all the seasons and include corrections for thermoregulation outside of the nest (which is different in summer and winter) and for the considerable costs of female gestation and lactation.

$$C.\ glareolus\ \text{DEB} = 2{\cdot}510\ W^{-0.50} \qquad (4)$$

$$A.\ flavicollis\ \text{DEB} = 2{\cdot}364\ W^{-0.50} \qquad (5)$$

These logarithmic functions express DEB in kcal/g-day, that is per unit of body weight. Experimental values in this form are convenient for calculation of respiration of cohorts, or the whole population, if only the sum of biomass-days of all animals is known.

The mean body weight in the population of voles and mice was determined for trappable animals and for all animals born. The weight of trappable voles in years of normal population density (1966–67) was on the average 18·5 g, but in the year of high density (1965) it was only 16·1 g. Similarly, the weight of trappable mice in the same years was 25·5 and 22·3 g respectively. The mean weights of all the animals born were proportionally lower: for voles 14·0 and 12·4 g and for mice 20·8 and 18·5 g. Costs of maintenance of the population were computed in relation to the biomass of trappable voles and mice.

When mean body weights of trappable voles are substituted into the equations (4 and 5) the calculated costs of maintenance becomes 0·521 kcal/g-day for voles and 0·484 kcal/g-day for mice.

In order to obtain cost of maintenance of a population the DEB value expressed in kcal/g-day is simply multiplied by the average biomass of the whole population $(\overline{NW})$, and in addition by duration of time (T = 365 days), or else by directly multiplying DEB and the numbers of biomass-days. These simple calculations are given below.

Cost of maintenance (respiration) of vole and mouse populations in kcal/ha-year:

	$\overline{NB}$ g		$M(DEB)$ kcal/g-day		T days		R kcal/ha-year
Bank vole (*C. glareolus*)	198	×	0·521	×	365	=	37,653·4
Field mouse (*A. flavicollis*)	175	×	0·484	×	365	=	30,915·5

Hence, the respiration of both populations reaches the average of 68,569 kcal/ha-year; varying in particular years from approximately 30,000—102,000 kcal/ha-year.

If detailed studies of daily metabolism of a particular species are not available, its DEB can be constructed on the basis of the intraspecific equations of ADMR/body size (Fig. 3.8). The formulae given in Fig. 3.8 were rounded off by forcing them to the exponent -0.50.

$$\text{Rodents} \qquad \text{ADMR}_{cc} \; O_2/g\text{-}hr = 19.94 \; W^{-0.50} \qquad (6)$$

$$\text{Insectivores ADMR}_{cc} \; O_2/g\text{-}hr = 26.80 \; W^{-0.50} \qquad (7)$$

Both equations should be re-written in terms of kcal and computed per whole day (by multiplying them by the caloric equivalent of 1 cc O_2 and by 24 hrs $= 0.0048 \times 24$):

$$\text{Rodents} \qquad \text{ADMR}_{\text{kcal/g-day}} = 2.297 \; W^{-0.50} \qquad (8)$$

$$\text{Insectivores ADMR}_{\text{kcal/g-day}} = 3.087 \; W^{-0.50} \qquad (9)$$

The corrections for thermoregulation and reproduction which were discussed earlier can be easily introduced to both equations (8) (9) by increasing only the intercept value 'a' (see equation (3)).

3.5 Assimilation and consumption

Assimilation (energy flow) is a simple sum of net production and respiration (1). These simple summations for vole and mouse populations are given below.

Energy flow (A) through vole and mouse populations in kcal/ha-year:

	P		R		A
Bank vole (*C. glareolus*)	918·6	+	37,653·4	=	38,572
Field mouse (*A. flavicollis*)	723·7	+	30,915·5	=	31,639

The total energy flow through voles and mice reaches 70,211 kcal/ha-year. In years with different population densities it ranged from about 31,000—105,000 kcal/ha-year. Even a very perfunctory examination of these figures shows that the share of production in the whole assimilation is very small.

Productivity of both predominating populations of voles and mice may also be completed by including the remaining species of less numerous rodents. In this case the average energy flow is 72,000 and consumption 86,000 kcal/ha-year. The total balance of energy flow through the populations of rodents in the beech forest were changing drastically in subsequent years following the fluctuations in numbers of voles and mice (Fig. 3.11).

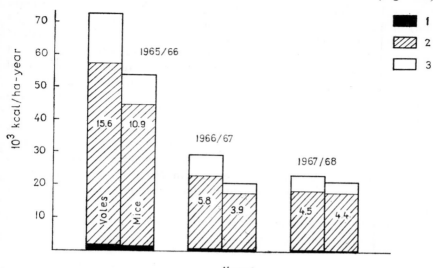

Fig. 3.11. Productivity of voles (*Clethrionomys glareolus*) and mice (*Apodemus flavicollis*) populations in the beech forest during three consecutive years. The mean numbers of voles and mice per 1 hectare in a given year are shown in the bars. 1—net production, 2— respiration (cost of maintenance), 3—rejecta (faeces and urine). 1+2 correspond to the energy flow (assimilation), and 1+2+3 represent the whole consumption. (After Grodziński *et al.*, 1970.)

Energetics of populations of small mammals were studied in different ecosystems including grasslands (Golley, 1960; Odum, Connell and Davenport, 1962; Grodzinski *et al.*, 1966), cultivated fields (Trojan, 1969), coniferous forests (Grodzinski, 1971; Hansson, 1971), deciduous forests (Grodzinski *et al.*, 1970; Bobek, 1971) and desert shrub communities (Chew and Chew, 1970). The comparison of assimilation and consumption of small mammals in these ecosystems (Table 3.2) indicates that the example of a beech forest discussed above is representative of forest ecosystems. Much higher energy flow was recorded only in small mammals inhabiting the cultivated fields and forest plantations.

Consumption (*C*) exceeds assimilation (*A*) by the amount of energy lost through rejecta (*FU*). This can be easily seen from the second bioenergetic equation (2). The assimilation and digestibility of the natural foods of small rodents was studied thoroughly by Drożdż (1967, 1968; see also Chapters 11A and 11E). Bank voles lose on average 13·2% of energy intake (consumption) in faeces and 3·9% in urine, and similarly field mice lose 11·2 and 2·1%, respectively. Digestibility of various natural foods in these animals is, on the average 86·8 and 88·2% whereas assimilation is 82·9 and 86·1% of energy taken with food (Fig. 3.12). The simplest way of estimating the total consumption of vole and mouse population is by adding 17·1 and 13·9% to their assimilation (energy flow) or by multiplying it by factors 1·206 and 1·161, respectively.

The total consumption of both populations reaches 83,420 kcal/ha-year, and in particular years fluctuated from 37,000 - 125,000 kcal/ha-year. Since consumption has been calculated, we can estimate how much food has been eaten by the rodents in a given ecosystem. However, we are not able to determine how much food was actually destroyed and dissipated by these animals.

Let us tabulate the values calculated so far, that is production, assimilation and consumption.

Productivity of vole and mouse populations in kcal/ha-year:

Av. standing crop ($\overline{NW}$)	541
Production (P)	1,642
Respiration (R)	68,569
Energy flow (A)	70,211
Consumption (C)	83,420

Golley (Davis and Golley, 1963) divided mammals into five feeding types which differed in efficiency of food digestibility and assimilation. A new survey of literature was recently completed and data on food utilization were found for 42 species of small mammals (Grodzinski and Wunder, in press). The mean coefficient of assimilation for grazing herbivore mammals is approximately 65%, for omnivores—75%, granivores—88%, for small insectivores and carnivores—nearly 90%. These approximate figures could also be used for estimating the consumption from known assimilation.

Table 3.2
Comparison of the primary net production, food available and consumption of small mammals in different forest ecosystems (After Grodzinski, 1971e).

Forest ecosystem	Principal small mammals	In 10³ kcal/ha-year			In per cent		Reference
		Primary net production[1] (P_p)	Food available to mammals (F_a)	Mammal consumption (C)	F_a/P_p	C/F_a	
Taiga forest (*Picea glauca*) College. Alaska, USA	*C. rutilus* *Tamiasciurus*	10,232	1,320 (1,320–1,630)	179 (47–494)	12.9	13.5 (3–38)	Grodzinski (1971e)
Pinewood (*Vaccinio-Pinetum*) Mazury lakeland, Poland	*C. glareolus*[2] *A. flavicollis*[2]	—	2,415–7,080[3]	21–42	—	0.6–1.2	Ryszkowski (1970)
Pinewood (*Cladonio-Pinetum*) Mazury lakeland, Poland	*C. glareolus* *A. flavicollis*	—	1,024[3]	20	—	1.9	Ryszkowski (1970)
Oak-pine forest (*Pino-Quercetum*) Mazury lakeland, Poland	*C. glareolus* *A. flavicollis*	—	13,040[3]	75–102	—	0.6–0.8	Ryszkowski (1970)
Mixed & deciduous Kompinos Forest, Poland	*C. glareolus* *A. flavicollis* *A. agrarius*	—	16,190[3]	105	—	0.6	Ryszkowski (1970)
Oak-hornbeam forest (*Querco-Carpinetum*) Cracow, Poland	*C. glareolus* *A. flavicollis*	—	2,050	95	—	4.6	recalculated from Grodziński (1961) Górecki & Gebczyńska (1962)

Table 3.2. (continued)
Comparison of the primary net production, food available and consumption of small mammals in different forest ecosystems.

Forest ecosystem	Principal small mammals[2]	In 10³ kcal/ha-year			In per cent		Reference
		Primary net production[1] (P_p)	Food available to mammals (F_a)	Mammal consumption (C)	F_a/P_p	C/F_a	
Beechwood (Fagetum carpaticum) Ojców, Poland	C. glareolus A. flavicollis	43,000	1,950	75 (45—129)	4.4	3.9 (2—7)	Grodziński et al. (1970) Drożdż (1968)
Spruce plantation (Picea abies) Björnstorp, South Sweden	M. agrestis	14,700— 19,200	nearly the same	332—699	—	1.5—2.8	Hansson (1971)
Forest plantation on peat-bog Augustów Forest, Poland	M. oeconomus	11,930	(6.650)[4]	375	(56)[4]	3.1 (5.6)[4]	Gębczyńska (1970)
Desert shrub (Larrea tridentata) San Simon Valley, Arizona, USA	Dipodomys merriami Lepus californicus	5,700	2,395	131	42.0	5.5	Chew & Chew (1970)

[1]Net above-ground primary production. [2]C—Clethrionomys, A—Apodemus. [3]All data from Ryszkowski (1970) are recalculated into 1 hectare; fraction which contains production of herb layer and fall of leaves and seeds (except coniferous needles). [4]Figures calculated roughly from data of Gębczyńska (1970).

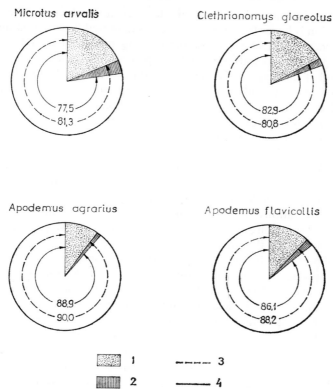

Fig. 3.12. Utilization of food energy by voles (*Microtus arvalis, Clethrionomys glareolus*) and field mice (*Apodemus agrarius, Apodemus flavicollis*). 1—energy loss in faeces, 2— energy loss in urine, 3—coefficient of assimilation (metabolizable energy), 4—coefficient of digestibility (digested energy). (From Droździ, 1968).

3.6 Production vs. Respiration

From a complete balance of energy flow and consumption for vole and mouse populations we can calculate the ratio of net production to respiration, to assimilation or to consumption. Net production in both populations is extremely low; on the average it is 2·3% of assimilation, and 2·1% of consumption. This means that in the energy flow through a rodent population the cost of maintenance entirely dominates over the production. The efficiency of production is lower for all homoiotherms than poikilotherms (Englemann, 1966; Turner, 1970). Analysis of the existing published data for 44 small mammal populations indicates that the production efficiency of rodents is 2·28%, and of small shrews is only 0·66% (Grodzinski and French, in press). Hence, both rodents and shrews show extremely low production efficiencies. This is due to the high respiratory cost of homoiothermy in these small animals. For this reason we will probably never eat mouse meat. It simply would be too expensive a dish.

The relationship of production and respiration is however quite stable and therefore can be used for predicting production from respiration. McNeill and Lawton (1970) were first to develop such functions for both homoiotherms and poikilotherms. Recent analysis of production and respiration in 44 populations of small mammals from Europe and North America leads to the distinction of two separate regression relationships, one representing the rodents and the other insectivores (Fig. 3.13). The functions shown on this figure can be used as predictive equations for estimating small mammal production from the respiration of their populations.

A total energy flow through a population of small mammals can be roughly estimated from field determinations of numbers and population biomass. The cost of maintenance can be estimated from general functions (8) or (9). Afterwards, the production can be estimated from using the functions given in Fig. 3.13. This does not imply however that good ecologists should limit themselves only to estimating numbers and biomass in the field!

Finally, it is important to realize what determines the accuracy of computations of the energy flow through vertebrate populations. In other words, which parameters are critical for the whole computation. A simple analysis of sensitivity (French, Grant and Grodzinsky, msc) indicates that accuracy depends first of all on the estimation of population numbers itself, and next on the measurement of food utilization and the metabolic rate under laboratory conditions. The latter measurements determine the daily energy budgets, and consequently cost of maintenance. Errors and some difficulties associated with estimating net production itself will have little influence on the whole balance. Nevertheless, it is obvious that for some purposes, e.g. energy flow to the higher trophic level in an ecosystem, an accurate calculation of net production is also important. Consequently there is an urgent need to measure the metabolic rate under field conditions and to check the computations of daily energy budgets. This should soon become possible using some recently described procedures, namely double labelled water method (Mullen, 1973) and biotelemetry (Morhardt and Morhardt, 1971).

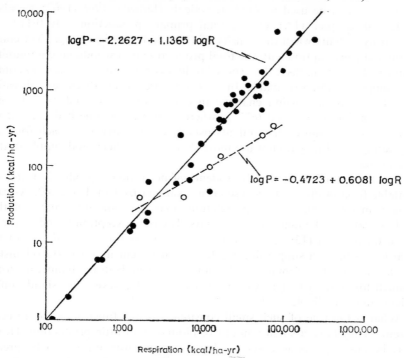

Fig. 3.13. Production (P) as a function of respiration (R) in rodent (solid line) and insectivore (broken line) populations. (After Grodziński and French, msc).

3.7 Consumption and available food

The last question concerns the relationship of the total consumption of the rodent populations to the net primary production of the forest, or more specifically to the food available. The food available to rodents can be defined as that food which is easy to find and is being chosen and eaten by these animals (Grodziński, 1968).

The total net production of the Ojców beech forest was found to be 10·3 metric tons/ha-year (Medwecka–Kornas and Lomnicki, 1967), which corresponds to approximately 44,000,000 kcal/ha-year (Drozdz, 1966). Yet out of this enormous amount only 2 million/ha-year, or 4·4% of the primary production, can be used as food by rodents (Drozdz, 1966, 1967; see also Table 1 in chapter 11A). Of the total primary production only 0·17% is utilized by rodents when the density of voles and mice is average. Yet comparing the consumption with the total production in a woodland community is an ecological nonsense; the ratio of rodent consumption to their potential food supply is more reasonable. In the beech forest being discussed this ratio is 3·94%. This proportion changed in consecutive years and with a high population density reached 6·77%, whereas with a low one it decreased to 2·39%. Even assuming that rodents usually destroy twice as much food as they can eat, still they do not use more than 10% of their food supply in the beech forest.

The degrees of utilization by rodents of net primary production and of available food in different ecosystems are summarized in Table 3.2. As can be seen in this table, rodent populations may consume 0·6—5·5% of the food available in the majority of habitats. The only exception found so far is in the taiga forest (13·5%). It seems that, in most forests, food cannot be regarded as being a simple limiting factor for rodent numbers (Grodzinski, 1963). As a rule the primary production of grasslands in the temperate zone is much lower than that in various woodlands. However, grasslands offer rodents more available foods.

What is the role of such small consumers in the functioning of an ecosystem? They have a small amount of biomass and little production. However, because metabolic costs are high, consumption must also be great. Being very efficient energy transformers, small mammals accelerate material cycling and energy flow within ecosystems.

References

ALLEN K.R. (1950) The computation of production in fish populations. *New Zealand Sci. Rev.* **8**, 89–101.

BOBEK B. (1969) Survival, turnover and production of small rodents in a beech forest. *Acta theriol.* **14**, 191–210.

BOBEK B. (1971) Influence of population density upon rodent production in a deciduous forest. *Ann. Zool. Fennici* **8**, 137–144.

BUECHNER H.K. & GOLLEY F.B. (1967) Preliminary estimation of energy flow in Uganda kob (*Adenota kob thomasi* Neumann). In Petrusewicz K. (ed.) *Secondary Productivity of Terrestrial Ecosystems.* pp. 243–254. Warszawa, Kraków.

CHEW R.M. & CHEW A.E. (1970) Energy relationships of the mammals of a desert shrub (*Larrea tridentata*) community. *Ecol. Monogr.* **40**, 1–21.

DAVIS D.E. & GOLLEY F.B. (1963) *Principles in mammalogy.* Reinhold Publ. Corp. New York. p. 335.

DROŻDŻ A. (1966) Food habits and food supply of rodents in the beech forest. *Acta theriol.* **11**, 363–384.

DROŻDŻ A. (1967) Food preference, food digestibility and the natural food supply of small rodents. In Petrusewicz K. (ed.) *Secondary Productivity of Terrestrial Ecosystems.* pp. 323–330. Warszawa, Kraków.

DROŻDŻ A. (1968) Digestibility and assimilation of natural foods in small rodents. *Acta theriol.* **13**, 367–389.

DROŻDŻ A., GÓRECKI A., GRODZIŃSKI W. & PELIKÁN J. (1971) Bioenergetics of water voles (*Arvicola terrestris* L.) from southern Moravia. *Ann. Zool. Fennici* **8**, 97–103.

ENGELMANN M.D. (1966) Energetics, terrestrial field studies, and animal productivity. In Cragg J. B. (ed.) *Advances in Ecological Research.* Vol. 3, 73–115. Academic Press, London–New York.

FRENCH N.R., GRANT W. & GRODZIŃSKI W. (msc) Small mammal energetics in grassland ecosystems.

GĘBCZYŃSKI M. (1966) The daily energy requirement of the yellow-necked field mouse in different seasons. *Acta theriol.* **11**, 391–398.

GESSAMAN J.A. (1973) Methods of estimating the energy cost of free existence. In Gessaman J. A. (ed.) *Ecological Energetics of Homeotherms— a View Compatible with Ecological Modeling.* pp. 3–31. Utah State University Press, Logan, Utah.

GOLLEY F.B. (1960) Energy dynamics of a food chain of an old-field community. *Ecol. Monogr.* **30**, 187–206.

GOLLEY F.B. (1962) *Mammals of Georgia—A Study of their Distribution and Functional Role in the Ecosystem.* Georgia Univ. Press, Athens, Georgia. p. 218.

GOLLEY F.B. (1967) Methods of measuring secondary productivity in terrestrial vertebrate populations. In Petrusewicz K. (ed.) *Secondary Productivity of Terrestrial Ecosystems.* pp. 99–124. Warszawa, Kraków.

GOLLEY F.B. & BUECHNER H.K. (Editors). (1969) A practical guide to the study of the productivity of large herboivores. IBP Handbook No. 7., Blackwell Sci. Publ., Oxford and Edinburgh. p. 320.

GÓRECKI A. (1965) Energy values of body in small mammals. *Acta theriol.* **10**, 333–352.

GÓRECKI A. (1967) Caloric values of the body in small rodents. In Petrusewicz K. (ed.) *Secondary Productivity of Terrestrial Ecosystems.* pp. 315–321. Warszawa, Kraków.

GÓRECKI A. (1968) Metabolic rate and energy budget in the bank vole. *Acta theriol.* **13**, 341–365.

GÓRECKI A. (1971) Metabolism and energy budget in the harvest mouse. *Acta theriol.* **16**, 213–220.

GÓRECKI A. & GĘBCZYŃSKI Z. (1962) Food conditions for small rodents in a deciduous forest. *Acta theriol.*, **6**, 275–295.

GRODZIŃSKI W. (1963) Can food control the numbers of small rodents in the deciduous forest. *Proc. XVI Int. Congr. Zoology*, Vol, 1, 257. Washington, D.C.

GRODZIŃSKI W. (1966) Bioenergetics of small mammals from Alaskan taiga forest. *Lynx* 6, 51–55.

GRODZIŃSKI W. (1968) Energy flow through a vertebrate population. In Grodziński W. & Klekowski R. Z. (eds.) *Methods of Ecological Bioenergetics.* pp. 239–252. Warszawa, Kraków.

GRODZIŃSKI W. (1969) [Two measures of metabolic rate in Common voles, *Microtus arvalis* (Pall,)]. In Blaxter K. L., Thorbek G. & Kielanowski J. (eds.) *Energy Metabolism of Farm Animals.* EAAP. Publ. No. 12, 399–400. Oriel Press Ltd. Newcastle upon Tyne.

GRODZIŃSKI W. (1971a) Energy flow through populations of small mammals in the Alaskan taiga forest. *Acta theriol.* 16, 231–275.

GRODZIŃSKI W. (1971b) Food consumption of small mammals in the Alaskan taiga forest. *Ann. Zool. Fennici* 8, 133–1336.

GRODZIŃSKI W., BOBEK B., DROŻDŻ A. & GÓRECKI A. (1970) Energy flow through small rodent populations in a beech forest. In Petrusewicz K. & Ryszkowski L. (eds.) *Energy Flow through Small Mammal Populations.* pp. 291–298. PWN. Warszawa.

GRODZIŃSKI W. & GÓRECKI A. (1967) Daily energy budgets of small rodents. In: Petrusewicz K. (ed.) *Secondary Productivity of Terrestrial Ecosystems.* pp. 295–314. Warszawa, Kraków.

GRODZIŃSKI W., GÓRECKI A., JANAS K. & MIGULA P. (1966a) Effect of rodents on the primary productivity of alpine meadows in Bieszczady Mountains. *Acta theriol.* 11, 419–431.

GRODZIŃSKI W., PUCEK Z. & RYSZKOWSKI L. (1966b) Estimation of rodent numbers by means of prebaiting and intensive removal. *Acta theriol.* 11, 297–314.

GRODZIŃSKI W. & WUNDER B.A. (in press) Ecological energetics of small mammals. In Petrusewicz K. (ed.) *Productivity Studies of Small Mammals.* Cambridge Univ. Press, Cambridge.

HANSSON L. (1971) Estimates of the productivity of small mammals in a South Swedish spruce plantation. *Ann. Zool. Fennici* 8, 118–126.

HANSSON L. (1972) Estimation of reproduction and production in small mammals. Small Mammal Newsletters 6, 139–155.

HANSSON L. & GRODZIŃSKI W. (1970) Bioenergetic parameters of the field vole *Microtus agrestis* L. *Oikos* 21, 76–82.

HART J.S. (1971) Rodents. In Whittow G.C. (ed.) *Comparative Physiology of Thermoregulation.* Vol. II. Mammals. pp. 1–149. Academic Press, New York and London.

HERRIED C.F. II & KESSEL B. (1967) Thermal conductance in birds and mammals. *Comp. Biochem. Physiol.* 21, 405–414.

HOLMES R.T. & STURGES F.W. (1973) Annual expenditure by the avifauna of a northern hardwoods ecosystem. *Oikos* 24, 24–30.

KACZMARSKI F. (1966) Bioenergetics of pregnancy and lactation in the bank vole. *Acta theriol.* 11, 409–417.

KALE H.W. II (1965) Ecology and bioenergetics of the Long-billed marsh wren in Georgia salt marshes. Publ. Nuttall Ornithol. Club. No. 5, Cambridge, Mass. p. 142.

KLEIBER M. (1961) *The fire of life—an introduction to animal energetics.* J. Wiley Sons, Inc. New York–London. p. 454.

MANN K.H. (1965) Energy transformation by a population of fish in the river Thames. *J. Anim. Ecol.* **34**, 253–275.

McNAB B.K. (1963) A model of the energy budget of a wild mouse. *Ecology* **44**, 521–532.

McNEILL S. & LAWTON J.H. (1970) Annual production and respiration in animal populations. *Nature* **225**, 5231, 472–474.

MEDWECKA-KORNAŚ A. & LOMNICKI A. (1967) Discussion of the results of ecological investigations in the Ojców National Park. *Studia Naturae s.A.* **1**, 199–213.

MIGULA P. (1969) Bioenergetics of pregnancy and lactation in European common vole. *Acta theriol.* **14**, 167–179.

MORHARDT J.E. & MORHARDT S.S. (1971) Correlations between heart rate and oxygen consumption in rodents. *Amer. J. Physiol.* **221**, 1580–1586.

MORRISON P.R. & Ryser F.A. (1951) Temperature and metabolism in some Wisconsin mammals. *Federation Proc.* **10**, 93–94.

MULLEN R.K. (1973) The D_2 ^{18}O method of measuring the energy metabolism of free-living animals. In Gessaman J. A. (ed.) *Ecological Energetics of Homeotherms—a View Compatible with Ecological Modeling.* pp. 32–43. Utah State Univ. Press, Logan. Utah.

MYRCHA A., RYSZKOWSKI L. & WALKOWA W. (1969) Bioenergetics of pregnancy and lactation in white mouse. *Acta theriol.* **14**, 161–166.

ODUM E.P., CONNELL C.E. & DAVENPORT L.B. (1962). Population energy flow of three primary consumer components of old-field ecosystems. *Ecology* **43**, 88–96.

PEARSON O.P. (1964) Carnivore-mouse predation: an example of its intensity and bioenergetics. *J. Mammal.* **45**, 177–188.

PETRIDES G.A. & SWANK W.G. (1966) Estimating the productivity and energy relations of an African elephant population. *Proc. Ninth Inter. Grassland Congr.* pp. 831–842. São Paulo, Brazil.

PETRUSEWICZ K. (1966) Production vs. turnover of biomass and individuals. *Bull. Acad. Pol. Sci., Cl.II.* **9**, 621–625.

PETRUSEWICZ K. (1967) Concepts in studies on the secondary productivity of terrestrial ecosystems. In Petrusewicz K. (ed.) *Secondary Productivity of Terrestrial Ecosystems.* pp. 17–49. Warszawa, Kraków.

PETRUSEWICZ K. & MACFADYEN A. (1970) *Productivity of terrestrial animals—principles and methods.* IBP Handbook No. 13., Blackwell Sci. Publ., Oxford and Edinburgh. p. 190.

PINOWSKI J. (1968) Fecundity, mortality, numbers and biomass dynamics of a population of the tree sparrow (*Passer m. montanus* L.). *Ekol. Pol. s.A.* **16**, 1–58.

SAWICKA-KAPUSTA K. (1968) Annual fat cycle of field mice, *Apodemus flavicollis* (Melchoir, 1834) *Acta theriol.* **13**, 329–339.

TROJAN P. (1969) Energy flow through a population of *Microtus arvalis* (Pall.) in an agrocenosis during a period of mass occurrence. In Petrusewicz K. & Ryszkowski L. (eds.) *Energy Flow through Small Mammal Populations.* pp. 267–279. PWN, Warszawa.

TROJAN P. & WOJCIECHOWSKA B. (1967) Resting metabolism rate during pregnancy and lactation in the European common vole—*Microtus arvalis* (Pall.). *Ekol. Pol. s.A.* **15,** 811–817.

TURNER F.B. (1970) The ecological efficiency of consumer populations. *Ecology* **51,** 741–742.

ZIPPIN C. (1956) An evaluation of the removal method of estimating animal populaions. *Biometrics* **12,** 163–189.

Invertebrate Bioenergetics

4

Parameters of an Energy Budget

A. DUNCAN AND R.Z. KLEKOWSKI

4.1 Introduction

Clearly how many of the parameters of an energy budget can be measured in a study depends on the number of workers available as well as the main emphasis of the study. Where the main interest lies in the inter-relationship between trophic levels, then both consumption and respiration are important parameters in addition to production. Measurement of the former involves not only quantitative estimates of how much energy is being removed from the previous trophic level but also the identification of which species or parts of species are being selected as food and whether there exists any kind of seasonal variation; whereas respiration is a measure of how much of the assimilated energy must be utilised in order to maintain life and how much is left over for production which, eventually, passes on to other trophic levels. On the other hand, where the main interest lies in the growth and production of one species-population only, then the only parameters that need be measured are those associated with the determination of population numbers, growth rates, mortality and birth rates.

In this section, certain of the budget parameters are dealt with in detail, namely, consumption (C), egestion (F) and respiration (R) because these are indispensable to an energy budget and a review of the methods involved in their measurement was thought to be useful. Excretion (U or in FU, the Gjecta) is dealt with in the section on methods for assessment of assimilation. errowth forms the basis of the theoretical concept of production (P) and requires a more extensive and complex discussion of theoretical ideas and their practical application than is possible in this handbook. The discussion in this book is therefore confined to a consideration of the relationship between R:P (Engelmann, 1966), of various growth efficiencies (K_1, K_2; Ivlev, 1939a, b, 1945, 1966; Odum 1959) and the inter-relationship between growth metabolism and temperature and what light this sheds upon population production and biomass (Winberg, 1962, 1966, 1968). For theoretical discussions on growth, the following books are recommended: Brody (1945), Bertalanffy (1964), Winberg (1968), Ricker (1971 ed.) Petrusewicz and Macfadyen (1970).

4.2 Consumption and assimilation

4.2.1 Nature of food

There already exists a very rich literature on the nature of the food of animals, particularly in relation to man, domestic animals (Blaxter, 1965; Crampton, 1965; Crampton and Lloyd, 1960; Morrison, 1956; Blaxter, 1967) and insects (Roeder, 1953). For some time the main emphasis was on the energy requirements of humans or domestic animals and their food was analysed mainly for their 'energy' content with the implicit assumption that the higher the calorific content the better. However, it soon became clear that not only is the calorific value important but also the nature of the food, namely, in what proportions carbohydrates, fats and proteins occur and also whether certain accessory but 'essential' substances are present, namely, vitamins, mineral salts, water and certain 'essential amino acids and fatty acids.' These are termed 'essential' when they are necessary to the animal but it itself cannot manufacture them. It is important to be aware of the existence of such substances as they may become 'limiting' in long term experiments where the animal is provided with only one type of food. Fraenkel and Blewett (1947) give an example of the effect of lipids and vitamin deficiency on the growth of caterpillars of *Ephestia*, and Moss (1967) of the selection of the tips of heather plants which are rich in nitrogen and phosphorus by the red grouse (*Lagopus lagopus scoticus*). It is in fact rather rare for an animal species to feed upon the same food throughout its life cycle or even throughout one developmental stage and it is more usual for animals to be polyphagous. Even monophagous species such as the wax moth (*Galleria mellonella—* Lepidoptera) feeding on the honeycomb is able to digest proteins (Duspiva, 1936). Since animals need to ingest daily sufficient protein and fat to provide all the essential amino and fatty acids or water in adequate quantities, this need may be the explanation for the occurrence in overcrowded populations of cannibalism of eggs by adults and larvae in such monophagous species of the flour beetle (*Tribolium* sp.) (Park, 1934, Rich 1956, Sonleitner 1961, Park *et al.*, 1965, Prus 1968).

In the desert environment, where scarcity of water is the major problem, food consumption is related not so much to the energy needs of animals as to their daily water requirements. Thus the amount of food consumed is governed by how much metabolic water may be obtained from the metabolism of carbohydrates and/or fats (Schultz, 1930) or by how much 'free' water is associated with the food available; the many behavioural and physiological adaptations associated with this problem are reviewed in (Mellanby, 1939), however, the mechanisms involved occur in temperate terrestrial insect species, as shown by Fraenkel and Blewett (1944).

4.2.2 Time of retention of food in the digestive tract (t_d).

The time of retention of food in the digestive tract (t_d) is related in general to the character of the food and feeding and, in particular, to the rate of ingestion of food, the size of the meals being ingested and the rate at which digestion takes place. The form or nature of the food greatly influences the feeding mechanisms of animals, according to whether the food consists of small particles, such as algae, protozoa, detrital particles suspended in water or scraped from some solid surface, or of larger masses, such as deposit, detrital masses, parts of living macrophytes, dead or living prey organisms, or is a fluid e.g. plant fluids, blood (Yonge, 1928, 1937). Most particulate feeders are aquatic and filter off the small particles by some filtering mechanisms. Such filter-feeding is usually a fairly continuous process as is also the feeding in molluscs which scrape solid surfaces or the feeding in deposit-feeders such as earthworms or in herbivores such as caterpillars or grasshoppers. Whereas other animals feed at regular or irregular intervals, for example, various predators or bloodsuckers such as leeches or mosquitoes. Superimposed upon the continuous or discontinuous nature of the feeding process may also be a periodic pattern of rhythmical behaviour associated with circadian or other rhythms.

The form and function of the digestive tract is also correlated with the kind of food normally eaten. The functional regions of the digestive tract may be briefly outline as follows: (1) the region of reception, consisting of the mouth with associated appendages, sense organs, secretory organs and buccal cavity; (2) the region of conduction and storage, consisting of the oesophagus and crop; (3) the region of internal trituration and early digestion, consisting of gizzards, gastric mills, stomachs or other grinding mechanisms, designed to reduce the size of food particle so that mixture with digestive enzymes is facilitated; (4) the region of final digestion and assimilation, consisting of anterior intestine or posterior insect mid-gut; (5) the region of faeces formation, consisting of the insect hind-gut or the vertebrate colon. The relative importance of these various regions varies in different species according to their needs. Thus storage crops are important in leeches and mosquitoes, the triturating region in animals feeding on larger food masses and most terrestrial animals have relatively long hind-guts where water ansorption takes place.

Food is propelled along the digestive tract at various speeds either by ciliary action (in aquatic ectoprocts, entoprocts as well as in molluscs) or by muscular action. The muscles involved may be the somatic muscles of the body also associated with locomotion as in annelids, or the visceral musculature of the gut itself. The food is moved along the gut and at the same time is further triturated and mixed with the digestive enzymes.

Thus the time (t_d) of the retention of food in the digestive tract may be as much as several months in the medicinal leech, or several weeks in the bug *Rhodnius* (Wigglesworth, 1943) or 9 to 33 hours in *Blatta* sp. (Chauvin, 1956) or rather short as in caterpillars, 2 to 3 hours in silkworms (Shinoda, 1931); $3\frac{1}{4}$ hours in *Prodenia eriderma* (Crowell, 1941) or in grasshoppers or locusts, 1 hour at 32°C in *Locusta*, (Nenjukov and Parfentiev, 1929); or 2 to 3 hours in *Callipttamus* (Tareev and Nenjukov, 1931); 4 to 6 hours in *Tribolium castaneum* Hbst. (Żyromska-Rudzka pers. comm). In order to assess the length of this time t_d, it is useful to know the feeding biology of the species being investigated, for example to know:

1. whether the feeding is continuous, discontinuous or periodic for a period of time such as 24 hours or better the feeding cycle;
2. whether the ingestion rate during active feeding is variable and how such variation affects the values of consumption per 24 hours or per feeding cycle;
3. whether within the digestive tract a period of storage takes place, how extensive is the mixing of food with previously or subsequently ingested food and how quickly does digestion proceed.
4. what effect the size of the meal has on the retention time. Several techniques are available for determining t_d, all of them involving marking food so that the marked food or faeces may be detected. Phillipson's 'direct method' of determining assimilation efficiency, described in the section of gravimetric methods, is one technique which relies on natural coloured food species as markers and may be extended to other feeders than carnivores; this technique is described in Phillipson (1960a, b) as well as in another section of this handbook (pp. 236-247).

The other two methods described here involve food marked with radioactive isotopes; in one method where the animal is fed for a long period of time on labelled food until some state of equilibrium is reached and then the animal is transferred into unlabelled food during which time the faeces are separated off at regular intervals. When the radioactivity of the faeces drops to background level, then t_d is the time from transfer to unlabelled food to the first period of production of unlabelled faeces. The second radioactive method for determining t_d involves short term feeding; the animal is fed on labelled food for a few minutes only, just long enough to obtain a 'signal' amount of labelled food. The animal is then transferred to unlabelled food and faeces are regularly collected and their radioactivity measured until the labelled faeces are produced; the time between the 'signal' feeding and the production of radioactive faeces represents time t_d. In both methods, it is essential that the faeces are separable from the food by the experimentor; in the former method, it is important that any radioactivity which may be excreted in the nitrogenous waste does not contaminate the faeces; in the latter method, extensive mixture of food in the gut may invalidate the method.

4.2.3 Period of the feeding cycle (t_C) and defaecation cycle (t_F)

As can be seen from Table 7A.2. (pp. 249-257) information about t_d is essential for many of the methods described there for determining the levels of consumption and assimilation and especially in those animals whose faeces are inseparable from their food. However, for the non-radioactive marker methods, e.g. using platinum or ash as markers, where measurement of consumption is based on the collection of faeces, what is necessary is some information on the normal defaecation cycle. It is possible to determine the period of this cycle (t_F) by collecting and weighing faeces at regular short intervals of time and then summing the faecal weights for ever-longer, over-lapping periods (as shown in Fig. 4.1) until a period of time is found during which the quantity of faeces produced which is more or less constant, irrespective of when the period of collection was started; Phillipson obtained such results in his 1960a paper for *Mitopus mori* and this is illustrated in Fig. 4.2. This period of time, t_F, represents the defaecation cycle time and it is essential to know it for methods W4b and M4b (in Table 7A.2. p. 239) where it is assumed to be equivalent to the feeding cycle period, t_C.

Figure 4.1 is an attempt to illustrate the hypothetical relationship between t_d, t_C and t_F in animals with different patterns of feeding-defaecation cycles.

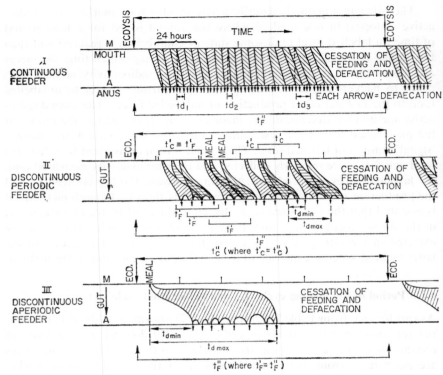

Fig. 4.1. Hypothetical relationship between t_d, t_C and t_F in animals with different patterns of feeding-defaecation cycles.

Fig. 4.1A—Patterns of food retention in the gut showing the relationship between t_d, t_C and t_F.

For simplicity, only three types of feeding-defaecation cycles are considered, namely for species which can be termed continuous, discontinuous periodic and discontinuous aperiodic feeders. However, because Arthropoda form a very high proportion of any community and characteristically undergo periods of ecdysis at genetically controlled intervals during which feeding and defaecation ceases, these three cycles are illustrated for arthropodan species.

Thus Fig. 4.1IA presents the feeding-defaecation cycle for more or less continuous feeders, i.e. for relatively long periods of time, such as caterpillars or deposit-feeders or animals such as *Tribolium* which live in their food medium, which also defaecate regularly and whose time of retention of food in the gut, t_d, varies only slightly so that any mixing of food does not seriously affect the measurement of the ingested and egested energy. The method suggested above for determining t_F is applicable to this type of feeding-defaecation cycle only for periods with constant t_d, although uniform rates of feeding and defaecation may appear for quite short periods of time. In arthropodan species, before ecdysis, feeding ceases but the animal continues to require energy for maintenance. Although our animals feed continuously during the time they do feed, it is very likely that their rate of consumption will vary, which in turn influences both the time t_d because of the *size* of the meal, and the rate of defaecation; the resulting changes in the instantaneous consumption and defaecation rates (in calories and assuming a 50% assimilation efficiency) throughout one instar period are shown in Fig. 4.1.IB and the resulting consumption and defaecation (in calories) cumulated for different moments and up to the end of the instar period are shown in Fig. 4.1.IC. It is interesting to note that, although we assumed a uniform 50% assimilation of consumed food throughout the instar period, simultaneous measurement of C and F for short Δt, e.g. 2 hrs at different times during the instar period would give quite different and erroneous values for assimilation efficiency (even $> 100\%$).

Figure 4.1.IIA presents the feeding-defaecation cycle for the discontinuous periodic feeders, such as forest or litter predators, with variable times of retention and considerable mixing of food in the gut; notice that faeces produced at any one moment may originate from quite different meals and that one meal may contribute to faeces egested at different times; if the time of retention (t_d) were considerably longer, this mixing of food may even span the feeding cycles. Again, in arthropodan species, feeding and defaecation ceases before ecdysis. Thus, in this type of feeding-defaecation cycle, there are two distinguishable feeding cycles, each incorporating a non-feeding, non-defaecating interval, one with a short-term periodicity (often circadian) which may be termed the 'first order or daily cycle (t_C and t_E)' which the whole population probably exhibits and the other which may be termed the 'second order or instar cycle (t_C and t_F)' which is regular for any one species but which may be staggered for different individuals in a population composed of different ages. These two cycles are demonstrated in Fig. 4.1.IIA. Figure 4.1.IIB presents the instantaneous consumption and defaecation rates for such discontinuous periodic feeders throughout one instar and Fig. 4.1.IIB the cumulated consumption and defaecation for one instar period, assuming

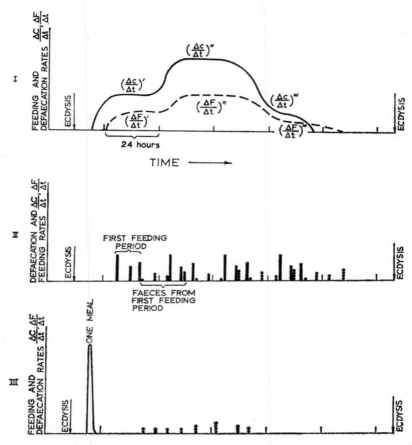

Fig. 4.1B—Consumption and defaecation rate through one instar period.

two levels of assimilation efficiency, 75% and 25%. This method suggested above for determining t_F is also applicable to this type of feeding-defaecation cycle (as is demonstrated in Fig. 4.2 for *Mitopus mori;* Phillipson 1960a) and, where t_C is not known and difficult to measure it can probably be assumed to be the same order of magnitude as t_F.

The third type of feeding-defaecation cycle is for the discontinuous aperiodic feeders, such as leeches which feed at very long intervals or the bug *Rhodnius* which feeds only once during an instar period but which may

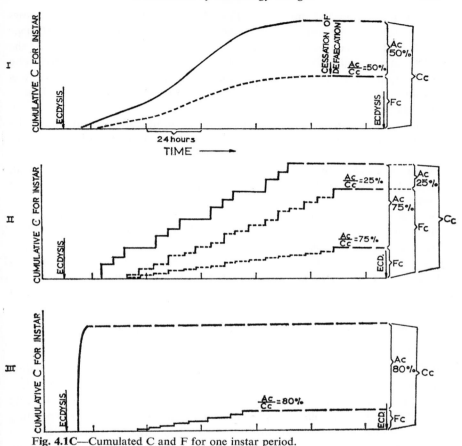

Fig. 4.1C—Cumulated C and F for one instar period.

produce faeces more regularly; this cycle is presented in Fig. 4.1.IIIA. Such animals have a great capacity for storage of food so that the time of food retention is very long ($t_{d.min}$ to $t_{d.max}$). In such cases, it is only possible to measure the second order or instar feeding-defaecation cycle for which information about the feeding biology and developmental cycle is necessary, in order to decide upon an experimental time. Figure 4.1.IIIB presents the instantaneous consumption and defaecation rates, and Fig.1.IIIC the cumulated consumption and defaecation for one instar period of these discontinuous aperiodic feeders.

Figure 4.2, plotted similarly to Figs. 4.1.IA, 4.1.IIA and 4.1.IIIA, shows the pattern of ingestion-egestion cycles for male and female adult *Mitopus mori*, a litter predator, using the data, published by Phillipson (1960a); the information available was the period, t_d, the time of capture and how long after capture the first faecal pellet appeared. From this data, Fig. 4.2 was produced which is very similar to that of Fig. 4.1.IA for discontinuous periodic feeders.

It is probably always advisable to have some information about the feeding and defaecation cycles of the species being investigated, irrespective of which methods for estimation of C and A are being employed. This is because the experimental time, t, in a feeding experiment should be a function of t_C, the period of the species' feeding cycle, and the time during which faeces are collected should be a function of t_F, the period of the species' defaecation cycle. That the time spent in active feeding differs from the time spent actively defaecating in any one feeding-defaecation cycle as well as between different feeding-defaecation cycles is demonstrated in the hypothetical examples given in Fig. 4.1. What is the real relationship between t_C and t_F (and also t_d) does not seem to be known for many species. However, some relationship must exist, the basis of which is a balanced inflow and outflow of energy in the organism so that

energy ingested	=	energy assimilated	+	energy egested
during time t_C		during time $t_?$		during time t_F;

that is, t_C is in some way equivalent to t_F in a normally and regularly feeding, defaecating animal whose rates of energy inflow and outflow are balanced.

It is clear that in a bio-energetic study, the time during which no feeding or defaecation takes place must be considered since the animal is expending energy continuously for maintenance and may also be actively growing during this period. The proportion of time t_C which is spent in active feeding will vary with feeding type, being greater in continuous than in discontinuous periodic feeders and least in aperiodic feeders like leech and *Rhodnius;* it is difficult to generalise about the defaecation cycle. The amount of time lost for feeding due to ecdysis will vary from species to species. Thus, any average consumption or defaecation rates (per hour, per 24 hours or per any other unit of time for which an energy balance is required) must be calculated from a period of time which includes this non-feeding, non-defaecating time. It is suggested that the most convenient way is to determine the periods of the first order, e.g. daily cycle (t'_C and t'_F) and, for arthropods, the second order

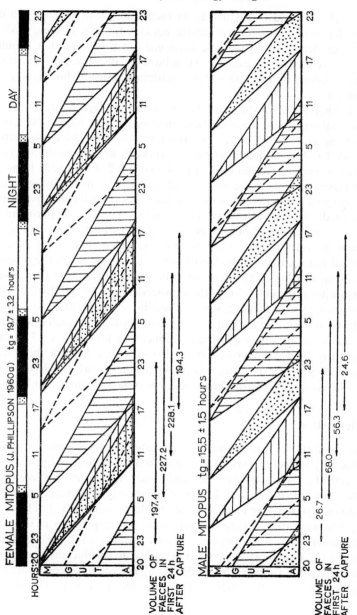

Fig. 4.2. Pattern of ingestion-egestion cycles for *Mitopus mori*. (After data in Phillipson, 1960a).

instar cycle (t''_C and t''_F) which is, in fact, the duration of the whole instar period. Knowing these characteristic periods for the species, it should be possible to decide upon realistic experimental times t which should be in some way related to t_C and t_F. An arbitrarily decided experimental time t will give a consumption rate which, multiplied up to 24 hours, is unlikely to be realistic.

In connection with the second order or instar feeding-defaecation cycle, certain problems may arise in those methods where t must be less than t_d because the faeces are inseparable from the food by the experimentor (see Table 7A.2.). An example is *Tribolium* whose t_d in flour is about 4 to 6 hours (Żyromska-Rudzka, unpubl.) so that t may be 2–3 hours. It is possible to obtain larvae of similar age and probably the same instar by collecting eggs laid during a known period; sufficient numbers for a feeding experiment can be obtained only when the period is 48 hours. Thus, the experimental larvae may differ in age by about 48 hours and experiments with coloured flour in which the gut contents were analysed revealed that about 20% of such larvae did not feed probably due to onset of ecdysis. Since the whole period of the instar feeding cycle (t''_C) cannot be employed, a realistic measurement of consumption which includes the non-feeders can be obtained in one of the two following ways. Either a series of feeding experiments 2 to 3 hours long must be carried out on the same larvae throughout a period equivalent to the instar duration or one experiment (replicated) lasting 2 to 3 hours using a sufficiently large number of larvae to ensure that 20% of them are in the non-feeding pre-ecdysis state. The main problem with the latter method is how to determine N, the minimum number of larvae which can be employed.

4.3 Metabolism

In aerobic animals, measurement of the rate of oxygen consumption gives a measure of the energy expenditure involved in the normal processes of the body and the metabolic rate per unit body weight of the intensity of its metabolism. This is not true, however for anaerobic animals in which rather the rate of glycolysis or fermentation should be determined. The amount of energy liberated during aerobic and anaerobic respiration is very different. During anaerobic respiration, only about 7% of the total free energy available in glucose (52 kcal/mol) is released whereas in aerobic respiration the first anaerobic stage is followed by oxidation by molecular oxygen which liberates a further 93% of the free energy of glucose (686 kcal/mol). This difference in the amount of energy released from one molecule of glucose emphasises the 'price' paid for the advantage of anaerobiosis which permits the survival without molecular oxygen in such habitats as the deeper layers of soil rich in organic compounds, in the bottom sediments of thermally stratified lakes and in gut parasites, all habitats with a super-abundance of food in common. Moreover, Beadle (1961) has suggested that those aquatic animals with a dependent respiration (their oxygen consumption declines with decreasing concentration of environmental oxygen) may partially satisfy their energy requirements in conditions of lowered oxygen conditions by anaerobic respiration.

Respiratory rate is a useful and sensitive measure of the organisms' daily expenditure of energy; in fact, the measurement of respiration can be so sensitive that it is modified by a great variety of factors, both internal and external, whose order of magnitude must be known before they can be ignored. The internal physiological factors which may affect the metabolism of animals are body size or age or developmental stage; reproductive state (whether bearing young or with developing gonads or just sexually mature); nutritional state (whether starving or well-fed; whether post-absorptive or just after ingestion); composition of diet; whether moulting or not; the state of acclimatisation to temperature, relative humidity or other environmental factors of significance to the species; state of stress (to immediate handling or other forms of disturbance); state of activity (whether moving or not). The external factors which affect metabolism are the diurnal and seasonal changes in temperature, relative humidity, changes in concentration of respiratory gases e.g. the level of carbon dioxide in respiratory chambers containing CO_2—absorbants, and any other significant environmental factor.

4.3.1 Types of metabolism

In order to measure respiration, one or more animals are placed in respirometers of various kinds, some of which are described in the present handbook, others in general reviews by Dixon (1952), Umbreit *et al.* (1964), Kleiber (1961), Kleinzeller *et al.*, (1954) and others. It is necessary to consider what kind of metabolism is being measured in such respirometers and whether it is possible to establish some kind of measurable metabolism relevant to ecological studies.

During the last century, comparative physiologists were searching for some sort of basal or 'zero' metabolism by means of which animals of different species could be compared. The first such basal metabolism was defined for homoiothermic animals by Brody (1945) using animals of known weight, age and developmental stage. This was *basal metabolism* which is the resting energy metabolism in a thermo-neutral environment in a post-absorptive state but without psychic or sexual stress; that is, the metabolism of animals in a state of minimal activity, not requiring energy for maintenance of body temperature, neither feeding nor digesting actively but not starving and inactive but not asleep. Krogh (1914, 1916, 1941) attempted to define a comparable condition of minimal activity for invertebrates and poikilotherms, which he termed *standard metabolism*.

Neither of these metabolisms are directly relevant for ecological studies on animals under field conditions but they are useful as a measure of the minimal energy requirements of a minimally active animal. Fry (1947, 1957) has suggested that there are only three levels of metabolism which can be usefully distinguished (in fish), namely, *standard metabolism*, that of minimal activity; *maximum active metabolism*, that of sustained maximal activity; *routine metabolism;* that when the animal exhibits spontaneous rather than directed movements. The upper limit for maximal active metabolism is probably physiological e.g. the area of respiratory membrane in poikilotherms or the store of creatine PO_4 in homoiotherms per unit body weight, so that maximal level measurable will depend on both duration and degree of activity. Fry (1947) points out that the difference between the standard and active metabolic rates provides some idea of the 'scope for activity' of the organism. Winberg (1956) considers that, although many authors have preferred to use Krogh's standard metabolism because of the difficulties in measuring the basal, resting metabolism, this latter term has a clear theoretical meaning and is worth retaining. He thinks that most published measurements have been made on animals which can move, which give rates higher than the basal level; these he terms 'ordinary metabolism' which is equivalent to Fry's routine metabolism and retains terms 'active metabolism' for when the organism is actively moving.

For characterising the normal metabolic rate of fish in nature, Winberg (1956) suggests that this can be taken to be twice the level of this 'ordinary metabolism' as measured by laboratory respirometry, which Mann terms 'the routine metabolism of a resting fish'; Mann (1969) discusses this hypothesis together with some evidence which supports it. Very little published information exists on the 'scope for activity' of animals other than fish (Fry 1947, 1957) and *Gammarus oceanicus* (Halcrow and Boyd, 1967).

Recently, Newell (1969) has pointed out that continuous recording of oxygen consumption provides data on respiratory levels of animals in various states of activity. Thus, when these are plotted logarithmically against body size of a large number of animals of different size, the upper limit represents the maximal rate and the lower the minimal rate of the possible levels of oxygen consumption; regression lines can be fitted to these maximal and minimal rates, which vary with body size in a way described in a later section (p. 123). If such measurements are made at a variety of temperatures, information is obtained on both the 'scope for activity' and the 'potential temperature range of activity'. Morrison and Grodziński (p. 75) have employed and 'average daily metabolic rate' (ADMR) as the best measure of metabolism for constructing daily energy budgets for small mammals. This ADMR represents the summed metabolism over 24 hours of an animal kept in a respirometer large enough to allow normal freedom of movement and feeding at a temperature similar to its natural habitat. Phillipson (1967) stresses that 'all the year round respirometric studies of all life stages, at all times of the diel should be made if one is to even approach a reasonably accurate estimate of the annual respiratory metabolism of a species population'. This author has developed a short-cut, 'best estimate' method for determining the annual population metabolism, which is described in detail in Phillipson (1970).

4.3.2 Factors affecting metabolic rates

What kind of metabolism and how to measure it for laboratory bioenergetic, i.e. physiological, studies of a species and what kind metabolism and how to measure it for field investigations on the energy flow and productivity of a natural population of this species are very different aims, despite the fact that very often it is the laboratory respiratory data which is applied to the field census or biomass data in order to estimate the population metabolism.

In bioenergetic studies, the metabolism can be measured in considerable detail as a function of the most significant factors, both internal (such as age, size, physiological state, degree of activity etc.) and external (such as temperature, relative humidity, concentration of respiratory gases, the nature and availability of food etc.). Thus, animals of known age, developmental stage and feeding condition can be used for respiratory measurements. The possibility of examining the metabolic responses under two or more environmental variables acting in combination (e.g. McLeese, 1956) is particularly valuable. Such studies can provide good information about the species' potentialities under different conditions. The main difficulty is in maintaining the laboratory culture in a realistic environment and in providing it with the same quality and quantity of food as in the field; cultures tend to produce either abnormally large, well-fed individuals or rather poor, ill specimens. Another difficulty is to ensure during the actual measurement of respiration reasonably realistic conditions for the species. Here, the relevant aspects are the form of the respiratory chamber (its shape, whether 'closed' or 'open' to the environment); whether the respiratory gases or the relative humidity are present in normal concentrations (high concentrations of CO_2–absorbants reduce the level of both CO_2 and water vapour, both abnormal conditions for soil or litter animals); whether temperature and light conditions are constant or variable throughout a 24 hour period; whether respiration is recorded continuously in order to encompass all normal levels of metabolism. MacFadyen (1961, 1967) discusses some of these problems and describes (1961) a continuously recording respirometer for small air-breathing invertebrates which will operate under varying temperatures. In general, it would appear that continuous flow, open-type respirometers, although more complex, provide better possibilities for maintaining controlled environmental conditions than closed-type respirometers.

In closed-type respirometers, where the depletion of oxygen and the accumulation of egested and excreted material may become a problem after some time, measurements cannot last too long. This is not serious in those species not exhibiting circadian-type rhythms of activity. However, such rhythms are often present in terrestrial species, either as inherent cycles or in response to diurnal changes in environmental temperature or relative humidity, and may be reflected in variation of metabolic rate over a 24 hour period. Then, either measurements must last long enough to include the whole period of the cycle (the average daily metabolic rate of Morrison and Grodziński p. 73; Phillipson, 1967) or, where this is not possible the R/t must be determined for a series of separate short periods during at least 24 hours. The period of this respiratory cycle (t_R) is likely to be related to the period of the feeding (t_C) and defaecation (t_F) cycles of a species which are discussed on p. 101.

The other cyclic event which affects respiration as an energetic parameter is the occurrence of ecdysis in arthropods; the metabolic intensity shows a regular 'instar rhythm', attaining peak values during the actual moult (Clarke 1957, Zwicky and Wigglesworth 1956, Gyllenberg 1969). For the calculation of energy budgets, the respiration should be summed for the whole instar period, including the period of the moult.

In terrestrial insects living in low humidities (Buck and Keister, 1955; Schneiderman and Williams, 1955) and in aestivating pulmonates (Klekowski, 1961), gas ventilation is not a continuous process. The relatively short period of gas inspiration and exspiration is interrupted by relatively long periods when the spiracles or pneumatostome remain closed. This phenomenon is clearly revealed when single individuals are measured in micro-respirometers. In other animals (e.g. *Cecropia* pupae, Schneiderman and Williams 1955), delayed exspiration of carbon dioxide together with continuous inspiration of oxygen may cause the calculation of abnormal respiratory quotient.

The same problems arise in measurements of respiration for estimating energy flow in natural populations. A few attempts have been made to measure the metabolism of animals carrying out normal activities in the field. These have involved feeding either double-labelled water ($D_2^{18}O$) or ^{65}Zn as metabolic tags and re-capturing the animals after a period of freedom or using telemetric techniques for detecting the rates of breathing movements or heart beat of free animals. The main advantage of these techniques, which have been tried out on homoiotherms only, is that respiration is measured during normal activity, which is particularly difficult to achieve in most respirometers. There are very few published results of oxygen consumptions in poikilotherms during different, measured levels of locomotory activity (Fry, 1947, 1957; Blažka, 1960; Krogh and Weis-Fogh, 1951; Yurkiewicz and Smyth, 1966).

In those studies aiming at some estimate of population metabolism in which attempts have been made to determine invertebrate respiration, measurements have been carried out either on laboratory animals at several temperatures or on animals brought in fresh from the field at various times of the year and measured at field temperatures or at the mean temperature of the habitat. Workers adopting the former approach have used this laboratory data to establish an empirical relationship (Wiegert, 1965) or mathematical regressions (Berthet, 1964) of oxygen consumption with temperature which were then applied to field census data or the applicability of Krogh's (1914, 1916, 1941) normal curve has been accepted (Winberg 1956, 1968; Healey, 1967). Where field animals have been measured at field habitat temperature, respiratory rates have been used to calculate population metabolism without further temperature conversions; there is, however, the danger that such animals may show an initial heightened respiratory rate due to disturbance from handling or because they have not been given time to acclimatise to the mean temperature.

It may be that the only respiratory data available for the species being studied are original determinations or reliable published measurements for one life stage only, for one period of the year or for one temperature. In such circumstances, it is still possible to obtain some general idea of the order of magnitude of the annual population metabolism by calculating the respiration for the other developmental stages (provided the relationship between respiration and body weight is known for the species) and for different thermal conditions (from Krogh's normal curve) and applying these to population data. Such calculations will provide only very rough estimates of the annual population metabolism compared with the more detailed respiratory investigations outlined above since it assumes the validity of these two relationships in all circumstances.

4.3.3 Relationship of metabolism with body weight

This subject has a long past history and an extensive literature with many excellent reviews which should be consulted (Brody, 1945; Kleiber, 1947, 1961; Zeuthen, 1947, 1953, 1970; Winberg, 1956; Hemmingsen, 1960).

When the standard respiratory rate per individual is plotted logarithmically as the ordinate (y) against an abscissa (x) of body weight, an approximately linear relationship is revealed, irrespective of whether the data used is for individuals belonging to one species, but of different size (dogs—Rubner, 1883), to one taxonomic group (fish—Winberg, 1956; Ivlev, 1954; crustaceans—Winberg, 1950; molluscs—Winberg, 1959) or to one physiological type (homoiotherms—Kleiber, 1961), to a series of unrelated species (Zeuthen, 1947, 1953) or to the whole spectrum of living organisms (Hemmingsen, 1960). That is, individual respiratory rate appears to be related to some function of body weight, a relation describable by the exponential equation.

$$R = a \cdot W^b \tag{1}$$

or logarithmically,

$$\log R = \log a + b \cdot \log W, \tag{2}$$

where R is the respiratory rate per individual in volume or weight units of oxygen or in calories or some comparable units; where W is the fresh or dry weight of the individual, in gravimetric or some equivalent units; where

$a = R/W^b$, is approximately constant and represents the specific metabolism (Brody, 1945) or the respiratory rate per W^b/W^b = the metabolically effective body size (Brody, 1945) or the metabolic body size (Kleiber, 1961); where b is the regression coefficient defining the slope of the regression line on a log -log scale and is given by the tangent of the angle formed by the regression line with the absissa.

The value of 'a' depends on the units used to measure R and W whereas the value of the regression coefficient 'b' is independent of the units of R and W and is constant for a given regression. This is illustrated in Fig. 4.3 which

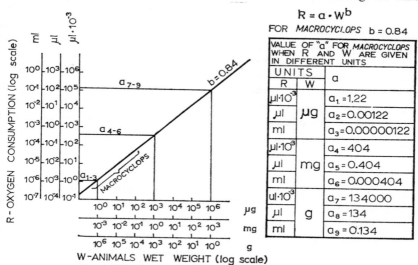

FOR *MACROCYCLOPS*	b = 0.84	

VALUE OF "a" FOR *MACROCYCLOPS* WHEN R AND W ARE GIVEN IN DIFFERENT UNITS		
UNITS		a
R	W	
μl·10^{-3}		$a_1 = 1.22$
μl	μg	$a_2 = 0.00122$
ml		$a_3 = 0.00000122$
μl·10^{-3}		$a_4 = 404$
μl	mg	$a_5 = 0.404$
ml		$a_6 = 0.000404$
ul·10^{-3}		$a_7 = 134000$
μl	g	$a_8 = 134$
ml		$a_9 = 0.134$

Fig. 4.3. Body weight-respiration relationship of *Macrocyclops albidus* (after Klekowski & Shushkina, 1966); value: R, W are given in different units.

shows the relationship between respiratory rate and body weight in *Macrocyclops albidus* Suv. (Klekowski and Shushkina, 1966); the relationship was $R = 1.22 . W^{0.84}$ where R was measured in μl·10^{-3}/ind .hr and W in μg wet weight. The accompanying table gives different values for the coefficient 'a' when R and W are expressed in other units.

For example, let us calculate the values of 'a' from the regression $R = 1.22$. $W^{0.84}$ (R, μl·10^{-3}/ind .hr; W, μg) for the following values of R and W:

1. The value of 'a_1' for R = μl·10^{-3}/ind.hr and W = μg
As 'a' = R when W = 1, 'a_1' = $1.22.(10^0)^{0.84} .10^0 = 1.22$

2. The value of 'a_6' when $R = $ ml $(=\mu l \cdot 10^{-3} \cdot 10^6)$/ind.hr and $W = $ mg $(= \mu g \cdot 10^3)$

As 'a' $= R$ when $W = 1$, 'a_6' $= 1 \cdot 22.(10^3)^{0.84} \cdot 10^{-6} = 1 \cdot 22 \cdot 10^{-3.48} = 0 \cdot 000404$. Thus, the formula is now $R = 0 \cdot 000404 \cdot W^{0.84}$; (R,ml/ind·hr;W,mg). By means of such re-calculations of the coefficient 'a', the respiratory results of different authors can be compared, even although R and W were measured in different units. Such a comparison is given later (Table 4.1).

The regression of individual respiratory rates against body weight can be readily obtained for any species by plotting the experimental data for as wide a range of body sizes as possible directly onto log-log paper rather than converting into logarithmic numbers and plotting this onto arithmetic paper. On log-log paper, the points will tend to form a straight line, through which, by eye, an approximate regression line can be drawn, for purposes of preliminary orientation. The approximate value of 'b' can then be determined as follows: lines parallel to the abscissa (x) and ordinate (y) are drawn and their absolute length (not logarithmic length) from their common meeting point to their points of contact with the regression line is measured in mm.

Then,

$$`b` = \frac{\text{length in mm of y}}{\text{length in mm of x}};$$

As $a = R$ when $W = 1$, the value of 'a' can be read off as the intercept of the regression line with a line parallel with the ordinate at a point on the abscissa where $W = 1$. The exponential equation is now approximately known. However, it is best to calculate the regression best fitting the experimental data, especially as this provides estimates of the errors of both 'b' and the intercept 'a' as well as some idea of the level of total variance. A description of the procedure of calculating a log-linear regression is given in Brody (1945—reprinted 1961, p. 398–401) although statistical textbooks such as Snedecor (1956) are better on errors. Data covering a short range of body weights may give erroneous slopes, especially if very variable.

The above regression can be used to express the relationship of the metabolic rate of weight specific respiration (R/W) with body weight;

$$\frac{R}{W} = \frac{a W^b}{W} = aW^{(b-1)} \qquad (3)$$

or logarithmically,

$$\log \frac{R}{W} = \log a + (b-1) \log W. \qquad (4)$$

This regression also shows a linear relationship when plotted logarithmically.

As early as 1839, Sarrus and Rameau (1839) found that the metabolic rate in animals of different body weight does not increase directly with

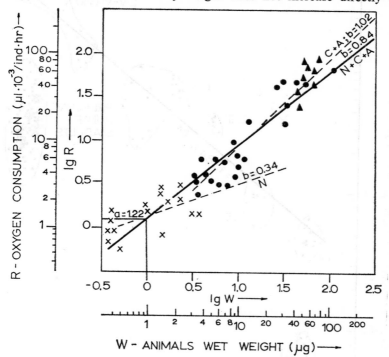

REGRESSION EQUATIONS:

for nauplii: $R = 1.25 \cdot W^{0.34}$

for copepodites and adults: $R = 0.76 \cdot W^{1.02}$

for all stages: $R = 1.22 \cdot W^{0.84}$

Fig. 4.4. Size dependence of respiratory rate of different developmental stages of *Macrocyclops albidus* Jur. (from, Klekowski & Shushkina, 1966; food conc. II=1mg/l).

increase in weight, when b would equal to 1·01 but rather to $W^{0.67}$ which represents the body surface area. This is the origin of the famous surface law of metabolism or law of Rubner (1883), where the exponent is 0·67 relative to absolute weight or −0·33 relative to unit weight. More recently, the value b = 0·75 has been adopted as more realistic by several authors (Brody, 1945;

Chapter 4

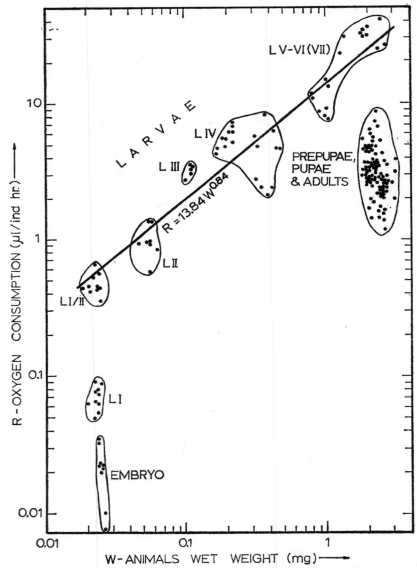

Fig. 4.5. Size dependence of respiratory rate of different developmental stages of *Tribolium castaneum*–cI. Regression equation for larvae II–VI (VII): R = 13.84·W$^{0.84}$. (After: Klekowski, Prus & Żyromska-Rudzka 1967).

Kleiber, 1961; Hemmingsen, 1960; Petrusewicz *et al.*, 1970). Published values for regression coefficients vary considerably but most lie between 0·6 and 0·8. Winberg (1950, 1956) gives 0·81 for aquatic crustaceans and 0·80 for freshwater temperate fish; Klekowski *et al.* (1972) obtained a value of 0·72 for soil inhabiting nematodes. Vernberg *et al.* (1970) present a table of b-values determined under different conditions which range from 0·423 to 1·05. Where authors have tested such differences in b-values, they have been found to be not significantly different, due to the variability of respiratory measurement (Schiemer *et al.*, 1974). Further examples and a fuller discussion can be found in Zeuthen (1947, 1953, 1970), Hemmingsen (1960) and Kleiber (1961).

This log-linear regression does not always describe adequately the respiration of a species throughout the whole of its life cycle. Thus, Fig. 4.4 shows regression equations which are different for the naupliar and older stages of *Macrocyclops albidus;* in this case, it may be that the naupliar regression covers too narrow a range of body weight to be reliable. But figure 5 demonstrates clearly that in *Tribolium* the calculated regression applies only to the larval stages which were intensively feeding and growing and not to the eggs, the non-feeding first larvae, the prepupae, the pupae and adults. Zeuthen (1970) suggests that in general the slope of the regression tends to decrease throughout development up to the stage in which growth ceases.

Activity in mammals causes an increase in the intensity of metabolic rate accompanied by a tendency for 'b' to decrease (Locker 1961;, Bertalanffy, 1964). In fasting or starving *Tenebrio* larvae, both the metabolic intensity and the regression slope decreases, from the non-fasting levels (Bertalanffy and Müller, 1941, from Bertalanffy, 1964) whereas in feeding larvae, the nature of the food eaten affects their metabolism (Ito and Fraenkel, 1966). Gyllenberg (1969) reports a gradual increase in both coefficients with temperature in grasshoppers acclimatised to four different temperatures.

4.3.4. Predictions and comparisons of metabolic rates

The term W^b is called the 'metabolic body size' by Kleiber (1961) and the 'metabolically effective body size' by Brody (1945) and represents the function of weight to which the animal's metabolic rate is proportional. When the generalised relationship between metabolism and individual body weight has been worked out for a homogeneous taxonomic group, it may be possible to use it to predict the approximate levels of metabolism of the animals belonging to the group on the basis of their weight by multiplying the metabolic body size (W^b) by the factor 'a' .For example, Brody (1945) states that the standard metabolic rate of mammals can be calculated from $70 \cdot kg^{3/4}$ kcal per day or from $3 \cdot kg^{3/4}$ kcal per hour. Winberg's published regressions for aquatic crustacea ($0 \cdot 165 \cdot g^{0.81}$, 1950) for molluscs ($0 \cdot 05$ to $0 \cdot 31) \cdot g^{0.75}$, (Winberg and Beliazkaya 1959) and for temperate freshwater fish ($36 \cdot g,^{0.8}$ 1956) can be used in the same way. Brody (1945) provides a table which gives the already calculated numerical values of W^b for values of W ranging from $0 \cdot 01$ to 2000 and for eight levels of the exponent; appendix 24 in Kleiber (1961) supplies similar information.

Metabolic body size also provides the best basis for comparison of the metabolic levels of different species. Kamler (1970) has published a useful comparison of the metabolism of seven arthropods (two insects, three crustacea, one pyriapod and one acarine) on this basis and this is reproduced, with some correction, in Table 4.1. In three species, the regression equation is based on respiratory studies for the whole life cycle but in others on parts of the life cycle only; the values for the coefficient 'a' (= respiratory rate per W^b) were converted to a standard temperature of 20°C to facilitate comparison. As can be seen in the table; values for '$a_{20°C}$' greater than $1 \cdot 0 \mu l$ oxygen/hr are not very frequent. For feeding larvae of *Tribolium castaneum* '$a_{20°}$' equals $6 \cdot 65 \mu l$ oxygen/hr which is unexpectedly high value but, as is shown in Table 4.2, is due to the intense metabolism of the larval stages (up to 15 μl oxygen/indiv.·hour) compared with the adult values between 1–2 μl oxygen/indiv.·hr. The larval stonefly, *Perlodes intricata*, also shows high '$a_{20°}$' values ($1 \cdot 6 \mu l$ oxygen/hr) but, as Kamler points out, this species normally lives at low temperatures, was measured at 10·5°C and 'a' was calculated only from the most metabolically intense stages of the life cycle. Notice that in four of the seven species, the exponent 'b' lies between 0·830 and 0·874 and in three of these the regression covers the whole life cycle.

Such comparisons can be made, although with caution, even when only a few respiratory rates are available for a species, provided that the metabolic body size (W^b) is known for the taxonomic group; the reason for caution is demonstrated in Fig. 4.5 where it is clear that the regression for *Tribolium* applies only to the actively growing larval stages and not to the eggs, larvae I, pupae and adults. An example of such comparison is given in Fig. 4.6 for several hymenopteran species and *Trioblium*. Very little information exists on hymenopteran respiratory rates, although Winberg (1965b) has published a regression ($R = 0 \cdot 240 \cdot W^{0.91}$) for the pupal stages of six species of ants.

Table 4.1. Comparison of the metabolism of seven arthropodan species (modified from Kamler, 1970).

Species	Range of live weights (W), mg	Experimental data O_2 consumption per indiv. ($=aW^b$), $\mu l/$ ind. hr	Calculated value of 'a', converted to 20°C, $t°C$ $\mu l/hr^*$		Period of life cycle measured	Authors
Macrocyclops albidus (Copepoda)	0.001–0.040	$0.404.W^{0.840}$	21	0.372	whole	Klekowski *et al*, 1966
Rhisoglyphus echinopus (Acarina)	0.001–0.050	$0.711.W^{0.830}$	24	0.510	whole	Stępień, unpubl.
Simocephalus vetulus (Cladocera)	0.017–0.350	$0.605.W^{0.874}$	22	0.512	whole	Klekowski, Ivanova, unpubl.
Tribolium castaneum (Coleoptera)	0.020–3.500	$13.840.W^{0.844}$	29	6.653	all feeding larvae	Klekowski *et al.*, 1967
Asellus aquaticus (Isopoda)	1.580–22.530	$0.450.W^{0.867}$	23	0.350	first 100 days	Prus, 1972
Polydesmus complanatus (Myriapoda)	5.00–90.00	$1.080.W^{0.680}$	23	0.841	2nd half of cycle	Stachurska unpubl.
Perlodes intricata (Plecoptera)	90.00–200.0	$0.631.W^{0.659}$	10.5	1.600	2nd half of larval life	Kamler unpubl.

*Properly: μl per animal of unit weight (mg) per hour.

Golley and Gentry (1964) have published respiratory rates for worker and soldier southern harvester ants (*Pogonomyrmex badius*) which are very similar although the soldier ants weighed about three times the weight of worker ants. This implies very intense metabolism in worker ants and may be one cause for the relatively high respiratory loss per unit production in this population which is the one invertebrate population which does not lie along Engelmann's (1966) regression line for log annual population production against log annual population respiration for invertebrates (shown in Fig. 4.7). There may be other causes (see Engelmann 1966; McNeill and Lawton, 1970) but here in Fig. 4.6. we attempt to compare the metabolic level per W^b,

Table 4.2. Respiratory rates of *Tribolium castaneum* throughout its life cycle (data from Klekowski Prus and Żyromska–Rudzka, 1967).

Stage	Age	Mean wt (fresh, mg)	Mean respiratory rate (μl/ indiv.hr)		Mean metabolic rate (μl O_2/mg. hr)
			Measured at 29°C	Calculated at 20°C	at 20°C (calculated from R/W)
L I	4	0.023	0.075	0.031	1.58
L I/II	6	0.022	0.480	0.031	10.48
L II	8	0.056	1.012	0.481	8.59
L III	10	0.117	3.451	1.660	14.19
L III	12	0.202	5.615	2.701	13.37
L IV	14	0.377	4.177	2.009	5.33
L V	16	1.003	10.405	5.005	4.99
L V	18	1.89	31.945	15.366	8.13
L VI/VII	20	2.26	23.541	11.323	5.01 and 1.14
L PP	22	2.338	2.771	1.333	0.57
P	24	2.595	2.536	1.220	0.47
P	28	2.498	1.869	0.899	0.37
P	31	2.461	2.865	1.378	0.56
A M&F	33	2.131	2.969	1.428	0.67
A M&F	35	2.035	3.765	1.811	0.89
A M&F	44	2.032	4.859	2.337	1.15
A M&F	57	2.101	5.328	2.563	1.22

using Winberg's $W^{0.91}$ of these two adult ant castes with some unpublished data on Polish *Apis mellifica* (Hymenoptera) and the various developmental stages of *Tribolium castaneum* (Coleoptera; Klekowski *et al.*, 1967).

In order to plot the R-W regression lines for *P. badius* in Fig. 4.8, their $W^{0.91}$ was calculated for the fresh dry weights of soldiers and worker ants quoted by Golley and Gentry (1964) and the value for '$a_{20°}$' by dividing their individual respiratory rates by their $W^{0.91}$ (i.e. a = $R/W^{0.91}$); the resulting regression is shown in Fig. 4.6. Ten over-wintering adult *Apis mellifica* were confined individually in 15 ml respiratory vessels in March 1969 and their oxygen consumption was measured at 22°C in a constant pressure respirometer: their mean oxygen consumption was 152 μl oxygen·hr for individuals weighing 110–115 mg fresh weight; these data were used to determine the level for '$a_{22°}$' and a regression for various weights plotted in Fig. 4.6. The *Tribolium* regression was plotted from the data given in Table 4.2, where

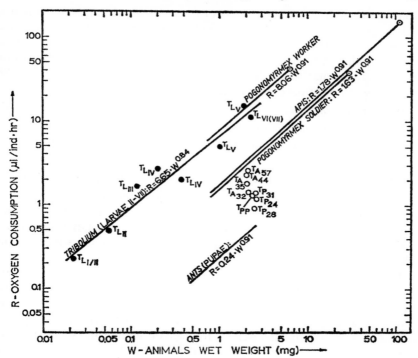

Fig. 4.6. Size dependence of respiratory rate. *Tribolium*: Regression for all feeding larvae–T_L full circles, empty circles–other stages: PP—prepupae, P—pupae, A_{57}—adults 57 days old (after: Klekowski, Prus, Żyromska-Rudzka, 1967). Ants (Pupae)—regression, from Winberg (1956b). *Pogonomyrmex*—data from Golley & Gentry (1964). *Apis*—unpubl. data. Regressions for *Pogonomyrmex* and *Apis* calculated taking b = 0.91, from Winberg (1956b).

individual rates measured at 29°C were re-calculated for 20°C, using a method described in the next section, and plotted against their mean fresh weights. The pupal ant regression was plotted from Winberg's $0.240 \cdot W^{0.91}$.

Figure 4.6 is quite illuminating and demonstrates that, provided that Golley and Gentry's respiratory measurements are reliable, that worker ants of *P. badius* have a more intense metabolism then either soldiers ants of *P. badius*, adult *Apis millifica* and adult *Tribolium*, whereas all these respire more intensely than pupal ants or pupal *Tribolium*. What is striking is that the metabolic level of adult worker ants of *P. badius* is comparable with that of actively growing *Tribolium* larvae, which themselves have an intense insect metabolism (see Table 4.2) It would be very informative to check the respiration of the adult *P. badius* as well as to measure the oxygen consumption of the other developmental stages. The comparable oxygen consumptions per W^b for the species mentioned above are therefore, in order of magnitude:

Worker of *P. badius*	8·06 μl oxygen/hr
Growing larvae of *T. castaneum*	6·65 μl oxygen/hr
Adult over-wintering *A. mellifica*	1·78 μl oxygen/hr
Soldier of *P. badius*	1·63 μl oxygen/hr
Pupae of several ant species	0·24 μl oxygen/hr

It is, of course, necessary to be cautious about predicting the metabolic level of active summer individuals of various weights from oxygen measurements of over-wintering individuals; the difference in metabolism between 'resting', diapausing, dormant animals and active ones can be as much as one order of magnitude.

There is also the possibility, mentioned earlier, of calculating the approximate order of magnitude of the metabolism of a field population from the respiratory-body weight regression for the species or even for the taxonomic group. Good population data is needed for this (numerical density, size analysis and mean individual weights for each developmental stage) collected either at frequent intervals or less frequently but with some information on growth, reproductive and mortality rates. It is important to recognise that only the individual body weight (usually the mean value) of an instar or age class can be substituted into the formula $a \cdot W^b$ in order to predict the individual respiratory rate of that stage, which is then multiplied by the numerical density in the field to give the total respiratory rate of the stage per unit area. Any calculation taking the form $a(N \cdot W)^b$ or $a(B)^b$ instead of $N \cdot aW^b$ will give erroneous results since the exponent is applied to N as well as W; for rough estimates of population respiration based upon biomass rather than numerical data, see the 'best estimate' method described by Phillipson 1970.

An example of the above is Mann's work (1964, 1965, and reviewed in 1969). He calculated the respiratory requirements for several species of fish in the River Thames from Winberg's regression for cyprinid fish ($R = 0·336 \cdot W^{0.8}$ at 20°C, where R is in ml oxygen per hour and W is in g) with appropriate temperature corrections for the average temperature of the river in each month (predicted from Krogh's 1916 curve) and using the value of W appropriate to the average weight of fish in each age class during the year. Later (1965), by assuming that the whole of the year's growth and mortality occurred at a constant relative rate during the months April to September, he was able to include in his calculations the decline in numbers and increase in weight of the fish during the course of the year. Before starting, Mann established experimentally that the temperature function of respiratory rate for the fish was adequately described by Krogh's curve and that Winberg's cyprinid regression was applicable to the Thames fish; the metabolic rate of fish in nature was taken to be about twice the routine level given by the regression, again from Winberg (1956).

4.3.5. Temperature

Most organisms will tolerate a certain range of temperature, above and below which temperature becomes lethal or sub-lethal, the organism becomes 'dis-orientated', normal activities are inhibited and it dies after a shorter or longer period of time. Within the tolerable temperature range of Fry's (1947) potential temperature range of activity, the organism can carry out normal activities but the level of environmental temperature exerts a major influence on the speed of various biological processes. Some quantitative expression of this influence would be useful in energetic studies.

One of the most widely adopted methods for comparing the magnitude of the effect of temperature on the velocity of different rate processes such as chemical reactions or physical or biological processes is Van't Hoff's Q_{10} approximation which is the factor by which the velocity of a rate process is increased for a rise in temperature of 10°C. Thus,

$$Q_{10} = \left(\frac{V_2}{V_1}\right)^{\frac{10}{t_2 - t_1}} \tag{5}$$

where V_1 and V_2 are the velocities of the process at temperature t_1 and t_2, respectively. From which:

$$\frac{V_2}{V_1} = Q_{10}^{\frac{t_2 - t_1}{10}} \tag{6}$$

and

$$V_2 = V_1 \cdot Q_{10}^{\frac{t_2 - t_1}{10}} \tag{7}$$

Various physical properties and simple chemical reactions tend to have characteristic and uniform values of Q_{10} for various ranges of temperature but this is not true of biological processes in whole organisms, such as respiration, growth and feeding. Moreover, the temperature responses of such processes in whole animals is much less regular than those for enzymatic reactions for which the Q_{10} approximation was first designed. This is probably because whole organisms respond to environmental stimulation in an integrated way and have the capacity to regulate their metabolism. Thus the phenomena of acclimatisation in poikilotherms and thermo-regulation in homoiotherms complicate the situation.

Although it has proved practically impossible to discover some simple relationship between respiratory rate and temperature in homoiotherms because of the heavy energy demands of thermo-regulation, the situation is somewhat better in poikilotherms. In 1914, Krogh (1914, 1916, 1941) produced an empirical curve relating respiratory rate with environmental temperature based on measurements of the resting metabolism (Krogh's standard metabolism) of a series of poikilotherms (a frog, a goldfish, *Tribolium* pupae) and a curarised dog. By plotting these measurements so that the maximal and minimal respiratory levels for each species coincided, he obtained a simple empirical relationship of increase in metabolism with increase in temperature which fitted the temperature responses of all these species. This curve has been called 'Krogh's normal curve' and since then has been the subject of much discussion as well as the basis for comparison with new respiratory measurements for other species. A detailed discussion on the biological significance of Krogh's curve can be found in Winberg (1956, 1968). This author considers that, for practical applications, it is the best statement, relating temperature with respiration and growth which is available at present, and has been applied to both aquatic and terrestrial poikilotherms.

One practical application of Krogh's curve is the possibility of calculating an unknown rate R_2 at a desired temperature t_2 from a rate R_1 measured at the temperature t_1. It is possible to insert the known values into the equation (7) but this calculation is complicated by the fact that the Q_{10} of biological processes varies with temperature (as is shown by the shape of Krogh's empirical curve) so that several values of Q_{10} may have to be applied for the temperature range between t_1 and t_2. Winberg (1956, 1968) has published the values of Q_{10} for various temperature intervals taken from Krogh's curve and these are given below:

Temperature interval: 5–10°C 10–15°C 15–20°C 20–25°C 25–30°C
Q_{10} values 3·5 2·9 2·5 2·3 2·2

In order to simplify the labour involved in temperature conversions of respiratory or growth rates, Winberg also published a table of 'correction coefficients' $\left(q = Q_{10} \dfrac{20-t}{10} \right)$ for each range of temperature between the measured one and an arbitrary standard temperature of 20°C (Table 4.3).

Table 4.3. Temperature corrections (q) for converting respiratory rates measured at temperature t to 20°C, according to the 'normal curve' (Winberg 1956, 1968).

t°C	q	t°C	q	t°C	q	t°C	q
5	5.19	11	2.40	17	1.31	23	0.779
6	4.55	12	2.16	18	1.20	24	0.717
7	3.98	13	1.94	19	1.09	25	0.659
8	3.48	14	1.74	20	1.00	26	0.609
9	3.05	15	1.57	21	0.920	27	0.563
10	2.67	16	1.43	22	0.847	28	0.520
						29	0.481
						30	0.444

With the aid of this table it is possible to 'correct' all respiratory measurements made at different temperatures (t) to one arbitrary temperature of 20°C using the appropriate values of q:

$$R_{20°} = R_t \cdot q \tag{8}$$

It is also possible to convert respiratory rates (R_t) measured at one temperature (t') into the unknown respiratory rate ($R_{t'}$) at another temperature (t'') from the ratio of q' and q'':

$$R_{t''} = R_{t'} \cdot \frac{q'}{q''} \tag{9}$$

Where: q'—is the coefficient for converting respiratory rates from t' to 20°C, and q''—from t'' to 20°C.

Thus, in a practical example, the respiratory rate of the third larvae of *Tribolium castaneum*, measured at 29°C, was 3·451 μl oxygen/ind·hr. The respiratory rate at 20°C can be calculated from equation 8, applying the value q $= 0.481$; $R_{20°} = 3.451 \cdot 0.481 = 1.66\mu$l oxygen /ind·hr. The respiratory rate at 23°C can be converted from $R_{29°}$ from equation (9), applying the q values 0·481 and 0·779: $R_{23°} = 3.451 \cdot \dfrac{0.481}{0.779} = 2.14\mu$l oxygen /ind·hr.

In those cases where a series of respiratory measurements have been carried out on a species at different temperatures, it is also possible to obtain approximate values of Q_{10} by a graphical method. A semi-log plot of respiration against temperature will reveal whether the relationship is linear for any part of the temperature range and Q_{10} values can be read off directly for any 10°C interval.

Any comparison of respiratory data, one's own or from the literature has to take into account the temperature at which the measurements were made. Equation (8) provides a quick way of converting the measurements to comparable rates at one temperature. Such a conversion assumes that the temperature influence on respiration is adequately described by Krogh's curve under all circumstances, which may not always be true.

Temperate species with fairly long life cycles complete their development under thermal conditions which vary throughout the year. Where it is not possible to measure the respiratory rate of the various life stages at their field temperature, laboratory measurements made at one or several temperatures can be converted to rates for the temperatures at which each stage normally lives in nature. Again, this assumes the validity of Krogh's curve at all seasons of the year and for all developmental stages.

In some cases, the respiratory data may be a little thin, for one size group only or may have been taken from the literature. For very rough estimations of the population metabolism, it may be possible to calculate the respiratory rates for the remaining sizes of the species from a respiration—body weight relationship above and earlier. Sometimes, even such a great approximation may be useful.

Obviously, Krogh's curve greatly over-simplifies the effects of temperature on the complex metabolism of whole organisms. Winberg (1956) himself suggests that the most reliable approximations are obtained only within the rather narrow middle region of a species range of tolerated temperatures. He further emphasises that great care must be taken not only at temperatures approaching the critical maximal and minimal sub-lethal levels but also with species living in habitats with very low or high temperatures. In those species where acclimatisation to low or high temperatures cause increased or decreased metabolism relative to the metabolic levels predicted by Krogh's curve for these environmental temperatures (Scholander, Flagg, Walter and Irving, 1953), application of 'corrections' will produce erroneous results.

Winberg's recommendation that the middle region of a species' range of tolerated temperature is where Krogh's curve is most reliably applicable is disputed by Stroganov (1956, 1962). On the basis of a series of experiments on temperature acclimatisation in fish (*Gambusia*) this author states that in the zone of temperature to which an animal has been acclimated or in which it normally lives, the respiration-temperature curve flattens, that is, Q_{10}s are normally low or zero, so that 'corrections' for temperature should not be needed.

Figure 4.7 represents an attempt to produce a synthetic diagram relating metabolism of an aquatic poikilotherm with environmental temperature, locomotory activity and state of nutrition, based upon the work of various authors. On the ordinate is heat production or oxygen uptake and along the abscissa the range of environmental temperature tolerated by a species, that is between the upper and lower lethal levels. The controversial zone is that labelled 'range of relative temperature independance' or 'zone of relative thermoneutrality' which we conjecture as a characteristic phenomenon in anyway aquatic poikilotherms. The curve of standard metabolism with its flat plateau in the region of acclimatisation temperature is taken from Stroganov (1956, 1962) and for other temperatures from Krogh's curve (1914, 1916, 1941). The relationship between the various levels of standard, routine, active and starvation metabolisms come from Fry (1947, 1957), Fry and Hart (1948) and Kausch (1969). The increased oxygen consumption in actively feeding animals, which Brody (1945) names 'specific dynamic-action' is taken from Conover (1966) in copepods, Karpevich (1957) in fish, and Kausch (1969) in young carp. Graham (1949 in Precht, 1955) demonstrates that the temperature at which Precht's 'turning-point' occurs is lower in active fish than in inactive fish. That the maximal standard metabolism at the maximum temperature reaches a similar level (apart from losses in efficiency due to thermal losses) to that of maximal active metabolism within the range of relative thermal independence is pure speculation, as also is the coincidence of the species' maximal possible active metabolism with the upper region of the species' range of relative thermal independence. The narrowing of the range of temperature independence together with increase in slope of the respiration-temperature curve with increased activity is also conjectural.

K

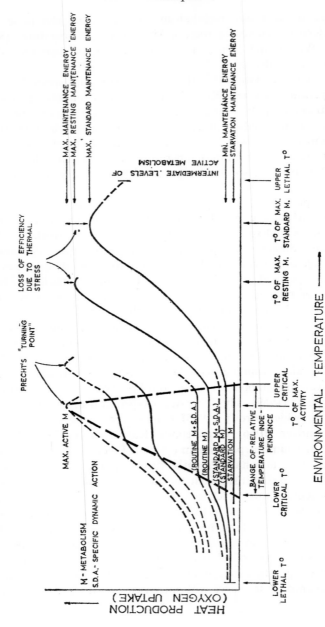

Fig. 4.7. Compiling diagram relating metabolism of aquatic poikilotherms with environmental temperature, locomotory activity and state of nutrition.

Newell (1969) has reviewed some recent evidence which seems to support some aspects of this 'model' and his distinction between the effects of short-term and long-term temperature fluctuations on metabolism and activity have clarified the general picture. He suggests that the active metabolism of some inter-tidal and freshwater invertebrates varies markedly with short-term fluctuations in temperature whereas their standard metabolism remains relatively independent of such temperature changes over the normal seasonal temperature prevailing in their habitats. The latter conclusion is based on the shape of the respiration-temperature curve for the standard metabolism of whole *Gammarus oceanicus* (Halcrow and Boyd, 1967), on the results of regression analysis in which the line relating the minimal rate of respiration (Newell and Northcroft, 1965, 1967; Newell and Pye, unpublished) and on observations on the mean rate of respiration of intact, relatively quiescent limpets (Davies, 1966, 1967) and freshwater *Trichoptera* and gammarids (Collardeau, 1961; Collardeau-Roux, 1966; Roux and Roux, 1967). Longer-term or seasonal fluctuations of temperature, Newell suggests profoundly influence the standard rate of metabolism in that the temperature-independent region of the respiration-temperature curve becomes appropriate to the temperature prevailing in the environment (Newell, 1969).

4.4 Discussion on population efficiencies, ratios and rough predictions of population production.

A considerable literature already exists on ecological efficiencies (Lindeman, 1942; Kozlovsky, 1968; MacFadyen, 1948, 1963, 1967; Odum, 1959; Patten, 1959; Slobodkin, 1960, 1962; Wiegert, 1964; Winberg, 1967, 1962; Ivlev, 1939). Recently Kozlovsky (1968) has reviewed the literature on ecological efficiencies and defined these as non-dimensional ratios of various parameters of energy flow in or between the trophic levels of a natural community, in or between populations of organisms or in or between organisms; usually the energy flow within a single organism is considered as a physiological phenomenon unless used in an ecological context. In section 2.4, efficiencies derived from instantaneous and cumulated energy budgets for individual organism were considered; here, it is mainly population efficiencies which are considered, their significance and how they may be used to predict population productions.

Chapter 4

Table 4.4 presents the main ecological efficiencies employed by various authors and analysed by Kozlovsky (1968); in this table, the efficiency is given in symbolic form, verbally and according to the terminology used by various other authors. They are divided into two groups, those concerned

Table 4.4. Some ecological efficiencies (modified from Kozlovsky, 1968).

Symbols	Meaning	Equivalent terminology and authors
1.1. Efficiencies within one trophic level (n)		
$\dfrac{A}{C}$	$\dfrac{\text{assimilation at n}}{\text{consumption at n}}$	:assimilation efficiency (Clarke, 1946; Odum 1957, 1959; Gerking, 1962; Winberg 1962, 1968) :efficiency of digestion (Engelmann, 1966)
$\dfrac{NP}{C}$	$\dfrac{\text{net production at n}}{\text{consumption at n}}$	:ecological growth efficiency (Odum, 1957, 1959; Gerking, 1962) :energy coefficient of growth of first order (Ivlev, 1939) :gross efficiency of growth (Richman, 1958) :efficiency of assimilation (Patten, 1959) :coefficient of utilization of consumed energy for growth (Winberg, 1962, 1968)
$\dfrac{NP}{A}$	$\dfrac{\text{net production at n}}{\text{assimilation at n}}$	:growth efficiency (Clarke 1946; Odum & Smalley, 1959) :tissue growth efficiency (Odum, 1957; Gerking, 1962) :net efficiency of growth (Richman, 1958) :energy coefficient of growth of second order (Ivlev, 1939) :coefficient of utilization of assimilated energy for growth (Winberg ,1962)
$\dfrac{R}{NP}$	$\dfrac{\text{respiration at n}}{\text{net production at n}}$	:respiratory coefficient (Lindeman, 1942)
2. Efficiencies between trophic levels (n, n−1)		
$\dfrac{C_n}{C_{n-1}}$	$\dfrac{\text{consumption at n}}{\text{consumption at n}-1}$	:Lindeman's efficiency, ratio of intake of trophic levels (Odum, 1957) :ecological efficiency (Slobodkin, 1960, 1962; MacFadyen, 1963) :gross efficiency of yield to ingestion (Wiegert, 1964) :efficiency of transfer to next higher level in terms of ingested energy (Patten, 1959)
$\dfrac{A_n}{A_{n-1}}$	$\dfrac{\text{assimilation at n}}{\text{assimilation at n}-1}$	:progressive efficiency (Lindeman, 1942) :Lindeman's efficiency, ratio of intakes of trophic levels (Odum, 1957) :biological efficiency, progressive relative efficiency at a given level in terms of relative productivities (Allee *et al.*, 1949) :biological efficiency (Dineen, 1953) :trophic level energy intake efficiency, Lindeman's

		efficiency, trophic level assimilation efficiency (Odum, 1959)
$\dfrac{P}{C}$	$\dfrac{\text{energy passed to } n+1}{\text{consumption at } n}$	:ecological efficiency (MacFadyen, 1963; Slobodkin, 1960, 1962) :gross efficiency of yield to ingestion (Wiegert, 1964) :efficiency of transfer to next level in terms of ingested energy (Patten, 1959)
$\dfrac{P}{NP}$	$\dfrac{\text{energy passed to } n+1}{\text{net production at } n}$	:efficiency of transfer to next level in terms of assimilated energy (Patten, 1959)

with energy flow within one trophic level and those with energy flow between trophic levels.

4.4.1. Assimilation efficiency $\left(\dfrac{A}{C} \cdot 100; U^{-1}\right)$

Published assimilation efficiencies for the mammalian vertebrates are usually higher (40–90%) than that of invertebrates (10–70%) whereas the proportion of assimilated energy being utilised for net production is consistently much lower in the mammals (less than 3%) than in invertebrates (up to 90%). The low production efficiency in mammals is clearly associated with their very high maintenance requirements as is shown in their extremely high R/P ratios (up to 90·0) compared with the poikilotherm invertebrates (less than 5·0).

The difference in assimilation efficiency of consumed food is also related to feeding type in invertebrates, or rather, to type of food associated with different feeding types. Species belonging to the detritus chain show the lowest assimilation efficiencies of 22–27% (Oribatei—Engelmann, 1961; *Oniscus asellus*—Phillipson and Watson, 1965) although Levanidov (1949) quotes a value of about 50% for *Asellus aquaticus*, which seems rather high for a detritus feeder, especially as Prus (1971) has obtained values between 26–35% which he believes to be the most reasonable estimates of percentage assimilation by *A. aquaticus*. However, in females of this species assimilation efficiency seems to be higher (41%) and especially in ovigerous females (44%) as compared with males (33%) Prus (1971). An assimilation efficiency of about 50% was also obtained by Ivlev (1945) for the aquatic worm *Tubifex tubifex* fed for 24 days on silt which came from a very polluted pond and with a high calorific value of 5·28 calories per mg. Evans and Goodliffe (1939) record an assimilation efficiency of 46% for the deposit-feeding

Tenebrio (which feeds upon the medium in which it lives). It appears that the amount of assimilable material in the food of detritus-deposit feeders may limit their assimilation efficiency. Herbivores belonging to the grazing chain are feeding on more nutritious food of higher calorific value but still with fairly high proportions of indigestible material; their assimilation efficiencies are somewhat higher than those recorded for detritus feeders but still low (37–40%; *Onychiurus* Healey, 1967; grasshoppers—Wiegert, 1965). Carnivores, feeding on animal food and able by their capture technique to select the most nutritious parts, also show high assimilation efficiencies of more than 50% (*Mitopus mori*—Phillipson, 1960a, b; *Lestes sponsa*—Fischer, 1967; *Macrocyclops albidus*—Klekowski and Shushkina, 1966). Whereas, the fluid food of *Philaenus* and blood suckers, containing the lowest proportions of indigestible matter, produces the very highest invertebrate assimilation efficiencies of 66% (*Philaenus* sp.—Wiegert, 1964).

The need for rough or short-cut methods for analysing the trophic-dynamic inter-relationships of complex communities have led several workers to search for relationships between energy parameters which may prove useful to predict total energy flow through populations belonging to different trophic levels. These attempts to establish such relationships are reviewed below.

4.4.2 Engelmann's relationship between annual production and annual respiration in animal populations.

Engelmann (1966) examined the relationship between log annual population production and log annual population metabolism obtained from fourteen published studies (see his Table 8). When the points for homoiothermic and poikilothermic species were separately considered, he found that the data for poikilotherms (9 populations) showed a linear relationship between population respiration and production, with the exception of one invertebrate population, that of the southern harvester ant (*Pogonomyrmex badius*—Golley and Gentry, 1964). The relationship for the poikilotherms is given by the equation $R = 4 \cdot 17 \cdot P^{0.86}$ and for the homoiotherms, $R = 389 \cdot P^{1.75}$, where both P and R are in $kcal.m^{-2}.yr^{-1}$, demonstrating the very high respiratory cost of homoiothermic production. Englemann discusses these differences, suggesting that they distinguish thermoregulators from non-thermoregulators rather than vertebrates from invertebrates because of the high respiratory cost of homoiothermy; in general, poikilotherms are more efficient producers, although less efficient assimilators then homoiotherms whose production efficiencies vary between 1–3%. Engelmann suggests, as do McNeill and Lawton (1970), that these relationship may prove useful in predicting total energy flow through populations in which either P or R is known.

McNeill and Lawton (1970) have re-plotted this relationship between annual population production and annual population respiration, using a larger number of studies (53 species from 29 studies), including aquatic as well as terrestrial populations. On calculating regression lines separately for homoiotherms and poikilotherms, they found that their slopes to be almost parallel (0·9812 for homoiotherms; 1·0137 for poikilotherms; but neither differed significantly from 1·0), but the scatter was greater in the poikilotherm data. McNeill (1969) had already pointed out that the duration of the life cycle will affect the annual population P:R ratio, and McNeill and Lawton arbitrarily separated the poikilotherms into populations of species known to be long-lived (in which a proportion of individuals were known to exceed two years of age) from species with shorter generation times (less than two years). When a further regression line was calculated for the comparatively short-lived poikilotherms only, the slope of this regression was 1·1740 and did differ significantly from 1·0. A consequence of this was that the mean net population production efficiencies, calculated from the regression equation, increased markedly at the lower production levels. Most of the comparatively short-lived poikilotherms, below log P = O, overwinter as eggs so that respiration during this resting period is reduced to a minimum. Most of the comparatively short-lived poikilotherms, above log P = O, overwinter as juveniles or adults and respiration continues at a relatively high level when production is virtually zero. Although a regression could not be calculated for the long-lived poikilotherms, it is clear that the respiratory costs at any level of production tended to be higher compared with the short-lived populations because a standing crop of older individuals is maintained from year to year.

A linear relationship between log cumulated production and log cumulated respiration can also be demonstrated for an average individual throughout its life cycle. This is illustrated in Figs. 4.8 and 4.9 in which the energy parameters have been cumulated for different intervals in the life cycle of an average individual from a laboratory population of *Tribolium castaneum* and for different sample dates from the field population of the annual *Oligolophus tridens*. The calculated slope varies in *Tribolium* from 0·738 for

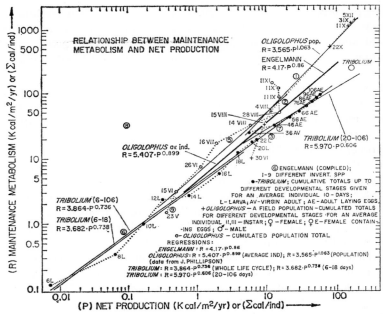

Fig. 4.8. Relationship between maintenance metabolism and net production.

growing larvae between 6–18 days old to 6·006 for late larvae, pupae, virgin and reproducing adults, ranging in age from 20–106 days and was 0·736 for the whole life cycle from 6–106 days. The life cycle of this species has two periods of active production, larval growth (6–18 days) and egg production (46–106 days) interrupted by a non-producing period from 20–36 days during which pupation occurs and the virgin adult emerges; Fig. 4.8. shows this period as a break in the otherwise linear regression for the whole life cycle. The calculated slope of the regression line for the growing average *Oligolophus* individual (from sample date 23.V to 28.VIII) was 0·899 but was much steeper (1·063) when calculated for the whole period when the species was actively present in the field. Up to 28·VII, the relationship between log cumulated production and log cumulated respiration was approximately linear, bearing in mind that these are based upon field samples but after 28.VIII no production occurs and only non-producing males and females are present with ever higher respiratory rates as they become more senile. It seems therefore that the linear relationship between log cumulated production

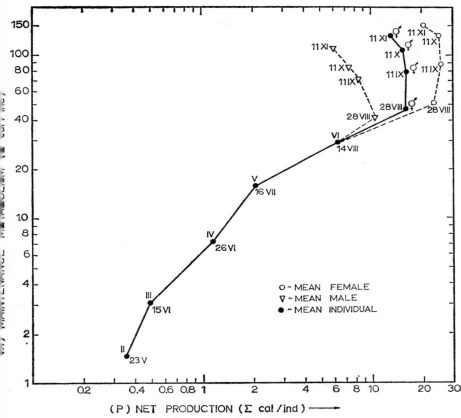

Fig. 4.9. Relationship between maintenance metabolism and net production in mean individual of *Oligolophus tridens* from a field population (data: from J. Phillipson). Empty circles—mean females, triangles—mean males, full circles—mean individuals. Dates near each point: end of instar periods.

and log cumulated respiration of individuals holds only for actively producing periods in the life cycle. There is an interesting parallel here with the linear regression between log individual respiratory rate and log body weight; sometimes the slope of the regression differs for different stages or a regression may be calculable only for the growing stages in the life cycle, as in *Tribolium* (Fig. 4.5). A plot of cumulated parameters probably reveal the relationship more truly than instantaneous ones.

4.4.3. Coefficients K_1, and ratios P/B and 'turnover rate'.

(a) K_1 and K_2

Ivlev (1939b, 1945, 1966) developed theoretically the concepts of food assimilability ($U^{-1} = \dfrac{A}{C}$, where A = (P + R)), the coefficient of utilisation of consumed food for growth (K_1) and the coefficient of utilisation of assimilated food for growth (K_2). He pointed out that assimilability of food depends on the chemical nature of the food material, the specific characteristics of the consumer's digestive system as well as, to some extent, on external environmental conditions, all of which are being confirmed as further studies are being completed. The coefficient K_1 is the ratio between production and food consumed $\left(\dfrac{P}{C}\right)$ and is the same as Odum's 'ecological growth efficiency' and K_2 is the ratio between production and assimilation $\left(\dfrac{P}{A} \text{ or } \dfrac{P}{P+R}\right)$ and has also been called the 'growth efficiency', 'tissue growth efficiency' or 'net efficiency of growth' by various other workers (see Table 4.4). These two growth coefficients were considered by Ivlev (1938, 1939b) as quantitative indices of changes in energy balance in an organism, that is, changes in the proportion of consumed or assimilated energy that was accumulated as growth energy. As indices, they were simple, easy to use, were very stable (as they indicated physiological processes) and were relatively unaffected by external environmental conditions. However, changes in energy balance occurred throughout the life cycle of an animal and he suggests that the growth coefficients have their maximal values in the early stages of development and show a regular reduction with age down to zero at the moment when growth ceases. This is illustrated in Fig. 4.10 which reproduces Ivlev's (1966) diagram of changes in the energy balance throughout the life cycle of a hypothetical animal; notice that U^{-1} also is maximal in young animals and decreases with age. Such changes in the values of the instantaneous growth coefficients can be summarised by calculating them in a cumulative form, provided the duration of each developmental stage is known; actual examples of such cumulative energy budgets and coefficients for various named species are given in an earlier section (pp. 32-52). Table 4.5 shows some values of K_2 calculated from the population quoted in McNeill

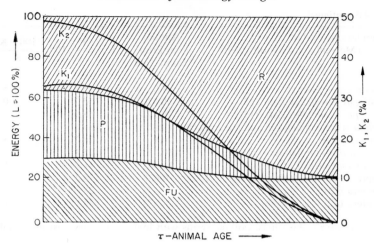

Fig. 4.10. Changes in the energy balance throughout the life cycle of hypothetical animal. Food ration (consumption—C) = 100%.

Table 4·5. Some efficiencies and ratios

A. Populations/kcal.m^{-2}.yr^{-1} (McNeill and Lawton 1970)	Annual Pop. Prod. /kcal.m^{-2}.yr	$K_2 = \dfrac{P}{A} \cdot 100$	$\dfrac{100 - K_2}{K_2} = \dfrac{R}{P}$
Homoiotherms			
Adenota kob thomasi Neumann (1)	0.81	1.30	76.1
Microtus pennsylvanicus p. Ord. (2)	0.517	2.95	32.9
Mustela rixosa allegheniensis (2)	0.013	2.34	41.8
Clethrionomys rutlus dawsoni Merriam (3)	0.00024	1.89	51.9
Tamiasciurus hudsoni prebli Howell (3)	0.00020	1.37	72.0
Microtus oeconomus macfarlani Merriam (3)	0.00024	1.82	54.0
Glaucomys sabrinus yukonensis Osgood (3)	0.00002	1.18	84.0
Pitymys subterraneus (De Selys Long) (4)	0.0955	1.57	60.0
Microtus agrestis (L.)			
Peromyscus polionotus Osgood (5)	0.12	1.79	54.7
Passerculus sandwichensis (Gmelin) (5)	0.04	1.11	88.7
Loxodonta africana (6)	0.34	1.46	67.6
Poikilotherms			
(a) Comparatively short-lived poikilotherms with low respiratory cost resting stages			
Leptoterna dolobrata (L.) (12)	0.064	50.8	0.97
Leptoterna dolobrata (L.) (12)	0.140	58.1	0.72
Leptoterna dolobrata (L.) (12)	0.096	52.3	0.92

Table 4.5 continued

A. Populations/kcal.m^{-2}.yr^{-1} (McNeill and Lawton 1970)	Annual Pop. Prod. /kcal.m^{-2}.yr	$K_2 = \dfrac{P}{A} \cdot 100$	$\dfrac{100 - K_2}{K_2} = \dfrac{R}{P}$
3 Orthoptera species (13)	4.0	15.6	5.40
Chorthippus parallellus Zett. (15)	0.420	55.3	0.81
Chorthippus parallellus Zett. (15)	0.273	58.9	0.70
Orchelimum fidicinium Rehn and Hebard (18)	10.8	36.7	1.72
Trichoptera (19)	20.9	23.7	3.23
Calopsetra dives (Johannsen) (19)	130.7	25.1	2.98
Philaenus spumarius L. (21)	15.963	41.2	1.42
Orthoptera: chiefly *Melanopus* spp.(22)	0.638	36.3	1.75
Orthoptera (22)	0.373	38.5	1.60
Orthoptera (22)	4.573	34.4	1.90

(b) Comparatively short-lived poikilotherms with high respiratory cost non-productive stages.

Pteromalus puparum (L.) larvae (4)	0.074	56.9	0.76
Pimpla instigator (Fabr.) larvae (4)	0.017	65.4	0.53
Oribatid mites (5)	0.43	21.4	3.67
Asellus aquaticus (6)	16.296	24.1	3.14
Onychiurus procampatus Gisin (7)	2.61	46.2	1.16
Pyrrhosoma nymphula (10)	4.436	58.2	0.71
Pyrrhosoma nymphula (10)	3.889	51.3	0.94
Ligidium japonica (16)	3.5	18.0	4.57
Japonaria laminata armigera (17)	26.9	46.5	1.15
Japonaria laminata armigera (17)	1.5	50.0	1.07
Limnodrilus hoffmeisteri claraparede (19)	160.5	24.9	3.01
Asellus miliaris Hay (19)	108.3	18.2	4.49
Pisidium virginicum (Gmelin) (19)	81.8	47.3	1.11
Anatopynia dyari (19)	37.4	15.2	5.59
Gammarus pseudolimneus Bousfield (20)	183.0	15.0	5.67
Physa integra Haldane (20)	170.6	27.9	2.58

c. Long-lived poikilotherms.

Allolobophora rosea (Sav).	2.717	17.58	4.68
Scrobicularia plana	17.5	24.3	3.12
Scrobicularia plana	124.0	20.65	3.84
Modiolus demissus Dillwyn	16.7	30.0	2.33
Rutilus rutilus (L.)	10.7	8.3	11.6
Alburnus alburnus (L.)	30.5	8.8	10.3
Leuciscus leuciscus (L.)	1.5	5.96	15.8
Perca fluviatilis (L.)	1.8	7.2	12.9
Gobio gobio (L.)	2.6	6.6	14.2

Littorina irrorata (Say)	40.6	14.0	6.1
Lithobius fortificatus (L.)	0.085	4.08	23.5
Lithobius crassipes (L. Koch)	0.074	5.51	17.1
Cherax albidua Clark	158.0		

(d) Cumulated individual budgets (see Section 2.3.5)	K_{2c}	$\dfrac{100-K}{K_{2c}}=\dfrac{P}{R}$	
Asellus aquaticus L. (Prus, 1972) (0–116 days)	18.2	4.49	
Lestes sponsa L. (Fischer, 1967) (larvae 0–37 days)	64.0	0.56	
Simocephalus vetulus O. F. Müller (Klekowski &			
Ivanova, in lit) ♀♀ immature 0–7 days	73.1	0.37	
♀♀ reproducing 0–20 days	64.7	0.54	
Ctenopharyngodon idella Val.			
on plant food (6 months, W=20–70g.) (Fischer, 1970)	14.5	5.90	
on animal food (6 months, W=40–120g) (Fischer, 1972b)	40.4	1.47	
Rhisoglyphus echinopus F. et R. (Klekowski & Stępień, in lit.)			
intra-generation budget, see section 2.3.5			
larva active (0–11 days)	35.2	1.84	
larva resting (0–13 days)	18.7	4.35	
protonympha active (0–16 days)	59.1	0.69	
protonympha resting (0–17 days)	49.5	1.02	
trichonympha active (0–20 days)	62.6	0.60	
trichonympha resting (0–21 days)	55.6	0.80	
average new hatched adult (0–22) days	60.0	0.67	
average adult + eggs (0–72 days)	70.0	0.43	
Tribolium castaneum Hbst. (Klekowski *et al.*, in press)			
prepupa (0–24 days)	29.4	2.40	
new hatched average adults (0–36 days)	20.8	3.80	
average adult + eggs (0–116 days)	44.2	1.26	

and Lawton (1970) together with some values of K_{2c} calculated from the cumulative individual budgets described earlier (p. 32); it is striking that for comparable species, the population K_2 shows similar levels with individual K_{2c}, the latter being less time-consuming to obtain.

In 1962, Winberg published an important paper in which he discusses the possibility of characterising the properties of ecosystems, namely their cycling of matter and transformation of energy, on the basis of certain general quantitative relationships of the organisms forming the ecosystem's component trophical levels. Important among these quantitative relationships are Ivlev's (1939a, b) indices (U^{-1}, K_1 and K_2) and the exponential equation relating resting metabolism and body weight, since these reflect the properties involved in the effectiveness of utilisation of matter and energy. His argument is summarised as follows: It is clear that,

$$C = P + R + FU \tag{10}$$

or, more usefully,

$$C = U (P + R), \tag{11}$$

where $P + R$ is the food energy assimilated and U is ratio of food energy consumed to that assimilated, i.e. $= \dfrac{C}{A} = \dfrac{1}{U^{-1}}$, where U^{-1} is the fraction of consumed food that is assimilated. Moreover, from Ivlev (1939b)

$$K_1 = \frac{P}{C} \tag{12}$$

From these three equations (10, 11, 12), incorporating six variables, three unknown variables may be calculated, provided that three others have been measured and that each parameter is expressed in calories for time^{-1}. However, according to Winberg, where the energy equivalent of a unit weight is known, it is possible and may be convenient to express parameters in terms of unit weight per unit time; thus, respiratory energy may be expressed in terms of the equivalent loss in body weight. Moreover, it may be ecologically more useful to quantify the food energy consumed as the food that is consumed per unit production, which is given by the 'food coefficient' or $\dfrac{1}{K_1}$ $\left(= \dfrac{C}{P} \right)$ although, where this has been calculated in weight units, it is necessary to multiply $\dfrac{1}{K_1}$ by the ratio of the calorific values of the food and the growth increment.

It may be preferable to utilise the coefficient K_2, thus adding a seventh variable, because this is a more stable index, according to Winberg, which is less affected than K_1 by the abundance and nature of the food available, its digestibility and other external environmental conditions. The product of $\dfrac{1}{U^{-1}} \cdot K_1 = K_2$ (Ivlev's second growth coefficient). The maximal known values for K_2 are between 0·7—0·8, achieved by micro-organisms under optimal conditions for growth or in embryonic growth of invertebrates and vertebrates. However, both K_1 and K_2 approach zero when growth ceases in adults; in field populations, the values of K_1 and K_2 will depend upon the age composition of the population. Ivlev's original definition of K_2 is expressed by

$$K_2 = \frac{P}{P + R} \qquad (13)$$

From these four equations (10, 11, 12, 13) and seven variables, it is only necessary to determine three of these in order to compute the remainder, as for example,

$$R = P \cdot \frac{(1 - U \cdot K_1)}{U \cdot K_1} \text{ or } FU = P \cdot \frac{(U - 1)}{U \cdot K_1} \qquad (14)$$

Winberg points out that it may even be possible to calculate very roughly the unknown variables of a simple ecosystem consisting of four trophic levels, on the basis of some reasonably realistic value for U^{-1} and K_1, taking the daily production of the animals occupying the fourth trophic level to be 1000 kcal.day and assuming, as do Lindeman (1942) and Engelmann (1966), that the production of each level is fully utilised as food by the animals of the next level, i.e. $P_1 = C_2$, $P_2 = C_3$, $P_3 = C_4$. Thus Table 1 in Winberg (1962), see below (Table 4.6).

Table 4·6. An example of the computation of the energy balance in hypothetical ecosystem, assuming as known the values marked*. Energy units per units time^{-1}: e.g. kcal per day.

Trophic level	K_1	1/U	R	FU	P	C
4	0	0.90*	900	100	0*	1000*
3	0.30*	0.90*	2000	333	1000	3333
2	0.25*	0.85*	8000	2000	3333	13333
1						
Total			10900	2433		

(Table 1 in Winberg, 1962).

Some rough idea of how large a biomass (B) of organisms could be responsible for the values of production calculated in the above table can be estimated in various ways.

One approach follows form formula 5, namely that,

$$P = R \cdot \frac{(U \cdot K_1)}{1 - U \cdot K_1} = R \cdot \frac{K_2}{1 - K_2} \qquad (15)$$

and so, dividing both sides by B,

$$\frac{P}{B} = \frac{R}{B} \cdot \frac{(U \cdot K_1)}{1 - U \cdot K_1} = \frac{R}{B} \cdot \frac{K_2}{1 - K_2} \qquad (16)$$

That is, for given values of K_1 and U, the production (growth and reproduction) per unit biomass (P/B) is proportional to the respiration per unit biomass (R/B). Knowing the values for P and R/B per unit time and the ratios K_1 and U (or K_2), some estimate for the biomass responsible for that particular value of P can be obtained.

Another approach involves the use of the inverse relationship between metabolism and body weight, expressed by the exponential equation, $Q = aW^b$; Winberg quotes some established equations which may prove useful, all recalculated as cal.day and for individual weights in grams:—

$Q = 422 \cdot 1 \, W^{0.75}$ (mammalian basal metabolism—Brody, 1945; Kleiber, 1961)

$Q = 19 \cdot 24 \, W^{0.81}$ (crustaceans at 20°C—Winberg, 1950)

$Q = 16 \cdot 54 \, W^{0.75}$ ('standard' metabolism of poikilothermic metazoan animals of various systematic groups at 20°C; 'active' metabolism can be taken to be twice this value—Hemmingsen, 1960)

$Q = 2.02 \, W^{0.755}$ (protozoa at 20°C—Hemmingsen, 1960)

$Q = 35 \cdot 8 \, W^{0.81}$ (resting metabolism of temperate fish at 20°C; under natural conditions, the 'active' metabolic rate can be taken to be twice this value in temperate fish and $1 \cdot 5$ times in tropical fish;—Winberg, 1956a, 1961).

As an example of the calculation involved in this method of estimating the numbers and biomass of organisms occupying the various trophic levels of the hypothetical ecosystem mentioned above, Winberg (1962) gives the following table (Table 2 from his paper, Table 4.7 below).

Thus, assuming that the fourth trophic level of the hypothetical ecosystem mentioned earlier is occupied by one mature non-growing mammal, whose

Table 4·7. An example of the computation of the number of individuals, and their biomasses, occupying four trophic levels of a hypothetical ecosystem (* values assumed to be known).

Trophic level	Organism	R Respir. of whole trophic level	w wt. of 1 indiv. in g	Q respir. of 1 indiv. cal. day	N number of individuals	B total biomass in kg
4	mammal	900	17.5×10^3	900.10^3	1	17.5
3	fish	2000	1	35.8	55.9×10^3	55.9
2	crustaceans	8000	10^{-4}	16.6×10^3	482×10^8	48.2
1	*Chlorella*		4×10^{-11}		267×10^{12}	10.7

daily respiration was 900 kcal and whose active metabolism was 1·5 times its standard metabolism described by $R = 422·1 \ W^{0.75}$, it can be calculated that its body weight is 17·5 kg. Further assuming that the third trophic level is occupied by tropical fish whose average body weight is, say, 1g and whose individual respiratory rate at 20°C is 35·8 cal.day, the number of fish occupying this level is given by the respiration of the whole trophic level ($R_3 = 2.10^6$ cal.day) divided by 35·8 cal.day and equals 55900 individuals and their biomass 55900 × 1g or 55·9 kg; had the average weight of each fish been 10g, the number of fish would be 8651 and the biomass 86·5 kg. These fish are postulated to feed on actively growing and reproducing crustaceans with an average individual body weight of 10^{-4} g and whose metabolism is 1·5 $(19·24W^{0.81})$ cal.day at 20°C which equals 0·01661 cal.day per individual at 20°C; the number of individuals is calculated from $\dfrac{8·10^6}{0·01661} = 482·10^6$ individuals or 48·2 kg.

The proportional relationship between the population production per unit biomass (P/B) and the populatilon respiration per unit biomass (R/B), for given values of K_2, as demonstrated in equation (16), can be used to predict the approximate values of one, knowing the level of the other. However, a more convenient expression of the metabolic intensity of a species or group is ml oxygen consumption per 1 g dry weight per hour: (Q/W). In order to compute the metabolic intensity Q/W represented by R/B the respiration R, cal.day, must be divided by the oxycalorific coefficient,

4·86 cal.ml oxygen consumed, and 24 hours and the biomass B, calories, by the calorific equivalent of 1g dry weight (c), so that

$$\frac{Q}{W} = \frac{R}{B} \cdot \frac{c}{4 \cdot 86 \times 24} \qquad (17)$$

or, substituting in equation (16),

$$\frac{Q}{W} = \frac{P}{B} \cdot \frac{c}{4 \cdot 86 \times 24} \cdot \frac{1 - K_2}{K_2} \qquad (18)$$

Using this equation, Winberg computed the following table of metabolic intensities for various values of daily or annual P/B coefficients and three levels of K_2, assuming that the biomass had a calorific value of 600 cal·g (Table 3 in Winberg, 1962 ; Table 4.8 below)

Table 4·8. Computed values of metabolic rates for various P/B coefficients and at three levels of K_2, assuming a calorific value of 600 cal/g.

P/B coefficients		Metabolic rate /ml O_2 per gram per hour/		
per day	per year	$K_2 = 0.5$	$K_2 = 0.25$	$K_2 = 0.1$
3	1095	15	45	135
1	365	5	15	45
0.3	109.5	1.5	4.5	13.5
0.1	36.5	0.5	1.5	4.5
0.03	10.95	0.15	0.45	1.35
0.01	3.65	0.05	0.15	0.45
0.003	1.10	0.015	0.045	0.135
0.001	0.365	0.005	0.015	0.045

At this stage in the development of the energetic approach to the study of ecosystems, such calculated relationship provide some basis for assessing how realistic are measured values, particularly of the neglected groups of micro-organisms.

b. P/B

In a later work (1968), where Winberg re-emphasises the usefulness of the P/B coefficient as an index of productivity or as a rate of production characteristic of a specific population under a given set of conditions, the author also points out that the coefficient will have a constant value, and so be meaningful, only when the population is in a stationary state, that is has an unchanging age structure and a constant level of biomass: for non-stationary populations, it will vary. It can therefore be calculated only for those periods of time when the population is reasonably stationary; such time intervals may be 24 hours, a week, a month or a whole year and then the production for this time interval is divided by the mean biomass for the same period of time. The coefficient must be labelled appropriately, the daily, weekly, monthly or annual P/B coefficient. Sometimes, it may be useful to calculate an annual P/B coefficient which represents the ratio of the annual production to the maximal or minimal biomass of that year; this should be denoted as the P/B_{max} or the P/B_{min} coefficient.

Winberg also points out that the commonly used American 'turnover rate' is analogous to the P/B coefficient. In steady state populations, the production will renew the biomass once every generation period; that is, during a period of time that equals the average life span of the members of the population, $P/B = 1$. In such a situation, the P/B coefficient per unit time^{-1}, e.g. per day, will equal the turnover rate, $1/L$ per unit time,$^{-1}$ e.g. per day, provided the time units are the same (where L is the life span in days). Thus, the turnover rate represents that proportion of the life span which is spent every day whereas the daily P/B coefficient represents that portion of the mean biomass that is produced daily. Both coefficients can be used to determine production, provided that the assumption of a nearly stationary population can be made.

c. The American 'turnover rate'

Edmondson (1970) suggests that the concept of turnover time and turnover rate can be applied to any component of the population, including numbers as well as chemical components. When dealing with numbers of a population of a steady size, the turnover rate is equal to the finite death rate or birth rate and the turnover time is the mean length of life. However, calculation of numerical turnover in a changing population is difficult on a finite basis because it is complicated by the fact that the birth rate and death rate are unequal. It is possible, however, to employ instantaneous coefficients for expressing numerical turnover in a changing population where the birth rate and death rate are unequal and the rate of loss or gain is proportional to the size of the population. Thus the replacement rate coefficient can be calculated from the fraction of the population that dies each day, which will be a constant fraction in a population growing in logarithmic phase.

References (see end of Chapter 2)

5

Calorimetry and Body Composition

5A MEASUREMENT OF CALORIFIC VALUE USING PHILLIPSON MICROBOMB CALORIMETER

T. PRUS

5A.1. Introduction.

In bioenergetic investigations it is most often necessary to know the calorific values of animal's body, of its food, and of unassimilated matter, i.e. faeces. These values can be obtained by burning the above mentioned materials in a calorimeter and measuring the amounts of heat produced. The heat produced at burning is equivalent to calorific content of the oxidized material. Calorimetry has been in use by many investigators dealing with bioenergetic studies (Richman, 1958; Golley, 1960 a, b; Slobodkin and Richman, 1960; Smalley, 1960; Comita and Schindler, 1963; Phillipson, 1964; Kendeigh and West, 1965; Nakamura, 1965; Klekowski *et al.*, 1967; Górecki, 1967; and many others). Various types of calorimeters have been employed by these authors.

When biological material is fairly abundant one can use macrocalorimetric technique since large amounts of dry matter are necessary for these measurements. For description of adiabatic bomb calorimeter as well as for general discussion of calorimetry in Ecological Studies see also Chapters 9A & 9B by Górecki. Some general account of calorific value of animals body, intra- and inter-species changes, the type of distribution of calorific value, and its biological consequences can be found elsewhere (Prus, 1970). The aquatic animals are mostly dealt with in that publication. However, when material is scant, it is most convenient to employ microcalorimeters in which one can burn small samples (less than 100 mg dry weight) with satisfactory accuracy.

Among the few types of microbomb calorimeters that are available at present (McEvan and Anderson, 1955; Phillipson, 1964), Phillipson's oxygen microbomb calorimeter seems to be most suitable for those who work with biological materials and have difficulty in collecting them in large quantities (Phillipson, 1964). The apparatus is of non-adiabatic type, excellent for measuring calorific value of organic substances. Simple construction, easy handling and operation, possibility to attain calorific value of small amount of material at disposal, low variation between replication samples and uniformity of results—all these connected with high accuracy of the measurements, are the advantages of this apparatus.

The calorimeter described below had been manufactured under the full consent of its constructor* in the workshops of the Nencki Institute of Experimental Biology, PAS, basing on the model of Phillipson miniature bomb calorimeter. However, several modifications have been introduced to the model in the process of manufacturing. Nevertheless the description of the apparatus, its calibration, operation and calculation of calorific values is quite similar to that in the Phillipson's (1964) publication.

5A.2 Description of the apparatus and accessory equipment

A complete set of the calorimetric system consists of the following: (1) microbomb, (2) stand with thermocouples and ring supporting the bomb, (3) polystyrene-steel jacket, (4) firing assembly, (5) recording potentiometer, and (6) accessory equipment.

(a) *Microbomb*, shown in Fig. 5A.1, is made of stainless steel, the wall thickness being of no less than 4 mm, with outer diameter of 34 mm, height of 86 mm. It consists of two parts which can be screwed together, the teflon washer (23), 1 mm thick, coming in between. In the upper part there are: inlet for fixing the oxygen filler (3) the needle valve (2) operated by key to pressurize bomb and cut off the oxygen supply, two ignition circuit terminals (5, 22) one insulated (22), the other one (5) in contact with the bomb mass, both connected by platinum fuse wire (21), $\phi - 0.15$ mm, which ignites the material; the sample holder (19) made of Pt sheet 0.5 mm thick, fixed to the metal sleeve (6) coating the uninsulated terminal and fastened to it with a screw; the Pt pan (20), 0.1 mm thick, $\phi- 6$ mm, height 2 mm, placed with pellet on the pan holder at each bomb operation, the Pt wire which presses down the pellet; a protective cap (24) similar in shape to the pan holder, but made of steel, attached to the top of terminal sleeve.

The lower part of the bomb is hollow. Outside at the bottom, there is a square notch to stabilize the bomb in the vice when tightening its two parts together. The notch is terminated with a short rod to join the bomb with

---

*We are greatly indebted to J. Phillipson for his allowance of construction as well as for supplying the model of the apparatus

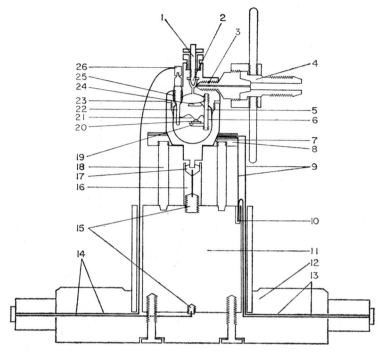

Fig. 5A.1. The bomb and stand, cross-sected. 1—needle valve rod; 2—pressurizing needle valve; 3—oxygen filler; 4—oxygen supply connective joint; 5—uninsulated terminal; 6—terminal sleeve; 7—thermocouple upper end; 8—textolite (turbax) ring; 9—thermocouple; 10—thermocouple lower end; 11—aluminium slug; 12—base; 13—thermocoupling leads to recording potentiometer, 14—ignition circuit leads to firing system; 15— brass connectives; 16— plexiglass coating rod and bowl; 17—mercury; 18—plexiglass rod supporting ring; 19—Pt sample holder; 20—Pt pan with sample; 21—Pt fuse wire; 22—insulated terminal; 23—teflon washer; 24—protective cap; 25—insulation coating; 26—brass cap.

mercury (17) in the stand. On the cone surface of the lower part, there is a crease to position the bomb in the ring (8) of the stand.

(b) *The stand* (Fig. 5A.1 and 5A.2) consists of the aluminium slug (11), 8 cm in diameter, 6 cm high, which is attached to the base (12). Over the aluminium slug, there is a textolite (turbax) ring (8), 6 cm in outer diameter, 1·8–2·3 inner diameters, 1 cm high. Twelve thermocouples (9) (copper and

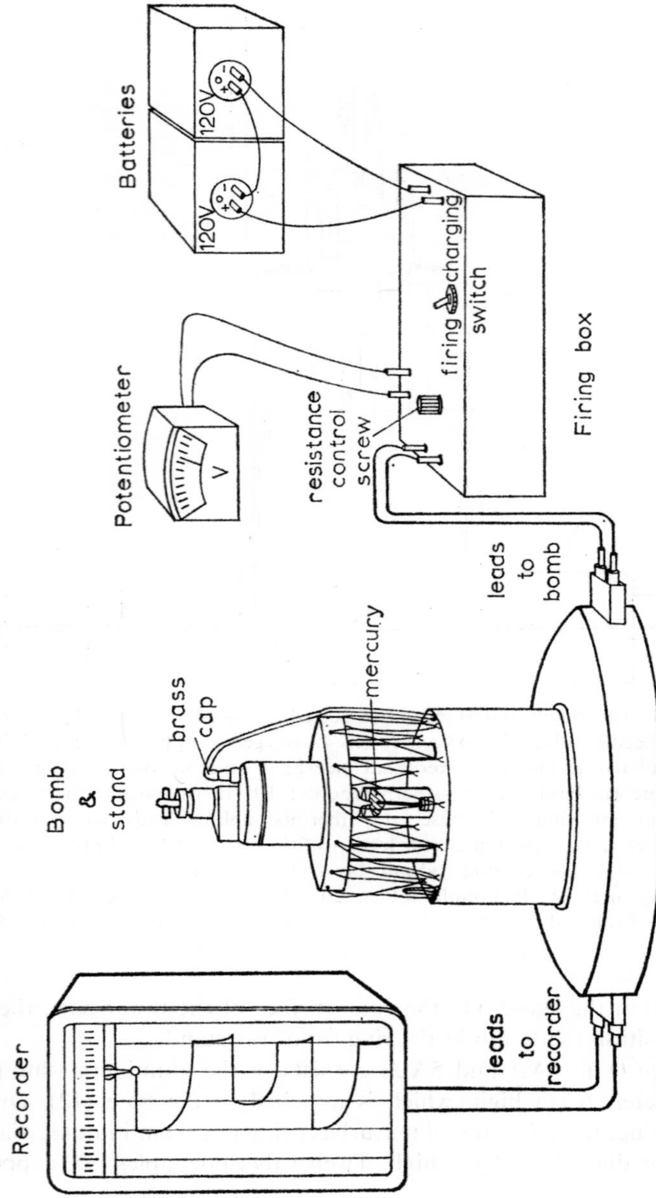

Fig. 5A.2. Calorimetric system assembled.

constantan), arranged in pairs, join the ring with the slug, being however insulated from both with a thin layer of polyethylene sheeting. Their upper ends (7) approach the inner cone surface of the ring, the lower ones (10) are mounted in the aluminium slug. Two main thermocoupling leads (13) to recording potentiometer end in the base, on its one side. The leads of the firing system (14) are on the opposite side of the base. One lead goes laterally through the aluminium slug, insulated from it, and terminates with a brass cap (26) fitted to the outer end of the insulated ignition terminal in the bomb. The other one, positioned centrally in the slug, ends with the lower screw (15) and starting from the upper screw (15) it goes farther through a plexi-glass rod (16) and ends in a plexiglass bowl filled with mercury (17). Mercury connects the firing circuit with uninsulated terminal in the bomb.

(c) *Polystyrene-steel jacket*, (not shown in the Figures) made of stainless steel sheeting, 1·6 mm thick, with polystyrene outer cover 4·5 mm in thickness, is placed over the bomb and the stand when operating the apparatus. It protects the whole calorimetric system (bomb, thermocouples, and aluminium slug) from changes in room temperature.

(d) *Firing system* is constructed according to the scheme presented in Fig. 5A.3. It consists of 2 batteries (Ba) (120 V each), a condenser (C), 2 resistors

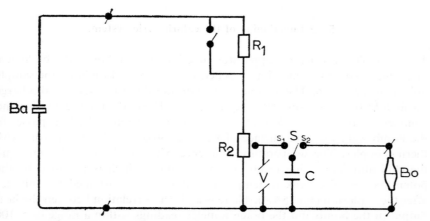

Fig. 5A.3. Ignition system, schematic. Ba—batteries, R_1 and R_2—resistors (47 and 50 kΩ), C—condenser (100 μF), V—potentiometer, S—operating switch, Bo—bomb. Condenser C is being charged when switch S is in charging position (S_1), when turned to firing (S_2), it closes the second circuit and the condenser discharges through the platinum wire, which ignites the pellet in the bomb.

(R_1 and R_2) (47 and 50 Ω), operating switch (S), two charging and firing circuits, and a potentiometer (V). On the firing box panel shown in Fig. 5A.2, there are sockets for the leads to the bomb, potentiometer and batteries. The panel holds also the resistance control screw to adjust voltage when charging condenser.

(e) *Potentiometer* is used for recording microvoltage generated in the thermocouples due to temperature difference between the bomb and the aluminium slug. Various apparatuses can be used, either galvanometers or potentiometers. In our laboratory we use a voltage recorder of German Democratic Republic make, MAW, type ek. N/T1 with a multiplier. Its full deflection amounts to 2 mV.

(f) *Additional equipment*. The calorimetric laboratory requires additional equipment as listed below: oxygen bottle with reducer and bomb filter, oxygen supply connective joint, platinum pans and wire, pellet forming tool, agate mortar (to homogenize material), vice, various keys, vacuum oven and desiccators, weighing containers, analytical balance with high accuracy at least of 0·01 mg, forceps, and chemicals: C_6H_5 COOH (for calibration), P_2O_5, $CaCl_2$ (desiccants), Hg (mercury contact), C_2H_5 OH (for washing), etc.

5A.3 Functioning of the calorimetric system.

The principle on which the apparatus functions is as follows: the biological material, dried and formed into a pellet, is placed with Pt pan on the sample holder in the bomb. The sample starts burning due to condenser discharge through Pt fuse wire which touches the pellet. The oxidation is explosive and complete so that the whole heat output is instantaneous. The heat passes to the bomb wall, the cone surface of which approaches closely the ring with thermocouples. The temperature difference between the heated bomb and the cold aluminium slug results in microvoltage which is recorded by the potentiometer. The peak in temperature difference is attained 6–7 minutes after the explosion (Fig. 5A.4). There is a direct correlation between the heat output in the bomb and the potentiometer readings within a range of 5–100 mg dry weight of benzoic acid burned in the bomb (Phillipson, 1964). Before pressurization, the bomb is washed with oxygen (3 times) to remove nitrogen from the bomb, making thus correction for nitrogen acids negligible.

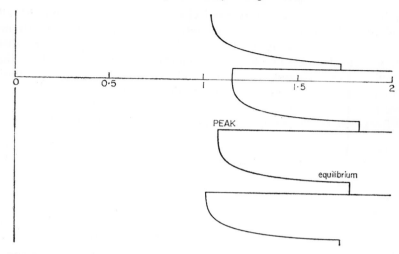

Fig. 5A.4. Original potentiometer record of a sample. The sampled record comprises the deflection between equilibrium point to the peak.

After the calibration of the bomb with material of a known calorific content, the calorific value of any biological sample burned in the bomb can be calculated. The calibration relates a known heat output to the potentiometer reading.

Finding the ash content in the sample by weighing the residual in the pan after burning, the ash-free calorific value of the material can be easily calculated.

5A.4 Operational instruction

To operate the bomb:
1. weigh pan, then pan and pellet; find weight of pellet; 2. switch on the recorder at least 15 min. before intended use; 3. fit 3–4 cm strip of Pt wire to bomb by fastening it to both terminals; 4. place pan with pellet in bomb and make sure that Pt wire touches pellet; 5. place lower half of bomb in vice, turn in upper part and tighten with a key; 6. fill bomb to 30 atm O_2, needle valve open then closed. To wash the bomb, release O_2 and refill for 3 times, then close needle valve and O_2 supply; 7. check bomb for leaks, cool in running, cold water; 8. place on stand, put contact cap on top of insulated

terminal; 9. place steel-polystyrene jacket over bomb and stand; 10. allow bomb to attain equilibrium ('zero') point, switch on recording tape; (to make sure the equilibrium is attained, check whether the line drawn is parallelled to tape vertical scale); 11. charge condenser then fire; 12. wait till maximum deflection is reached and line subsides; 13. count divisions from equilibrium point to peak; 14. depressurize bomb, disconnect its two parts, spare Pt wire; 15. weigh pan, find ash content; 16. calculate calorific values (total and ash free), acc. to the example given in data sheet included (Table 1).

Table 5A.1. Data sheet for calculation of calorific values (per mg dry weight and per mg ash-free dry weight), exemplified by *Tribolium castaneum*—CI females, 17 days after eclosion.

Sample No.	Material		Dry weight of sample (mg)	Ash-free calorific value of 1 mg (cal)	Calorific value of 1 mg (cal)
	Species	specification			
29	*Tribolium castaneum*	adult females 17 days after eclosion	21.79	6.6926	6.5329

Scale reading: 51.0 divisions	Calibrations: 1 div. −2.7912 cal

Calculations			Weight in mg	
I.	51.0 div.	— 142.3512 cal	pan + pellet	79.25
	21.79 mg	— 142.3512 cal	pan	57.46
	1 mg	— 6.5329 cal	pellet	21.79
			pan + ash	57.98
			pan	57.46
			ash	0.52
II.				
	21.27 mg	— 142.3512 cal	pellet	21.79
	1 mg ash-free	6.6926 cal	ash	0.52
			ash-free pellet	21.27

Date: 21 May, 1967

Notes:

5A.5 Calibration

The bomb is calibrated with benzoic acid of a known calorific value—6·332 cal/mg. The calibration follows in two steps. The first one is to check whether there is a direct linear dependence between mgs of the substance burned in the bomb and the recorder readings; the second one, to obtain calorific equivalent of one division on the recorder's scale. The first step involves burning benzoic acid samples ranging in weight from 5 to 100 mg and plotting the obtained deflection values against the weight of samples. The points should form a straight line on the graph, with a permissible variation. This procedure, unless done in the factory, does not need to be performed more than once for each apparatus.

The second step involves burning of 10 (or more) benzoic acid samples within a narrow weight range similar to that of material pellets whose calorific values are to be ascertained later. The calorific equivalent of one division of potentiometer reading is found for each sample, then the average, standard deviation, mean error, and coefficient of variation, are computed in order to estimate the accuracy of the apparatus (Table 5A.2).

Table 5A.2. Calibration 1 mg benzoic acid = 6.332 cal).

No. sample	Wt in mg	Calories	Scale reading, No. divisions	Cal. equivalent of 1 division
1	10.34	65.4729	22.7	2.8843
2	15.06	95.3599	36.0	2.6489
3	16.44	104.0981	37.0	2.8135
4	16.81	106.4409	38.8	2.8178
5	17.74	112.3297	40.3	2.7873
6	20.58	130.3156	47.0	2.7726
7	23.68	149.9418	54.0	2.7767
8	24.62	155.8938	56.5	2.7590
9	28.63	181.2852	67.5	2.6708
10	31.43	199.0148	74.3	2.6785

Mean: 1 div. = 2.7609 cal $\sim$ 20μV

It is advised to re-calibrate the bomb at certain intervals by burning 3–5 benzoic acid samples, especially after any change or repair has been made in the system.

To calibrate bomb, second step: 1. make 10 pellets using benzoic acid of a known calorific value with weight range, e.g. 15–20 mg; 2. calculate calorific content of each sample by multiplying mgs of weight by calories of 1 mg; 3. burn pellets and record for each the number of scale divisions; 4. find how many calories correspond to one division in the scale for each sample; 5. find average of the latter values and calculate S.D., M.E. and CV%, use this average for calculation of calorific value of biological samples (see Table 5A.1).

5A.6 Material drying and pellet forming techniques

It is absolutely necessary to have properly dried material before its calorific value can be measured in the bomb. When drying the material, precaution should be taken not to overheat the biological samples since that will cause underestimating the calorific value due to liberation of fatty acids from decomposing lipids. At the time of homogenizing and pellet forming procedure the dried up material will rapidly adsorb water. To avoid this error, it is necessary to dry anew the pellets before their weight for calorific calculation can be properly ascertained. If this recommendation is not fulfilled, exceptionally high, unexpected variation between samples of apparently homogeneous material may make the obtained results of little value.

To meet the above mentioned conditions, the following techniques can be recommended, based on our own experience with *Tribolium* material. 1. Collect uniform material (living animals, food or faeces) and find its actual (wet) weight. 2. Place material in vacuum oven with temp. of 60°C for 24–48 hrs, depending on the degree of hydration, size of objects etc. (for live material higher temperature can be applied at the beginning to kill the animals instantly). 3. Transfer the material in weighing containers to a vacuum desiccator with water absorbent and keep at room temperature. Weigh the material every 24 hr, unless constant weight is attained. 4. Homogenize material in an agate mortar and make pellets of adequate weight (e.g. 15–20 mg). 5. Dry the pellets again for at least additional 24 hr before their burning. 6. Calculate wet/dry weight ratio or percentage water content of the material.

More extended discussion of the drying techniques can be found also in Chapter 9B concerned mostly with larger objects, such as mammals and plant materials.

The largest volume of material burned hitherto in the apparatus at the disposal of the Department of Bioenergetics and Bioproductivity is that of the flour beetle *Tribolium castaneum* (Klekowski *et. al.*, 1967) the oribatid mite *Rhisoglyphus echinopus* (Klekowski and Stępień in litt.), aquatic isopod *Asellus aquaticus* (Prus, 1971, 1972) as well as their food items, and faeces of the latter.

References

COMITA G.W. & SCHINDLER D.W. (1963) Calorific values of microcrustacea. *Science* **140**, 1394–1396.

GOLLEY F.B. (1960a) Energy dynamics of a food chain of an old-field community. *Ecol. Monogr.* **30**, 187–206.

GOLLEY F.B. (1960b) Energy values of ecological materials *Ecology*, **42**, 581–584.

GÓRECKI A. (1967) Caloric values of the body in small rodents. In Petrusewicz K. (ed.) *Secondary Productivity of Terrestrial Ecosystems.* PWN, Warszawa–Kraków, **1**, 315–321.

KENDEIGH S.CH. & WEST G.C. (1965) Calorific values of plant seeds eaten by birds. *Ecology*, **46**, 553–555.

KLEKOWSKI R.Z., PRUS T. & ŻYROMSKA-RUDZKA H. (1967) Elements of energy budget of *Tribolium castaneum* (Hbst) in its developmental cycle. In Petrusewicz K. (ed.) *Secondary Productivity of Terrestrial Ecosystems.* PWN, Warszawa–Kraków, **2**, 859–879.

KLEKOWSKI R.Z. & STĘPIEŃ Z. (in litt.) Elements of energy budget of *Rhizoglyphus echinopus* (F.R R.).

MCEVAN W.S. & ANDERSON G.M. (1955) Miniature bomb calorimeter for determination of heats of combustion of samples of the order of 50 mg mass. *Rev. scient. Instrum.* **26**, 280–284.

NAKAMURA M. (1965) Bio-economics of some larval populations of pleurostict scarabaeidae on the flood plain of the River Tamagawa. *Jap. J. Ecol.* **15**, 1–18.

PHILLIPSON J. (1964) A miniature bomb calorimeter for small biological samples. *Oikos.* **15**, 130–139.

PRUS T. (1970) Calorific value as an element of bioenergetic investigations. *Pol. Arch. Hydrobiol.* **17**, 183–199.

PRUS T. (1971a). The assimilation efficiency of *Asellus aquaticus* L. (Crustacea, Isopoda). *Freshwat. Biol.*, **1**, 287–306.

PRUS T. (1972) Energy requirement, expenditure, and transformation efficiency during development of *Asellus aquaticus* L. (Crustacea, Isopoda). *Pol. Arch. Hydrobiol.*, **19**, 97–112.

RICHMAN S. (1958) The transformation of energy by *Daphnia pulex*. *Ecol. Monogr.* **28**, 273–291.

SLOBODKIN L.B. & RICHMAN S. (1960) The availability of a miniature bomb calorimeter
for ecology. *Ecology*, **41**, 784–785.
SMALLEY A. E. (1960) Energy flow of a salt marsh grasshopper population. *Ecology*
41,672–677.

5B. CHEMICAL COMPOSITION OF AN ANIMAL'S BODY AND OF ITS FOOD

A. DOWGIALLO

5B.1 Introduction

The only direct method of determining the calorific value of animal and
plant material is the physical, calorimetric one of bomb combustion. All
indirect methods of calorimetry are approximate only, giving values which
can be discussed only in relation to the physical ones. Some of these are based
on oxygen utilisation during the process of wet oxidation, assuming that $3\cdot4$
g-cal are liberated as heat per 1 mg O_2 consumed (Maciolek, 1962). Others are
based on the determination of chemical composition of the material being
examined. From the latter the calorific value may be computed, using
calorific equivalents of fats, proteins and carbohydrates, namely $9\cdot5$, $5\cdot7$, and
$4\cdot2$ kcal/g respectively (Bladergoen, 1955; MacFadyen, 1963; Winberg, 1971).
Computation from the chemical composition gives in many cases results of
low accuracy due to the complexity of biological material and hence the
difficulties in estimating the true level of total fat, protein and carbohydrate.
A complete estimation of calorific value of any material requires an additional
determination of dry weight and ash. The results obtained in our Laboratory
for various materials were usually lower than those of bomb combustion by
5–10%. This may be due partially to the improbability of a 100% recovery
of the examined material (Giese, 1967). On the other hand the bomb combustion of plant material with a high ash level may produce low results for
calorific value in comparison to that computed from chemical composition
(Szczepański, 1970).

The chemical method of determining calorific values of whole organisms
is then much more complex and time consuming than the calorimetric
method and is unlikely to replace it, unless the quantity of available material
is insufficient for bomb combustion. However, information about the chemical composition of bombed material may be required to interpret the calorimetric results, for example those reflecting the seasonal variability of food

resources or changes connected with the age and physiological state of the organism (Aleksiuk and Stewart, 1971; Barnes, 1972; Bernice, 1972; Boddington and Mettrick, 1971; Coles, 1970; Czeczuga, 1963; Lee *et al.*, 1970; Perkins and Dahlberg, 1971; Raymont *et al.*, 1964, 1968; Raymont *et al.*, 1971; Wissing and Hasler, 1971). Moreover, information about the relative proportions of the basic components in the food (along with the estimates of the respiratory quotient and nitrogenous excretion) provides the possibility of determining which food components are being respired and so contributing to energy transformations (Blažka, 1966 c; Conover and Corner, 1968; Corner *et al.*, 1967; Kleiber, 1961). Changes in the chemical composition of the animal body provide information on, for example, whether storage of fats is taking place or contrary whether the animal is starving and respiring its own body tissues (Zielińska, 1957).

Finally the comparison of chemical composition of food and faeces may provide some information on which food substances are being assimilated (Johannes and Satomi, 1966). The determination of the refractory organic fraction, such as cellulose, lignin, chitin, keratin etc. which are mostly indigestible by animals without the help of micro-organisms, makes possible an explanation of the differences between the expected and actual assimilation.

Since different kinds of biological material require special treatment and it is not possible to cover all variants, only a few analytical procedures are given below as examples, chosen for their simplicity from a great number of methods available. These examples are of an introductory character, this chapter being devoted for persons that have little or no experience in biochemical practice.

The proposed methods cover major groups of compounds rather than individual substances, and this is why they inevitably lack high accuracy. The sources of the possible errors are briefly discussed below, together with examples of particular analytical methods, the most common ones consisting of the application of inadequate standards and coefficients for calculation. However provided the level of the error is estimated and the results are critically evaluated, these methods are likely to be adequate for most bioenergetic studies.

To avoid important errors it would be advisable, when working with an unknown biological material, to check the methods of fractionation for completeness of the process and the methods of determination for the presence of interfering compounds. A balance sheet of recovery (in % fresh or dry

M

weight) assures the control whether or not considerable losses of the examined sample took place (Giese, 1967; Pourriot and Leborgne, 1970; Raymont *et al.*, 1964). If the material to be examined is particularly difficult to homogenise, at least triplicate portions of the same sample should be analysed and an average result calculated. Some useful introductory remarks to analytical work will be found in handbooks of analysis (e.g. Golterman, 1970; Harrow; *et al.*, 1958; Strickland and Parsons, 1968; Paech and Tracey, 1956), for statistical elaboration of results consult a specialised handbook (e.g. Bailey, 1959).

The storage of biological material for further analysis may constitute a source of erroneous results in comparison with the original one, due to autolysis, irreversible protein denaturation and oxidation of unsaturated fatty acids (making them less soluble), and error in estimating the dry weight because of loss of water etc. Excessive heating is particularly disadvantageous (Blažka, 1966 b; Paech and Tracey, 1956). For aquatic invertebrates the methods of preservation are discussed for example in papers (Boddington and Mettrick, 1971; Fudge, 1968; Pourriot and Leborgne, 1970; Raymont *et al.*, 1971) and in a review (Giese, 1967). According to the author's experience with *Tribolium* (unpublished) the least change in chemical constitution of preserved material compared with fresh was obtained by killing weighed insects by thermal shock (5 minutes at 65°C) to stop enzyme action, and then drying to a constant weight and further storing them in the same vacuum desiccator over P_2O_5 at room temperature until removed for analysis. They can also be stored in a deep freezer or in closed vessels over solid CO_2, but allowance should be made for losses of water. Lyophilisation of the ground material may result in loss of lipids. A safe storage period seems to be several months.

Some analytical methods referred to below were elaborated for aquatic organisms, but in most cases they may be directly applied to terrestrial material. An excellent review of these methods with suggestions on their choice in relation to various material was given by Giese (1967). A complete microanalytical scheme is given by Holland and Gabbott (1971). Directions on analytical methods may also be found in numerous papers and monographs referring to both aquatic (e.g. Beers, 1966; Bernice 1972; Blažka, 1966 a, b; Kosenko and Rashba, 1967; Luquet, 1970; Pourriot and Leborgne, 1970; Raymont *et al.*, 1964, 1968, 1971; Strickland and Parsons, 1968; Winberg, 1971) and terrestrial material (Karnavar and Nair, 1969; Lenartowicz *et al.*, 1967; Minderman and Bierling, 1968; Zaluska, 1959; Zielińska

1957). The methods applied were essentially analogical to those elaborated in general animal and plant biochemistry and quoted in numerous laboratory handbooks (e.g. Belozerskij and Proskurjakov, 1951; Colowick and Kaplan, 1957; Florkin and Scheer, 1967–1969; Hawk *et al.*, 1954; Lang and Lehnartz, 1953–1964; Official Methods, 1950; Paech and Tracey, 1955 –1964). The analytical procedures presented here were used to determine the chemical constituents of adult beetles *Tribolium castaneum* Herbst and their food, a 95:5 (by weight) mixture of wheat flour and dried ground baker's yeast. This species was chosen for analysis because it was also the subject of respirometric and calorimetric measurements during the IBP training course in 1968. However since then some work has been done on chemical analysis of aquatic invertebrates, of fish and of terrestrial plants, the practical remarks being included into this chapter. The valuable contribution of Mr. S. Kędzierski and Mrs. M. H. Watkowska in solving analytical problems is gratefully acknowledged.

5B.2 Basic equipment

Common laboratory equipment and glass-ware. Glass homogenisers with glass or teflon pestles, hand or electric motor-driven, after Potter and Elvehjem (1936). The homogeniser tube should be chosen to fit into a centrifuge in order to perform most operations in the same vessel, that is grinding, filling to volume with a solvent, heating in a bath and centrifuging, with a cap if the solvent is volatile. For example homogenisers of the total length 10 cm, marked at the volume 10 ml were applied by the author to a laboratory centrifuge (Fig. 5B.1). Water bath with tube or flask rack. Spectrophotometer. or filter photometer. Drying oven, muffle furnace, balances, vacuum desiccator, Parnas-Wagner apparatus for micro-determination of nitrogen. Laboratory centrifuge up to 6000 rpm. may be replaced in some instances by the equipment for vacuum filtration through sintered glass funnels or through glass fiber filters, for example the Millipore or Sartorius type glass filterholder with spring clamp for glass fiber filter paper. The glass fiber filter paper (e.g. Whatman GF/C, Nr. 83) circles may be easily obtained by pressing out of a sheet with the edge of a glass funnel against a hard surface. If no filter holder is available, the circle may be carefully and tightly placed on the surface of a sintered glass funnel of medium porosity.

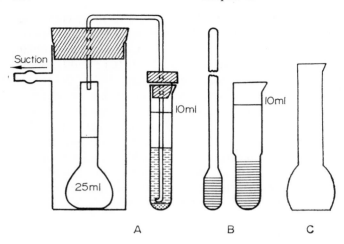

Suction

10ml

10ml

25ml

A B C

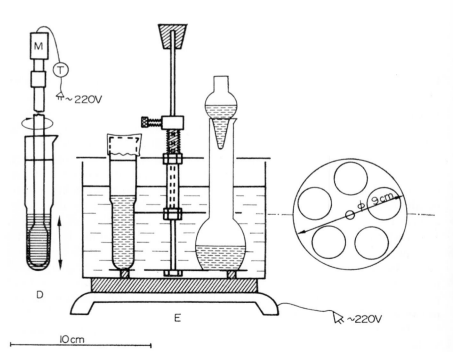

M

T

~220V

φ 9cm

~220V

D E

10cm

5B.3 Determination of dry weight and ash content

A weighed sample of fresh material is dried to constant weight at 60–100°C. With the lower temperature the constant weight is attained after a considerably longer drying and is usually higher by a few per cent from that at 100°C. In practice it is advisable to apply the lower temperature, if the material is dried prior to analysis or bomb combustion. With fresh material it is reasonable to use higher drying temperature and refer the results to dry weight, although losses of volatile matter are likely to occur.

The sample after dry weight determination may be used for ash determination by ignition at 450–500°C to constant weight, if drying was performed in tared platinum, nickel or heat-resistant glass crucible. The use of higher temperature up to 550°C is unsafe because of possible decomposition of calcium carbonate due to temperature inhomogeneity in various regions of a muffle furnace. A reconstitution of carbonates and of hydrated salts may be attained by wetting the ash with a 5% solution of ammonium carbonate and drying to constant weight at the temperature chosen for dry weight determination (Giese, 1967; Official Methods, 1950; Raymont *et al.*, 1964; Winberg, 1971).

Values obtained in the course of analysis for the level of particular components in the examined material as per cent of fresh weight, may be converted to dry weight per cent (to first decimal place) by multiplying them by the factor 100:D, where D = dry weight in per cent of fresh weight.

Fig. 5B.1. (left)   A. Device for decanting the supernatant after centrifugation (one of the possible ways). Glass thick-wall centrifuge tube, lipped, marked at 10 ml. or homogeniser tube. The shaped tube stopper of synthetical materials may be used also during centrifugation.

B. Potter-Elvehjem glass homogeniser tube with glass or teflon pestle, suitable for heating and centrifugation, marked at 10 ml (hatching indicates a ground surface).

C. Thin wall flask for heating in a bath, about 25 ml capacity.

D. Homogeniser fitted elastically to an electric motor (M) with speed regulation (T). During rotation of the pestle the tube is moved up and downwards.

E. Tube or flask rack in a water bath on a heating plate. Tube or flask capped with aluminium foil or equipped with a glass reflux condenser filled with cold water.

5B.4 Determination of total lipids

The lipids constitute a manifold group of natural compounds characterised by their solubility in non-polar solvents, like petroleum ether, benzene, ethyl ether, chloroform, acetone, alcohols etc., pure or in mixtures. This solubility is in accordance with their general hydrocarbon character, that in turn underlies their highest calorific value compared to the remaining constituents of organisms. The accurate determination of the lipid fraction of the examined material is particularly important, when the result must serve to calculate the total calorific value of the material, assuming a common calorific equivalent for the remaining non-lipid fraction.

No specific reaction exists for the determination of the whole group. The direct determination of the sum of lipids in the examined material requires extraction, purification of the extract, evaporation of the solvent and weighing of the residue. Some indirect methods of determination, also in micro scale, are based on oxidative procedures (Blažka, 1966 b; Bloor, 1947; Sperry, 1955). Other methods based on specific reactions of fatty acids give an estimate of milliequivalents (meq) of fatty acids, irrespective however of the chain length. The average chain length may be found acidimetrically by analysing a weighed sample of lipids isolated from the examined material (Entenman, 1957 b), or an empirical lipid equivalent of a known photometric standard may be established by comparing it with a weighed sample of isolated lipids (Blažka, 1966 b). More details on lipid extraction and determination may be found in papers and monographs (e.g. Bligh and Dyer, 1959; Chuecas and Riley, 1969; Entenman, 1957 a, b; Folch *et al.*, 1957; Giese, 1967; Hutchins and Martin, 1968; Niemierko *et al.*, 1952; Official Methods, 1950; Rouser and Fleischer, 1967). The chromatographic techniques were surveyed by Horning *et al.* (1964), Marinetti (1967–1969), Stahl (1969), Stein and Slawson (1966). It is recommended to determine the lipids in fresh material, unless the weighed samples were stored deep frozen with no access of oxygen. Heat drying results in decomposition and oxidation of lipids.

The method presented here is the indirect hydroxamate one of determination of esterified fatty acids, based upon the direct method of selective and quantitative extraction and weighing after evaporation of solvents.

Esters of fatty acids and aromatic acids react with hydroxylamine: $R-CO-OR_1 + NH_2OH = R-CO-NHOH + R_1-OH$ and the resulting hydroxamic acids produce with ferric salts red coloured compounds suitable for photometric determination at 520–540 nm. This reaction does not comprise the unsaponifiable fraction, e.g. cholesterol (Stern and Shapiro, 1953) which raises the value of the apparent equivalent weight of an ester group. No trace of acetone is admissible in the examined material.

5B.5 Reagents

Note. Unless otherwise stated, the term water means distilled water.

Alcohol-ether mixture 3:1 (Bloor, 1914, 1947; Sperry, 1955). Three volumes of ethanol 96% and one volume of ethyl ether, prepared daily from purified products. Ethanol is made free of acids and esters by first refluxing it for two hours with KOH pellets (ca 1 gram per litre) and then distilling. The first and last 10% are rejected. Ethyl ether is made free of peroxides by 15 min. shaking with 20% its volume of 10% ferrous sulphate, acidified with 2 ml concentrated sulphuric acid per litre. The ambient temperature should not exceed 18°C. The ether layer is separated in a separatory funnel, dried 24 hours over calcium chloride, distilled rejecting the first and last 10%, and stored in a dark cool place no more than 10 days. (Blažka, 1966 b). Caution: No open fire is admissible, ether vapour is highly inflammable.

Alcohol—diluted. One volume of ethanol 96% is mixed with one volume of water.

Two-molar hydroxylamine hydrochloride solution. Dissolve 14 g of $NH_2OH \cdot HCl$ in 100 ml of water. Equal volumes of this solution and of 3·5 n NaOH may be mixed immediately before use and 1·6 ml of the mixture used as below.

Alternatively $NH_2OH \cdot HCl$ solution is prepared in diluted alcohol.

Hydrochloric acid 4 n. One volume of concentrated HCl (specific gravity 1·19) is mixed with two volumes of water.

Ferric chloride 0·37 m in 0·1n HCl. Dissolve 10 g $FeCl_3 \cdot 7H_2O$ in 100 ml 0·1n HCl, allow to stand overnight and decant. Clear solution is used.

Standard solution. Methyl salicylate, e.g. 0·01m contains 152 mg this ester in 100 ml of alcohol 96% or alcohol-ether mixture. The weighing of the freshly distilled ester (boiling point 224°C at normal pressure) is best carried out in a tared glass capillary of inside diameter 0·05–0·3 mm, bent to shape of capital letter omega and suspended to a balance (Kędzierski, unpubl.). This concentrated solution can be stored up to 3 weeks at 0–4°C in a dark place in a flask with ground stopper. The standard is diluted ten times with alcohol-ether mixture shortly before use. Each millilitre of this solution corresponds to 0·3 mg (0·001 meq) of fat, assuming that the average molecular weight of animal lipids is 900 and that they consist mainly of triglycerides, then the equivalent weight of ester group averaging 300. The calibration curve is linear up to 0·002 ester-meq. per 1 ml of test solution.

Solvents for preparative lipid extraction. A mixture of 2 volumes of chloroform and 1 volume of methanol and a mixture of 1 volume of chloroform and 1 volume of hexane (or light petroleum ether). These solvents can not leave any weighable residue on evaporation.

Cotton wool washed several times with chloroform-methanol mixture, air dried and preserved from contamination.

5B.6 Procedure

A weighed sample (0·05–0·1 g fresh weight) of the examined material is homogenised with 4 ml of alcohol-ether mixture, heated for 2–3 minutes up to 60°C in a water bath, cooled and centrifuged for 5 minutes at 4–5000 r.p.m. The homogeniser should be capped e.g. with aluminium foil, the volume of solvent mixture exceeding that of water in the sample at least 20 times, to assure complete mixing. The clear extract is decanted into a calibrated flask or test with a ground stopper, the extraction with a fresh portion of solvent mixture repeated twice more for 10 minutes at 60°C in a water bath, using reflux condensers filled with water (Fig. 5B.1) followed by centrifuging and decanting. The combined extracts are made up to volume (e.g. 10 ml) with the solvent mixture.

For the colour reaction 5 ml of the lipid extract or of the diluted standard solution or 5 ml of the solvent mixture for reagent blank (zero optical density) are pipetted into separate flasks or test tubes with ground stoppers, to each one 0·8 ml of hydroxylamine hydrochloride and 0·8 ml of 3·5 n NaOH solutions are added, the content mixed and allowed to stand at room temperature for 20 minutes. After this time 1 ml of 4 n HCl is added and again mixed. Optical densities (OD) of standard and unknowns are read after 10 minutes up to 30 minutes, using a photometer, e.g. the Hilger BIO-absorptiometer with the filter Nr 52. In view of fading of the colour with time, the optical density of standard solution should be read at the beginning and at the end of a series, to enable a correction of results. The optical density of a pigment blank may be subtracted if necessary: extract with all reagents except for water instead of hydroxalamine hydrochloride solution, against a reagent blank with pure solvent mixture instead of extract and water instead of hydroxylamine as above.

With wheat flour, larger samples were needed (up to 0·5 g) because of the low lipid content of this material, other details of the procedure remaining unchanged. With animal material less suitable for extraction, e.g. preserved dry, the procedure may be modified in the way that first the sample is damped and extracted with cholroform-methanol mixture leaving it overnight and only then heating (Boddington and Mettrick, 1971). After repeated extraction, centrifugation, then evaporating the solvent mixture, the extracted lipids are dissolved in alcohol-ether mixture for further determination. For details see below 'Evaluation of the apparent equivalent weight'.

In the original method (Stern and Shapiro, 1953) the aqueous solution of hydroxylamine hydrochloride is used, particularly suitable for the determination of animal and plant-seed lipids. However with extracts of green parts of plants the addition of water in the pigment blank results in an increase of its optical density due to turbidity. This may be at least partially avoided using diluted alcohol instead of water in both hydroxylamine hydrochloride solution and the blanks. In this way a corrected reading of unknown optical density is obtained after the formula (see below for denotations):

$$D_u \text{ corrected} = D_u - y + m - n$$

where y, m, and n are optical densities measured according to the following scheme, which eliminates also the possible colour reaction of plant phenolic compounds with iron and the effect of solubility of mineral salts produced in the course of reaction. The presence of hydroxylamine hydrochloride in the mixture increases the solubility of plant lipids, thus lowering the turbidity. The determination of m and n can be omitted if they differ by less than $OD = 0·010$.

D_u—5 ml extract, 0·8 ml $NH_2OH \cdot HCl$ in diluted alcohol, 0·8 ml NaOH, 1 ml HCl, 0·8 ml $FeCl_3$

against blank—5 ml extract, 0·8 ml dil.alcohol, 0·8 ml NaOH, 1 ml HCl, 0·8 ml $FeCl_3$

y—5 ml alcohol-ether, 0·8 ml $NH_2OH \cdot HCl$ in dil.alcohol, 0·8 ml NaOH, 1 ml HCl, 0·8 ml $FeCl_3$

against blank—5 ml alcohol-ether, 0·8 ml dil.alcohol, 0·8 ml NaOH, 1 ml HCl, 0·8 ml $FeCl_3$

m—5 ml extract, 0·8 ml dil.alcohol, 0·8 ml NaOH, 1 ml HCl, 0·8 ml water
n—5 ml extract, 0·8 ml $NH_2OH \cdot HCl$ in dil. alcohol, 0·8 ml NaOH, 1 ml HCl, 0·8 ml water
m and n are both measured against distilled water as zero optical density
D_k—5 ml standard, 0·8 ml $NH_2OH \cdot HCl$ in dil.alcohol, 0·8 ml NaOH, 1 ml HCl, 0·8 ml $FeCl_3$
against blank—5 ml alcohol-ether, 0·8 ml $NH_2OH \cdot HCl$ in dil.alcohol, 0·8 ml NaOH, 1 ml HCl, 0·8 ml $FeCl_3$

5B.7 Evaluation of the apparent equivalent weight of lipid ester group

The lipids fraction comprises all compounds of the high calorific (hydrocarbon) character, although only a part of them contains an ester group. Consequently every ester bond represents in the colour reaction not only its own equivalent weight, but also a proportion of the non-ester lipid material. This augmented average equivalent weight of an ester group, denoted as the apparent equivalent weight, may vary considerably, thus strongly influencing the results of determination of total lipids. Its value has been found to be in insects 230 (*Calandra granaria*) and 303 (*Tribolium castaneum*), in fish 280 (*Perca fluviatilis*), in net plankton from African shelf-waters 510 (Klekowski and Wiktor, unpubl.), in Antarctic Amphipoda 190 (*Paramoera Walkeri*, Rakusa and Suszczewski, unpubl.), in wheat flour 560, and in dried fallen leaves (*Alnus glutinosa*) as high as 780. The mean value—400, for freshwater animals was found by Blažka (1966 b).

The determined ester equivalent weight is valid at least within the same developmental stage and for a given physiological state of an organism (Hutchins and Martin, 1968; Karnavar and Nair, 1969; Lee et al., 1970; Perkins and Dahlberg, 1971; Sulman, 1971), hence its determination is useful when more samples of a similar material are to be analysed by the photometric method. With only few samples the preparative isolation and weighing of lipids is more convenient. In both cases a complete and selective extraction of lipids is essential. This is obtained by first extracting with a warm solvent mixture containing a lower alcohol which provides dissociation of lipid-protein complexes, followed by re-extraction of the raw fat with a less polar anhydrous solvent mixture (Rouser and Fleischer, 1967).

The most efficient solvent mixtures for animal material are chloroform-methanol (2:1) (Folch et al., 1957) and for plant material ethanol-benzene

(1:1) (Minderman and Bierling, 1968; Paech and Tracey 1955-1964). With highly hydrated homogenates the mixture of chloroform and methanol of various ratios, i.e. 2:1, 1:1, 1:2, can be used for the first extraction step, of which 19, 9, and 4 volumes respectively are completely miscible with 1 volume of water in the homogenate (Bligh and Dyer, 1959). A selective re-extraction is best carried out with chloroform-hexane (or chloroform-petroleum ether) mixtures of a ratio 1:1 up to 1:2. It is important to store the preparation of isolated lipids at a low temperature either under nitrogen, or *in vacuo* in a small volume of solvent, that is evaporated before further treatment.

A sample of the examined material containing 20–25 mg of lipids is homogenised with 10 ml of chloroform-methanol mixture, warmed for 10 minutes in a 60–65°C water bath, cooled, centrifuged and the clear supernatant decanted into a clean wide neck flask. A new portion of solvent mixture is added to the residue, allowed to stand at room temperature for 1–2 hours, again warmed, cooled and centrifuged. A third extraction may be performed by leaving the residue with a new portion of solvent mixture till the next day (Drabikowski *et al.*, 1966; Perkins and Dahlberg, 1971).

The combined raw extracts are evaporated on a water bath under a mild stream of nitrogen and dried in a vacuum desiccator over KOH pellets. The residue on evaporation is re-extracted with chloroform-hexane mixture, with subsequent centrifugation of the extract, 3–4 times with a fresh portion of solvent mixture, leaving overnight with the last portion. The clear supernatants are collected in a tared, thin-walled wide-necked flask; if turbid, previously dried with anhydrous sodium sulphate, the solvent evaporated on a water bath under a stream of nitrogen, dried in a vacuum desiccator over solid KOH, and the preparation of isolated lipids is weighed.

The centrifugation of extracts may be replaced by filtration through a pad of cotton wool directly into the wide neck flask, extracting the lipids repeatedly with small portions of fresh solvent mixtures. Glass fibre filter paper can also be used.

For further colour reaction the lipid preparation is dissolved in 50 ml of alcohol-ether mixture in a calibrated flask, leaving it overnight to settle a possible turbidity. From this solution 5 ml are taken for the colour reaction. Some non-ester lipids may be insoluble in alcohol-ether mixture, but this does not influence the result, since the insoluble residue has been comprised in the weight of the lipid preparation.

5B.8 Calculation of results

A —lipids in per cent of fresh weight of the examined material
D_u —optical density of the examined lipid extract (unknown)
D_k —optical density of the standard ester solution
a —meq. of ester in 1 ml of standard (e.g. 0·0012)
b —mg of fresh material sample
f —ml of the final volume of the examined lipid extract (dilution factor)
X —lipid apparent equivalent weight
c —concentration of the known lipid solution (lipid preparation), in mg/l
d —weight of lipid preparation
Y —mg of fresh material for preparative lipid extraction
D_x —optical density of the known lipid solution

$$A = \frac{D_u \cdot a \cdot f \cdot X \cdot 100}{D_k \cdot b} \quad \text{where } X = \frac{D_k \cdot c}{D_x \cdot a} \text{ — for the photometric method}$$

$$A = \frac{d \cdot 100}{Y} \text{ — for the gravimetric method}$$

5B.9 Determination of protein

Among methods of total protein determination the most widespread is the indirect one, consisting in the determination of organic nitrogen by the Kjeldahl procedure, with and without deproteinising of the material. For details on this procedure see Appendix to this chapter. Since the per cent of nitrogen differs in various proteins, an empirical factor is used for calculating the amount of protein from the nitrogen level in the examined material. This factor varies from 4 to 8 for pure preparations of most common proteins, averaging 6·25 for animal tissues, 5·7 for wheat flour, 6·4 for casein, 4·4 for ashless matter of the fish scales (*Perca fluviatilis*), 6·0 for yeast (Hesse *et al.*, 1971) and 5·1 for leaf proteins calculated from total insoluble nitrogen (Kaushik and Hynes, 1971). All other methods are calibrated according to the determination of Kjeldahl nitrogen or to known amounts of pure proteins as standards (Ballentine, 1957; Belozerskij and Proskurjakov, 1951; Block,

1960; Boje, 1966; Giese, 1967; Hesse *et al.*, 1971; Layne, 1957; Official Methods, 1950; Winberg, 1971).

One of the direct methods is the gravimetric estimation of soluble proteins after precipitation by trichloroacetic acid (TCA) (Hoch and Vallee, 1953). Particular optical methods of determining proteins in solution are based on the existence of several functional groups in the protein molecule or on the formation of turbidity with protein precipitants (Hawk *et al.*, 1954; Layne, 1957). Insoluble proteins require solubilisation, usually in neutral or buffered salt solutions or in decidedly alkaline solutions (Blažka, 1966 a; Boje, 1966; Cleland and Slater, 1953; Helander, 1957; Jennings *et al.*, 1968; Layne, 1957; Lowry *et al.*, 1951; Pearse and Giese, 1966). These mild conditions are insufficient in many cases to solubilise all the proteins present in the examined material. On the other hand prolonged heating with concentrated alkalis leads to splitting of the protein molecules, that results in an evident increase of the non-protein nitrogen fraction and moreover losses of free ammonia are possible. To ensure completeness of extraction it is advisable to perform it stepwise, first in mild conditions and, after separation of the first solution, more drastically. The insoluble residue should contain chitin as a sole nitrogenous material. The completeness of extraction should be checked once for any particular examined material.

The separation of protein and nonprotein nitrogenous fraction may be achieved in various ways, but none of them is applicable universally (Bell, 1963). Some of the clarifying procedures remove from the solutions not only proteins and other high molecular nitrogenous or nitrogen-free substances, but also a large proportion of low molecular nitrogen compounds (Ashwell, 1957; Friedemann *et al.*, 1963; Winzler, 1955). Trichloroacetic acid and tungstic acid are known to precipitate quantitatively only true proteins, lower peptides being left in solution. (Haden, 1923; Hoch and Vallee, 1953; Ivanko, 1971; Kaushik and Hynes, 1968; Merrill, 1924; Raymont *et al.*, 1964). In contrast to this tannic acid removes from the solution also lower peptides (Belozerskij and Proskurjakov, 1951).

5B.10 The biuret method

For most purposes the biuret method of protein determination, as presented here, is applicable. Peptides containing at least two neighbouring groups

— CO — NH —

(as well as some similar groupings) form violet complexes with copper salts in alkaline solution (Official Methods, 1950). The colour intensity, proportional to the protein concentration, is measured photometrically at 540–550 nm (e.g. filter 55). For the sake of calibration of the photometric method the results are referred to the trichloroacetic acid insoluble (final concentration of TCA 4–10% weight per volume), or tungstic acid precipitable nitrogen content of the same protein solution. This calibration is worth-while doing when more samples of a similar material are to be analysed by photometric method; otherwise determination of Kjeldahl nitrogen fractions is more convenient and the protein nitrogen content multiplied by 6·25 to give the protein content in the solution. For routine work a photometric working standard solution of a pure protein preparation may be prepared daily, e.g. serum or egg albumin. Its optical density equivalent must be referred to Kjeldahl nitrogen content (protein-N) determined once for each new material to be examined (Blažka, 1966 a; Gornall *et al.*, 1949; Itzhaki and Gill, 1964; Layne, 1957). Using the biuret method for protein determination in fish homogenate a conversion factor (see Calculation) was found for the bovine albumin working standard, averaging 1.3.

Alkaline protein extract of some insects must be brown coloured due to their pigmentation, as happened with *Tribolium* beetles. In this case the biuret method cannot be applied, since the pigment increases the optical density in the useful range of wave lengths thus covering the colour reaction. Plant material containing pigments may be pre-extracted with warm mixtures of ethanol and ethyl ether, chloroform and methanol (see Extraction of Lipids) or with ethyl ether. Pre-extraction with ethanol alone is not recommended because certain plant proteins are alcohol-soluble (Bell, 1963; Belozerskij and Proskurjakov, 1951; Jennings *et al.*, 1968). The extraction of cereal proteins with the biuret reagent may be carried out in the presence of a drop of chloroform. A turbidity blank is determined by bleaching the blue colour with potassium cyanide (Maslowski and Skórko, 1966). For the extraction of nonprotein nitrogen from cereals and leaves, a higher concentration of TCA up to 10% must be used because at lower concentrations it dissolves a portion of plant proteins (Bell, 1963; Ivanko, 1971; Kaushik and Hynes, 1968).

If the protein cannot be dissolved without loss, as happened with *Tribolium* beetles, the protein nitrogen content may be found by subtracting the chitin nitrogen (see Determination of Carbohydrates) and the nonprotein soluble nitrogen from total nitrogen content in the material.

5B.11 Reagents

Note. Distilled water and all reagents must be free of ammonia.

Biuret reagent I. Aqueous solutions made in small portions of water of 80 g NaOH (112 g KOH), 15 g sodium-potassium tartrate ($NaKC_4H_4O_6 \cdot 4H_2O$— Seignette salt) and 3 g KI, are mixed and made up to 1 litre.

Biuret reagent II. 3·5 g $CuSO_4 \cdot 5 H_2O$ are dissolved in a water portion and diluted to 1 litre.

Biuret reagent III. Solutions I and II made in small volumes of water are mixed before diluting and only then made up to 1 litre.

The reagents I and III are stored in tight polythene bottles.

Alcohol-ether mixture—see Determination of Lipids.

Aqueous trichloroacetic acid (TCA) 50% w/v, sp. gr. = 1·24. A 5% solution is prepared by diluting 10 ml of 50% solution with water to 100 ml. Warning ! TCA solutions produce dangerous burns if brought into contact with the mucosa.

Sodium tungstate 10%. Ten grams $Na_2WO_4 \cdot 2H_2O$ are dissolved in water and made up to 100 ml.

Sulphuric acid 1 n.

Working standard. Dry bovine serum albumin is dissolved in biuret reagent I to provide concentration of 2 mg/ml. The calibration curve is linear up to 3·5 mg protein per 1 ml of solution.

5B.12 Procedure

A. *Animal material.* A weighed sample (0·1–0·2g) of material is homogenised for 1–2 min. with a few ml of biuret I reagent (alkali-tartrate-iodide), filled up to 10 ml with the reagent and allowed to stand in a stoppered centrifuge tube for 24–28 hours at room temperature. The residue after extraction of lipids may be used, but a fraction of nonprotein nitrogen is removed with the lipid extract. The homogenate is then mixed by swirling the tube or with a glass rod and centrifuged (4–5000 r.p.m.) for 10 minutes. If the sediment still contains protein, the tube should be heated at 100°C for 30 minutes prior to centrifugation. An aliquot of the clear extract is mixed with the same volume of biuret II reagent (copper sulphate) and after 30 minutes up to 48 hours the optical density is determined against a reagent blank, as zero reading of the photometer (equal volumes of biuret reagents I and II). Another aliquot is taken to determine the Kjeldahl total nitrogen (soluble).

The nonprotein nitrogen may be determined in a separate sample. A weighed sample is homogenized with 10 ml of 5% TCA, the tube covered with a glass condenser being heated for 10 minutes at 80°C and allowed to stand overnight at room temperature, centrifuged, and an aliquot of the clear supernatant mineralised for nitrogen determination. The nonprotein nitrogen may be determined also in an aliquot of solution used for optical density determination, after the protein has been precipitated with TCA (final concentration 5% w/v) with heating of the sample as above. The sample should be previously neutralised with sulphuric acid. The nonprotein N is subtracted from total N to obtain protein N.

Alternative technique. A sample of animal material is homogenised with 5 ml of biuret reagent III, then 5 ml of water are added (final volume 10 ml) and thoroughly mixed. The homogenate is allowed to stand at room temperature for 24–28 hours, centrifuged, and the clear liquid is read in the photometer against a reagent blank. If slight turbidity remains, the solution may be filtered through dry filter paper, or extracted with a few ml of ethyl ether by syphoning, whereafter the tube is heated up to 50–60°C (Gornall *et al.*, 1949; Layne, 1957).

B. *Plant material or animal material with a high lipid content.* A weighed sample (0·2–0·5 g) of fresh material is homogenised with a volume of alcohol-ether mixture (exceeding that of water in the material 20 times), centrifuged, and the supernatant discarded. If dry material is used, it is wetted with a few drops of water. The residue is heated with another volume of solvent for 10 minutes at 60°C, using reflux condensers (Fig. 5B.1), cooled, centrifuged, and the supernatant discarded. The extraction is repeated until the solvent is no more coloured. For the last extraction pure ether may be used. The tube is drained, heated for 10 minutes at 60°C to remove the solvents. The residue is mixed with 5 ml of biuret I reagent, left for 24 hours at room temperature, then heated for 30 minutes at 100°C, cooled, 5 ml of biuret II reagent are added, filled with water to exactly 10 ml and mixed. After 30 minutes—2 hours standing at room temperature the mixture is centrifuged and the optical density measured against the reagent blank.

Should the optical density be read the next day, the extract must be kept at 2–5°C to avoid copper reduction.

Certain plant materials, especially dried, produce brown alkaline extracts, unsuitable for photometry. The protein content may be then estimated from the difference between total and nonprotein nitrogen. The latter is determined as follows. The sample is homogenised with 5 ml of water and 0·5 ml of 1 n H_2SO_4, then 1 ml of 10% Na_2WO_4 is added and mixed, the pH of the mixture should be 4–5. After 1 hour the sample is heated for 5 minutes at 80–90°C, made up to 10 ml. mixed and centrifuged. An aliquot of the clear supernatant is taken for nonprotein nitrogen determination. If an alkaline protein extract is deproteinised with tungstic acid, it must first be neutralised with 1 n H_2SO_4.

To determine the protein-N content in autumn-shed leaves an alternative technique was applied (Kaushik and Hynes, 1968). The protein was extracted with 0·1 n NaOH, precipitated with TCA (final concentration 7–10%), centrifuged off and redissolved in 0·1 n NaOH for further determination of protein (Folin-positive material, Lowry *et al.*, 1951). Further extraction with 1 n NaOH doubled the results (Dowgiallo, in preparation).

Neither TCA nor tungstic acid were able to deproteinise completely both a homogenate of net plankton and an alkaline extract of this homogenate, thus giving too high results for nonprotein nitrogen. The best results were obtained in this case by extracting the lipids with a 20-fold excess of chloroform-methanol or ethanol-ether mixtures and determining the nitrogen content in the extract. The obtained value corresponds to nonprotein nitrogen content, found from total nitrogen, chitin-nitrogen and protein-nitrogen (Folin-positive) determinations in the homogenate.

A complete hydrolysis of the examined material followed by determination of amino acids is necessary to check the value of methods of fractionation into protein and nonprotein nitrogen, also if the examined protein is poorly soluble. This determination is particularly important for the evaluation of nutritive quality of foods. Chromatographic and photometric techniques are both widely applied for this purpose (Boyd, 1970; Boyd and Goodyear, 1971; Cowey and Corner, 1963; MacGrath, 1972; Manoukas, 1972; Piez and Saroff, 1964; Walker, 1972; Yemm and Cocking, 1955).

The colour reaction using the Folin-Ciocalteau reagent is about 100 times more sensitive as compared to the biuret one, although susceptible to interference from the presence of phenolic substances and purines in the protein solution. The protein to be tested must be moreover readily soluble in water or diluted alkali and free of pigments. The method is particularly suitable for protein determination in serum as well as in cultures of microorganisms. The following procedure is adapted to protein test solutions in 0·1 n NaOH (Drews, 1968; Layne, 1957; Lowry *et al.*, 1951; Winberg, 1971).

N

5B.13 Reagents

Regaent A—30 g Na_2CO_3 in water, made up to 500 ml, stored in polyethylene bottle.

Reagent B—0·1 g $CuSO_4 \cdot 5H_2O$ and 0·2 g sodium-potassium tartrate are dissolved in water and made up to 500 ml.

Reagent C—equal volumes of A and B are mixed. Prepared daily.

Reagent F—Folin reagent diluted with water to make it 1 n in acid. A portion of the concentrated reagent, prepared according to the original procedure (Layne, 1957; Lowry, 1951; Winberg, 1971) diluted 20 times with water is titrated with 0·5 n NaOH to phenolphtalein as indicator.

Standard protein solution —5–25 mg of crystalline serum albumin are dissolved in 100 ml of 0·1 n NaOH. This solution can be stored at $+4°C$ for 6 months, if protected from evaporation. The calibration curve is not linear, the readings are performed at any chosen wavelength from 600 to 750 nm.

5B.14 Procedure—animal material

A weighed sample (0·05–0·1 g fresh weight) of material is homogenised with NaOH solution to make the final concentration in 10 ml of the homogenate 0·5 n NaOH. This may be left from 15 minutes to 24 hours at room temperature and centrifuged. The residue is mixed with 5 ml 1 n NaOH, left overnight at room temperature and centrifuged. If the material is difficult to extract the centrifugation is preceded by heating the tube for 30 minutes at 100°C. The residue is washed with a portion of water, centrifuged, and the combined supernatants are made up to 100 ml, giving a protein solution in 0·1 n NaOH.

Duplicate portions of 1 ml of this solution or of working standard solution and 1 ml of 0·1 n NaOH for reagent blank (zero optical density) are pipetted into thin wall flasks (Fig. 5B.1), followed by 5 ml of reagent C. The contents are mixed and allowed to stand for 20 minutes at room temperature. Then to each flask 0·5 ml of reagent F is added and the content immediately mixed. The flasks are left standing at room temperature for 2 hours protected from direct light, before the reading of the optical density, the colour development in 1 and 2 hours differing by 2% O.D. After that time they may be stored for further 24 hours at $+4°C$ with no marked fading of colour intensity (Dowgiallo and Watkowska, unpublished).

The pre-extraction of lipids is not recommended since the extract may contain a fraction of Folin-positive material, however a pigment blank may be determined if the alkaline protein solution is coloured or slightly turbid. For this purpose 5 ml of reagent C are mixed in a flask with 0·5 ml of reagent F and let stand at room temperature for 30 minutes. Then 1 ml of the examined solution is added, the flask content mixed and its optical density read. The found pigment O.D. is subtracted from the sample O.D. (Kleczkowski *et al.*, 1961).

5B.15 Calculation of results

A —protein in per cent of fresh weight of the examined material
D_u —optical density of the examined sample homogenate
D_k —optical density of the working standard protein solution
a —mg protein in 1 ml of working standard protein solution
b —mg of fresh material sample
f —ml of the final volume of sample homogenate (dilution factor)
X —conversion factor from working standard concentration to actual protein concentration in the homogenate
c —protein nitrogen in mg per 1 ml of sample homogenate
6·25—conversion factor from nitrogen to protein

$$A = \frac{D_u \cdot a \cdot f \cdot 100}{D_k \cdot b \cdot X} \quad \text{where } X = \frac{D_u \cdot a}{D_k \cdot c \cdot 6 \cdot 25}$$ —for the biuret method (alternative for the Folin-Lowry method)

$$A = \frac{c \cdot f \cdot 100 \cdot 6 \cdot 25}{b}$$ —for the Kjeldahl procedure

$$A = \frac{D_u \cdot a \cdot f \cdot 100}{D_k \cdot b}$$ —for the Folin-Lowry method (alternative for the biuret method)

5B.16 Determination of carbohydrates

This determination comprises total carbohydrates of animal material, the chitin being determined separately: by colorimetry (Krause, 1959; Strickland

and Parsons, 1968), by weighing or by calculation from Kjeldahl nitrogen. The chitin nitrogen content multiplied by 14·5 gives the amount of chitin. With plant material the sum of carbohydrates yielding to hydrolysis in diluted sulphuric acid may be determined in one operation and the weight of the insoluble residue may be considered as measure of holocellulose together with lignin (crude fibre). The fractionation of carbohydrates may be achieved into free (alcohol soluble) sugars, protein bound sugars and glycogen in animal material (Blažka, 1966 b; Carroll *et al.*, 1956; Giese, 1967; Lenartowicz *et al.*, 1967; Roe and Dailey, 1966; Seifter *et al.*, 1950), into different classes of carbohydrates in yeasts (Trevelyan and Harrison, 1952) and in plant material (Bachelard, 1968; Barashkov, 1963; Belozerskij and Proskurjakov, 1961; Blažka, 1966 b; Kosenko and Rashba, 1967; Shnyukova and Pirozhenko, 1972). For further details on carbohydrate fractionation and determination consult specialist monographs (Hassid and Abraham, 1957; Paech and Tracey, 1955–1964; Pigman ,1957; Strickland and Parsons, 1968; Whistler and Smart, 1953; Whistler and Wolfrom, 1962).

Hexoses (both mono- and polysaccharides) react with anthrone in concentrated sulphuric acid to form a blue-green coloured complex, suitable for photometric determination at 625 nm (Morris, 1948). Under the reaction conditions given below (Hilger photometer, filter No. 61), xylose at the same concentrations as glucose produces only 7–8% of optical density given by glucose. The determination of free pentoses with the orcinol–$FeCl_3$ reagent is more selective, glucose produces about 3% of xylose colour when present at the same concentration. The readings are performed at 670–700 nm against a reagent blank (Brown, 1946). The pentosans require not only solubilization, but also complete hydrolysis prior to pentose determination (Minderman and Bierling, 1968). The reaction of sugars with phenol and sulphuric acid (Dubois *et al.*, 1956) is in common use in aquatic chemistry (Strickland and Parsons, 1968; Pourriot and Leborgne, 1970; Raymont *et al.*, 1964, 1971), however it has not been tested in this laboratory.

5B.17 Reagents

Anthrone reagent. 160 mg of anthrone are dissolved in 100 ml of sulphuric acid 84% (w/w,sp.gr. 1·78): 4 volumes of sulphuric acid (sp.gr. 1·84) are added to 1 volume of water. Sulphuric acid of lower concentration, down to 80% is also suitable. The reagent is prepared daily and filtered through a fine sintered glass filter.

Sulphuric acid 1 n.

Sulphuric acid 80% (w/w, sp.gr. 1·73): 3 volumes of sulphuric acid (sp.gr. 1·84) are added to 1 volume of water.

Ethanol 96%, purified after Blažka (1966 b) (see reagents for lipid determination).

Sodium or potassium hydroxide, 6 n.

Sodium sulphate, 2% aqueous solution.

Sugar stock solutions in saturated aqueous benzoic acid. One gram of anhydrous glucose is dissolved in 1 litre benzoic acid solution. One gram of anhydrous xylose is dissolved in 1 litre benzoic acid solution. Working standards containing 40 micrograms glucose per 1 ml or 10 micrograms xylose per 1 ml are prepared daily by diluting with water a portion of the stock solution.

Trichloroacetic acid 5% w/v.

Orcinol reagent. In 100 ml of concentrated hydrochloric acid (sp. gr. 1·19) 0·1 g $FeCl_3$ · $6H_2O$ and 0·2 g orcinol are dissolved. The reagent is prepared daily. The solution of $FeCl_3$ in concentrated HCl may be stored, in a portion of which orcinol is dissolved.

5B.18 Procedure

Basic procedure for hexoses. Of the solution tested 2 ml are pipetted into a test tube or a 25 ml thin wall flask, 10 ml anthrone reagent added, mixed, the tube is heated exactly for 5 minutes in a boiling water bath and cooled in tap water. Simultaneously 2 ml of distilled water (reagent blank) and 2 ml of working standard solution (duplicate) are treated in the same manner. Optical densities are read within 6 hours against the reagent blank as zero optical density. The readings should be corrected for the pigment blank of the examined solution. This is performed by mixing and heating 2 ml of the examined solution with 10 ml of sulphuric acid (sp.gr. 1·78) without anthrone (Hewitt, 1958; Strickland and Parsons, 1968). Duplicate determinations are recommended, the readings are performed against water. The concentration of carbohydrate should not exceed 0·1 mg per 1 ml of solution. The presence of TCA in the test solution up to 5% and the presence of H_2SO_4 up to 2 n do not interfere. Chlorides in the range of concentration 0·5–1% Cl in the test solution increase the optical density of anthrone reaction by a constant value 11%.

Basic procedure for pentoses. A volume of 5 ml of the solution tested is pipetted into a tube or flask, 5 ml of orcinol reagent added, mixed and heated for 30 minutes in a boiling water bath, then cooled under tap water. Simultaneously 5 ml of distilled water (reagent blank) and 5 ml of working standard solution are treated in the same manner. The readings should be performed within 2 hours. Sulphuric acid in pentose (xylose) solution increases the optical density from 6% for 0·5 n H_2SO_4 to 45% for 2 n HSO_4.

A. *Alkali soluble polysaccharides in animal material.* A weighed sample (0·1–0·2 g) of fresh material is homogenised with 2·3 ml of 6 n alkali and heated for 20 minutes at 100°C. Subsequently 0·2 ml of sodium sulphate solution is added followed by alcohol to final volume 10 ml. The mixture is allowed to stand overnight at 4°C or it is heated up to 60°C for 5 minutes, but the latter procedure may not secure complete precipitation of the solubilised polysaccharides. After centrifugation for 15 minutes at 5000 r.p.m. the supernatant is discarded and the tube drained. The precipitate is hydrolysed for 30 minutes at 100°C with 5 ml 1 n H_2SO_4, diluted with water to 10 ml and centrifuged. The clear sugar solution (I) is decanted to a clean stoppered tube and preserved at 4°C for further sugar analysis up to 2–3 days.

The sugar solution I is diluted ten times for analysis.

The residue is heated with 3–4 ml 6 n KOH (NaOH) at 100°C for 20 minutes and, after addition of an equal volume of water, centrifuged, and the supernatant is discarded. The residue is heated again in 2–3 n KOH for 5–10 minutes, centrifuged and the supernatant discarded. The deproteinised residue is washed twice by heating it for 2–3 minutes with water, centrifuged, washed with alcohol, again centrifuged, finally dried in a desiccator. The obtained chitin preparation is preserved for nitrogen determination or weighed. The shells of crustacea or mollusca, before washing with alcohol, should be treated with 5 ml 0·5 n HCl for 15 minutes at 100°C and washed with water, to remove the carbonates.

The samples of mixed plankton for chitin determination should be treated with more diluted alkali (3 n) that is sufficient to dissolve all protein and eliminates the risk of loss of chitin (Krause, 1959).

B. *Carbohydrates TCA—soluble in animal material.* A weighed sample (0·025–0·050 g) of fresh material is homogenised with 8 ml of 5% TCA and heated for 15 minutes at 100°C in tubes covered with condensers. The tube content is filled up to 10 ml with water, mixed and centrifuged until clear. From this solution 2 ml are pipetted for hexose determination as above. The determination is short and it comprises both simple and condensed sugars, however the acid used does not solubilise quantitatively some glycoproteins, thus giving low results for total carbohydrates. It is important to use appropriate sugar standard solution containing e.g. galactose, if galactogen containing animals are examined (molluscs, parasitic worms). The results are then expressed in per cent of a particular sugar in the examined material. The use of cold TCA solution for carbohydrate extraction allows to fractionate the glycogen (Boddington and Mettrick, 1971).

C. *Carbohydrates hydrolysed by 1 n sulphuric acid in plant and animal material.* A weighed sample (0·1–0·2 g) of material cut into fine pieces is mixed with 3 ml of 1 n H_2SO_4 (homogenised if necessary) and hydrolysed for 30 minutes at 100°C, using condensers. The mixture is made up with water to 10 ml and centrifuged. The sugar solution is diluted 100 times for plant material and 10 times for animal material. From the diluted solution 2 ml are taken for analysis (basic procedure for hexoses) or with plant material 5 ml are taken for pentose determination as well, correcting the results for respective presence of sugars of the other group. Pentoses may constitute a considerable proportion of total carbohydrates also in animal material (Raymont *et al.*, 1968).

The residue on hydrolysis and centrifugation requires further treatment. If animal material is analysed (invertebrates) the subsequent digestion for 30 minutes with 6 n or 3 n KOH (NaOH) is essential to obtain pure chitin. The residue is then washed with water and alcohol as under A.

In plant material crude cellulose may be estimated by weighing the residue on hydrolysis, after it has been washed with distilled water, with water containing ammonia (e.g. 0·1%) to neutralize the sulphuric acid, and with alcohol, then dried at 90°C to constant weight. This residue after separation of easily hydrolysable carbohydrates (of the starch type) still contains cellulose, a part of hemicelluses, lignin, some protein and ash, and does not exactly correspond to crude fibre, which is the residue after acid and alkali treatment (Janas and Podkówka, 1972; Paech and Tracey 1955–1964; Strickland and Parsons, 1968). The use of detergents affords protein-free crude fibre (van Soest, 1963; Walicka, 1972). The carbohydrate part may be isolated from this material by further hydrolysis in the following way. The residue after mild hydrolysis, or a portion of fresh material, is mixed with 1 ml of 80% sulphuric acid and left at room temperature for 2·5 hours. Subsequently 14 ml of water are added, the tube covered with a condenser and heated in a boiling water bath for 5 hours, then centrifuged. Possible losses of water are refilled during heating. The whole supernatant is diluted to a known volume with water and analysed for hexoses and pentoses (basic procedures) (Belozerskij and Proskurjakov, 1951; Minderman and Bierling, 1968).

The lignin content can be evaluated as the difference between the weight of fresh material and the sum of weights of determined compounds, i.e. lipids, condensed pentoses and hexoses, protein, ash and water.

In samples containing no more than 0·2 mg of polysaccharide in the residue after hydrolysis (that accounts for not less than 10% of this residue), or in weighed portions of fresh material the hexoses may be determined directly. The sample should be first pre-extracted with alcohol-ether mixture (20-fold excess over the water volume in the sample), if it is strongly pigmented or it contains more fat than 5% in dry weight. Water is added to the dry residue or to sample homogenate to make 2 ml and 10 ml of anthrone reagent are added, according to the basic procedure for hexoses. If the solution is turbid after heating, it may be either centrifuged before reading in the photometer or filtered through glass fibre filter on a support of sintered glass.

For aquatic plants, as algae, submerged or floating plants, the method of acetylation of cellulose is recommended (Blažka, 1966 b).

5B.19 Calculation of results

A — carbohydrate in (condensed) glucose per cent of fresh weight of the examined material

D_u — optical density of the examined carbohydrate solution

D_k — optical density of glucose working standard solution

a — mg glucose in 1 ml of the working standard solution (e.g. 0·040)

b — mg of fresh material sample

0·9 — conversion factor from glucose to polysaccharide (condensed)

f — ml of the final volume of the sample (dilution factor)

$$A = \frac{D_u \cdot a \cdot 0\cdot9 \cdot f \cdot 100}{D_k \cdot b}$$

Table 5B.1. Example for a data sheet of chemical determination

Material	—*Tribolium castaneum* Herbst, adult 4 days, starved 4 hours
Sample fresh weight	—189 mg Conversion factor to % dry weight—1·87
Dry weight 100°C	—53·4%
Determination	—carbohydrate—polysaccharide
Method	—alkaline then acid hydrolysis, anthrone reagent
Apparatus	—Hilger BIO-absorptiometer, filter 6l, cell 1 cm
Standard	—glucose 0·040 mg/ml
Dilution factor	—100
Readings (against water)	—standard 0·423, 0·418, 0.419, mean 0·420 corrected 0·360
	—reagent blank 0·060
	—pigment blank 0·034
	—unknown 0·259, 0·265, — mean 0·262 corrected 0·168
Per cent content:	
of fresh weight	$- \;\; A = \dfrac{0 \cdot 168 \;\cdot\; 0 \cdot 040 \;\cdot\; 100 \;\cdot\; 100 \;\cdot\; 0 \cdot 9}{0 \cdot 360 \;\cdot\; 189} = 0 \cdot 9$
of dry weight	$- \;\; 0 \cdot 9 \;\cdot\; 1 \cdot 87 = 1 \cdot 7$
Remarks	—no correction for pentoses was necessary
Calculations	

References

See end of Chapter 5C.

5C. APPENDIX: Proposed methods for estimating the excreted non-protein nitrogenous waste products in mixed urinary-faecal material (rejecta).*

A. DOWGIALLO

5C.1 Introduction

According to Hoar (1966), 90% of the nitrogen excreted is accounted for by the catabolism of proteins and, with a few exceptions, this is mainly in the form of ammonia, urea or uric acid. Moreover, amino acids which are derived from the digestion of proteins are known to pass into the fuel system

---

*The cooperation of Dr. Annie Duncan in elaborating this Appendix is gratefully acknowledged.

if they are not required for growth, tissue repair or the production of enzymes or secretions of various kinds. How much passes along this pathway depends upon the amount of protein in the diet and whether other sources of respiratory substrate is available, among other factors. In starving animals, the cellular protein content declines, thus acting as a protein reserve which can be drawn upon, although protein is not stored in the same way as carbohydrates or fats. The significance of nitrogenous waste is that it is a form of chemical energy eliminated from the body of the organism whose energy budget is being studied and thus needs to be measured. If large amounts of nitrogenous substances are being excreted, it is likely that protein is being used as a respiratory substrate which may influence the availability of protein for growth, especially if the diet is low in nitrogen. Moreover, assuming that all nitrogenous waste derives from protein being used as a respiratory substrate, a knowledge of the respiratory quotient, the rate of oxygen consumption and carbon dioxide production allows one to calculate the nature of the substrate being catabolised, in what proportions and the amount of heat being produced. This technique is called indirect calorimetry and is described in Brody (1945) and Kleiber (1961).

In those taxonomic groups of animals which excrete their nitrogenous waste separately from their faeces, the amount of energy eliminated from the organism as urine is readily estimated, given the chemical expertise. However, in three groups (insects, birds and gastropod molluscs), the urine and faeces are eliminated together. Petrusewicz and Macfadyen (1970) have coined the term rejecta (FU) for this consumed but not utilised energy. It is a much more difficult problem technically to estimate the urinary nitrogen content of rejecta but the following method, which is the result of two years' testing in the laboratory, is offered as a possible solution.

5C.2 Basis of method

The main procedure is one of extraction in warm $O \cdot 1M$ lithium carbonate of the nitrogenous products of protein metabolism from complex mixture of substances contained in 'rejecta' (Jeżewska *et al.*, 1963). Usually not less than 90% of the excreted nitrogen is derived from protein metabolism (Hoar, 1966). The substances likely to be contained in such extract are urinary ammonium salts, urea, purines (including uric acid and guanine), creatinine and the faecal amino acids, proteins and peptides as well as some other nitrogenous and non-nitrogenous organic and inorganic compounds soluble in lithium carbonate (Campbell, 1970; Campbell and Goldstein, 1972; Jeżewska *et al.*, 1963; Kaplan, 1969; Nejedly, 1953; Potts, 1963; Prochazkova, 1964). Subsequently, any proteins and polypeptides present are precipitated by tungstic acid and the nitrogen content of the remaining substances is determined by the Kjeldahl method.

5C.3. Main procedure

a. Collection and preparation of material

Dry the egested material in an oven at 60°C for subsequent storage and later mass treatment or proceed with fresh material. The only possible losses due to drying at 60°C is free ammonia and this loss may be checked by comparing the total nitrogen content per mg dry weight of rejecta obtained from dry and from fresh material. Grind well the fresh or dry material into a uniform paste or powder from which a good representative sub-sample may be taken. If the material is not homogenous, use larger samples.

b. Determination of total nitrogen present in untreated sub-samples of rejecta.

Determine the total nitrogen in a sub-sample of the dried homogenised powder (about 25-50 mg) directly by means of the Kjeldahl method described below. This determination can be ommitted, should the information not be needed.

c. Extraction of soluble nitrogenous substances in warm lithium carbonate

Extract another sub-sample of about 25 mg of dried or 50 mg fresh homogenised material with 5 ml of warmed saturated solution of lithium carbonate (about 0·1 M). This is best carried out in a water bath at 80–90°C for 15–30 minutes with occasional stirring. Cool, centrifuge at about 4000 rpm and decant the clear supernatant. Add to the sediment a few drops of lithium carbonate plus 1–2 ml distilled water, stir, centrifuge and add the decanted solution to the first supernatant. Make up the volume of these combined supernatants to 50 ml with water; this is now solution I and contains the total soluble nitrogen present in all the compounds listed in paragraph 1. In an aliquot of solution I, determine the total soluble nitrogen by Kjeldahl method, expressing the results as nitrogen content per 100 mg dry weight of rejecta. Check the completeness of nitrogen extraction once for each particular material by extracting the remaining sediment again with fresh lithium carbonate solution and determining the nitrogen content of the extract by the Kjeldahl method. Even after re-extraction, this sediment may still contain nitrogen in insoluble proteins, chitin etc.

d. Separation of proteins from the soluble nitrogenous substances

If solution I contains proteins from the faecal part of the rejecta, these must be removed by the following procedure. Take 5 ml of solution I, add 1 ml of a 10% solution of sodium tungstate (10 g $Na_2WO_4 \cdot 2H_2O$ made up to 100 ml with water); stir with a glass rod and add 0·50 ml of 1N H_2SO_4, stirring again; heat in a 80°C water bath for 10 minutes for the protein precipitate to settle. Test the clear supernatant for acidity (pH should be about 5); if the pH is more than 5 add sulphuric acid, drop by drop. Make up to about 10 ml with distilled water, mix well, centrifuge and decant the clear supernatant into a 25 ml flask, add 10 ml water to the sediment, heat at 80°C for 5 minutes, centrifuge, add the supernatant to the first one and make up to 25 ml (solution II). Repetition of this treatment will ensure the dissolving of all the uric acid in rejecta of high uric acid content (Bell, 1963; Haden, 1923; Hawk *et al.*, 1954). Solution II represents the non-protein soluble nitrogen in the rejecta and should contain most of the nitrogenous catabolic products plus any amino acids present (a very small proportion, e.g. in birds, Campbell, 1970). Take an aliquot of solution II to determine its nitrogen content per 100 mg dry weight of rejecta. If solutions I and II do not differ with respect to their nitrogen content or if no protein is precipitated with tungstic acid, then de-proteinisation can be ommitted.

e. Detection and estimation of ammonium compounds in fresh sub-samples of rejecta.

Some ammonium salts have been recorded in the rejecta of both birds and insects although usually not exceeding 10% of the total nitrogen excreted. Some of these may be lost during the initial drying at 60°C and some during the warm lithium carbonate extraction. The following method will detect the presence of such salts and can possibly be used to estimate their concentration in fresh undried rejecta. Mix a sub-sample of fresh rejecta (e.g. 10

mg) with 100 ml distilled water and either centrifuge or allow to settle; decant 50 ml of the clear supernatant. To this add 0·2 ml saturated sodium-potassium tartrate (Seignette salt) and 0·8 ml 27% NaOH, mix, add 0·1 ml Nessler reagent, mix again and determine the optical density as described below in the section 'nitrogen micro-determination'. The Nessler colour with NH_3 (like tea) appears immediately even with a concentration of as little as 0·1 μg NH_4-N/ml whereas urea does not produce such a colour. Details of the composition and preparation of the Nessler reagent are given below. In the presence of sulphides, a grey turbidity appears upon the addition of the reagents in which case the solution has to be acidified to pH 2–3 and aerated for 10 minutes first before adding these.

N.B. It is essential to use ammonia-free distilled water for the preparation of all the reagents and for dilutions of the extracts. This can be produced by either re-distilling distilled water after acidification with a few ml of phosphoric acid per litre or filtering distilled water through a column of Amberlite IR 120—acid form.

5C.4. The Kjeldahl method for determination of ammonia-nitrogen and organic nitrogen

The standard Kjeldahl method for the determination of ammonia-nitrogen and organic nitrogen has evolved during the century of its existence into literally thousands of variants, differing in more or less important details. The main stages in the Kjeldahl procedure are: 1. digestion (mineralisation) of the sample in sulphuric acid in order to convert organic nitrogen into ammonium sulphate; this conversion is quantitative, except for the oxygen compounds of nitrogen, 2. distillation of the ammonia, a stage which may be ommitted, 3. determination of ammonia either in the distillate or directly in the digestion solution. The procedure can be employed either in a macro or micro scale (Jacobs, 1965).

At each of these main stages, the procedure may be varied. The mineralisation stage may vary in the catalyst employed, according to the requirements of subsequent analysis; distillation may be performed in a variety of types of apparatus, of which the Parnas-Wagner and the Markham steam distillation apparatus are the most common; the Conway method utilises diffusion of the ammonia without heating. The greatest variation occurs in the methods used for determination of the final ammonia. The most widely employed methods are: titrimetric (acidimetric or iodometric) or photometric, based either on Nessler or on the formation of a dye of the indophenol-blue type, and others.

The basic methods for the determination of nitrogen are given in most laboratory handbooks (e.g. Ballentine, 1957; Block, 1960; Golterman, 1970; Hawk *et al.*, 1954; Strickland and Parsons, 1968; Official Methods, 1950; etc.). Some analytical procedures can be recommended because of their simplicity, for example, the Nessler reagent used with a photometric determination of the ammonia colour directly in the digestion solution (Burck, 1960; Ivanko, 1971; Lang, 1958; Polley, 1954) or in water (Mackereth, 1963). A special Nessler reagent which produced no turbidity in high concentrations of ammonia was used without distillation by Williams (1964). It is also possible to produce the indophenol-blue directly in the digestion mixture (Blachère and Ferry, 1957; Glebko *et al.*, 1967) as well as in natural waters (Newell, 1967; Riley 1953; Tetlow and Wilson 1964). Newell (1967) used chloramine T instead of hypochlorite with subsequent extraction with an organic solvent of its reaction products with ammonia. In a method similar to that of Tetlow and Wilson (1964), Kaplan (1969), and Stegemann and Loescheke (1962) suggest that it may be possible to ommit the extraction stage, thus simplifying the procedure. Prochazkova (1964) used the bispyrazolone reagent with subsequent extraction of the dye with trichloroethylene.

5C.5. Micro-technique for determination of nitrogen

The following micro-scale technique for the determination of nitrogen is based upon methods described by Hrbaček (1962) and his co-workers for plankton, seston and water, supplemented after Rzymowska *et al.* (1953).

5C.5.1. Reagents

All reagents should be of the analytical grade; only ammonia-free distilled water should be used (see Section 5C.3e).
1. 10% NaCl in water (w/v).
2. Concentrated sulphuric acid (ammonia-free).
3. 60% KOH in water (w/v), stored in a polythene bottle.
4. 27% NaOH in water (w/v), stored in a polythene bottle.

5. Nessler reagent: to 23·88 g KI (containing no free iodine) dissolved in 30 ml H_2O add 36·40 g HgI_2, stir until dissolved and make up to 100 ml. Filter through a hard filter paper and store in a dark bottle.
6. A standard stock solution of ammonium chloride containing 3·818 g NH_4Cl per litre of which 1 ml represents 1 mg NH_4-N. For calibration, diluted solutions should be prepared daily.
7. 2% $CuSO_4$: 4 g $CuSO_4 \cdot 5H_2O$ in 200ml water.
8. 30% H_2O_2 (Perhydrol), nitrogen- free.

5C.5.2. Procedure and calculation

A sub-sample of the material to be tested (not exceeding 0·2 g and containing up to 2 mg N) is rinsed down a Kjeldahl flask (volume 50–100 ml) with water, 1 ml 10% NaCl, 2 ml concentrated H_2SO_4 and 0·2 ml 2% $CuSO_4$ are added. Place the flask in an inclined position on an electric plate. N.B. gas burners cannot be used, cigarettes should not be smoked nor should bottles of ammonia be opened in the room during a Kjeldahl procedure.

The flask is heated and, after the water has evaporated, white fumes appear in the neck of the flask and the liquid becomes clear and colourless within 3–4 hours (if tungstic acid is present a precipitate is produced); if de-colourisation takes place with difficulty, add 0·5 g of potassium sulphate or 1–2 drops of H_2O_2 directly into the liquid and re-heat. The final colour should be that of $CuSO_4$ in H_2SO_4 (a pale blue-green). After de-colourisation heat one hour extra. A control containing only the reagents for mineralisation (H_2SO_4, $CuSO_4$, NaCl, K_2SO_4 or H_2O_2) should be treated along with the samples. The mineralised samples may be stored before distillation in a desiccator over concentrated H_2SO_4 with a ground glass surface lubricated with H_3PO_4 to protect them from atmospheric NH_3.

For distillation of ammonia, dilute the mineralised sample with water up to 100 ml, put an aliquot containing no more than 100 μg N (e.g. 5 ml, dilution factor 20) into the distillation apparatus, rinse the funnel with water, add 2 ml 60% KOH to make the liquid strongly alkaline, rinse again, close immediately and distil, collecting exactly 50 ml of the distillate in a volumetric flask within 10–15 minutes. This may be kept until the next day if tightly stoppered (test solution).

Add to the flask 0·8 ml 27% NaOH, mix, then add 0·1 ml Nessler reagent, stopper flask and mix thoroughly. Read off the optical density at 428–430 nm in a 1–5 cm cuvette against a reagent blank (i.e. NaOH plus Nessler in distilled water); this must be done within 30 minutes to 2 hours. Distil the mineralised control, read off its optical density and subtract it from the sample optical density reading.

Add NaOH and Nessler reagent to a series of diluted standards (containing, for example, 30, 60 and 90 μg N per 50 ml) in a 50 ml flask; determine their optical densities and plot a calibration graph of optical densities against concentration of nitrogen. Apply the sample optical densities to the graph and read off the sample nitrogen concentrations. If the calibration reveals a straight line relationship, calculate the nitrogen concentration from

$$A = \frac{E_u \cdot a \cdot 100 \cdot f}{E_k \cdot b}$$

where

A —nitrogen as percentage of the examined material (mg per 100 mg dry wt),

E_u —optical density of the test solution,

E_k —optical density of the diluted standard ammonia solution (taking the average of two replications),

a —mg $NH_4 - N$ per 1 ml of diluted standard (e.g. 0·001 mg.),

b —mg of the examined material processed,

f —dilution factor.

Check the reproducibility of the method by using for each series only one of the diluted standards, replicated twice. It is advisable to run one control determination for each batch of reagents; add 2 ml of ammonia stock solution together with all the other reagents in appropriate amounts to a Kjeldahl flask and proceed as above.

5C.6 References

ALEKSIUK M. & STEWART K.W. (1971) Seasonal changes in the body composition of the garter snake (*Thamnophis sirtalis parietalis*) at northern lattitudes. *Ecology*, **52**, 485–490.

ASHWELL G. (1957) Colorimetric analysis of sugars. In Colowick S. P. and Laplan N. O. (eds), *Methods in enzymology*. Academic Press, New York, Vol. **3**, p. 73–105.

BACHELARD E.P. (1968) Interference of perchloric acid with the anthrone reaction for for carbohydrates. *Analytica chim. Acta*, **42**, 171–173.

BAILEY N.T.J. (1959) *Statistical methods in biology*. The English Univ. Press Ltd., London, p. 200.

BALLENTINE R. (1957) Determination of total nitrogen and ammonia. In Colowick S. P. and Kaplan N. O. (eds), *Methods in enzymology*. Academic Press, New York, Vol. 3, p. 984–995.

BARASHKOV G.K. (1963) *Khimija vodoroslej*. Chemistry of algae. *Izd. Akad. Nauk SSSR*, Moskva, p. 143 (Russian).

BARNES H. (1972) The seasonal changes in body weight and biochemical composition of the warm—temperate cirripede *Chthamalus stellatus* (Poli). *J. esp. mar. Biol. Ecol.*, **8**, 89–100.

BEERS J.R. (1966) Studies on the chemical composition of the major zooplankton groups in the Sargasso Sea off Bermuda. *Limnol. Oceanogr.*, **11**, 520–528.

BELL P.M. (1963) A critical study of methods for the determination of non-protein nitrogen. *Analyt. Biochem.*, **5**, 443–451.

BELOZERSKIJ A.N. & PROSKURJAKOV N.I. (1951) *Prakticheskoe rukovodstvo po biokhimii rastenij*. Practical manual of plant biochemistry. Izd. Sovetskaja Nauka, Moskva. p. 387. (Russian).

BERNICE R. (1972) Biochemical composition of *Streptocepholus dichotomus* Baird *Branchinella kugenumaensis* (Ishikawa) (Crustacea: Anostraca). *Hydrobiologia*, **39**, 155–164.

BLACHÈRE H., FERRY P., (1957) Dosage de l'azote dans les sols par microdiffusion. *Ann. Agron.*, **3**, 495–498.

BLADERGROEN W., (1955). Einführung in die Energetic und Kinetik biologischer Vorgänge. *Wepf a. Co. Verl.*, Basel. p. 368.

BLAZKA P., (1966a) Bestimmung der Proteine in Material aus Binnengewässern. *Limnologica*, **4**, 387–396.

BLAZKA P., (1966b) Bestimmung der Kohlenhydrate und Lipide. *Limnologica*, **4**, 403–418.

BLAZKA P., (1966c) The ratio of crude protein, glycogen and fat in the individual steps of the production chain. *Hydrobiol. Stud.*, **1**, 395–408.

BLIGH E.G., DYER W.J., (1959) A rapid method of total lipid extraction and purification. *Can. J. Biochem. Physiol.*, **37**, 911–917.

BLOCK R.J., (1960) Amino acid analysis of protein hydrolysates. In Alexander P., Block R. J., (eds), *Laboratory manual of analytical methods of protein chemistry*. Pergamon Press, Oxford, London, New York, Paris vol. **2**, p. 3–57.

BLOOR W.R., (1914) A method for the determination of fat in small amounts of blood. *J. Biol. Chem.*, **17**, 377–384.

BLOOR W.R. (1947) A colorimetric procedure for the determination of small amounts of fatty acid. *J. Biol. Chem.*, **170**, 671–674.

BODDINGTON M.J., METTRICK D.F. (1971) Seasonal changes in the chemical composition and food reserves of the freshwater triclad *Dugesia tigrina* (Platyhelmitesi Turbellaria). *J. Fish. Res. Bd Can.*, **28**, 7–14.

BOJE R. (1966) Proteine. *Limnologica*, **4**, 383–386.

BOYD C.E. (1970) Amino acid, protein, and coloric content of vascular aquatic macrophytes. *Ecology*, **51**, 902–906.

BOYD C.E., GOODYEAR C.P., (1971) Nutritive quality of food in ecological systems. *Arch. Hydrobiol.*, **69**, 257–270.

BRODY S. (1945) *Biogenetics and Growth*. New York. 1023 pp.

BROWN A.H., (1946) Determination of pentose in the presence of large quantities of glucose. *Arch. Biochem.*, **11**, 269–278.

BURCK H.C., (1960) Kolorimetrische Mikro-Kjeldahl-Methode mit direkter Nesslerisation zur routinemassigen Stickstoffbestimmung. *Microchim. Acta.* **2**, 200–203.

CAMPBELL J.W. (1970) *Comparative biochemistry of nitrogen metabolism*. (ed.) Academic Press, London, New York, vol. **1–2**, p. 45, 916.

CARROLL N.V., LONGLEY R.W., ROE J. H. (1956) The determination of glycogen in liver and muscle by use of anthrone reagent. *J. biol. Chem.*, **220**, 583–593.

CHUECAS L., RILEY J.P., (1969) Component fatty acids of total lipids of some marine phytoplankton. *J. mar. biol. Ass.* U.K., **49**, 97–116

CLELAND K.W., SLATER E.C., (1953) Respiratory granules of heart muscle. *Biochem. J.*, **53**, 547–556.

COLES G.C., (1970) Some biochemical adaptations of the swamp warm *Alma emini* to low oxygen levels in tropical swamps. *Comp. Biochem. Physiol.*, **34**, 481–489.

COLOWICK S.P., KAPLAN N.O., (1957) (eds.) *Methods in enzymology*. Academic Press, New York, vol. **3**, p. 1154.

CONOVER R.J., CORNER E.D.S. (1968) Respiration and nitrogen excretion by some marine zooplankton in relation to their life cycles. *J. mar. biol. Ass.* U.K., **48**, 49–75.

CORNER E.D.S., COWEY C.B., MARSHALL S.M., (1967) On the nutrition and metabolism of zooplankton V. Feeding of *Colamus finmarchicus*. *J. mar. biol. Ass.* U.K., **47**, 259–270.

COWEY C.B., CORNER E.D.S. (1963) Amino acids and some other nitrogenous compounds in *Colamus finmarchicus*. *J. mar. biol. Ass.* U.K., **43**, 485–493.

CZECZUGA B. (1963) Quantitative proportions of glycogen in certain species of the Tendipedidae (Siptera) larvae. *Hydrobiologia* (Haag), **22**, 92–110.

DRABIKOWSKI W., DOMINAS H., DĄBROWSKA M., (1966) Lipid patterns in microsomal fractions of rabbit skeletal muscle. *Acta Biochim. pol.*, **13**, 11–24.

DREWS G. (1968) Microbiologisches Praktikum für Naturwissenschaftler. Springer verl., Berlin, Heidelberg, New York, p. 214.

DUBOIS M., GILLES K.A., HAMILTON J.K., REBERS P.A., SMITH F. (1956) Colorimetric method for determination of sugars and related substances. *Analyt. chem.*, **28**, 350–356.

ENTENMAN C., (1957a) General procedures for separating lipid components of tissue. In Colowick S. P., Kaplan N. O. (eds), *Methods in enzymology*. Academic Press, New York, vol. **3**, p. 299–317.

ENTENMAN C. (1957b) Preparation and determination of higher fatty acids. In Colowick S. P., Kaplan N. O., (eds), *Methods in enzymology*. Academic Press, New York, vol. **3**, p. 317–328.

FLORKIN M., SCHEER B.Y., (1967–1968) (eds.) *Chemical zoology*, Academic Press, New York, vol. **1–4**,

FOLCH J., LEES M., SLOANE-STANLEY G.H. (1957) A simple method for the isolation and purification of total lipids from animal tissues. *J. biol. Chem.*, **226**, 497–509

FRIEDEMANN T.E., WEBER C.W., WITT N.F. (1963) Clarification of solution and removal of interfering substances in determination of reducing sugars. *Analyt. Biochem.*, **6**, 504–511.

FUDGE H. (1968) Biochemical analysis of preserved zooplankton. *Nature*, London, **219**, 380–381.

GIESE A.C. (1967) Some methods for study of the biochemical constitution of marine invertebrates. *Oceanogr. mar. Biol.*, **5**, 159–186.

GLEBKO L.I., ULKINA V.E., VASKOVSKY V.E. (1967) Spectrophotometrical method for determination of nitrogen in biological preparations based on thymol-hypobromite reaction. *Anal. Biochem.*, **20**, 16–23.

GOLTERMAN H.L., (1970) *Methods for chemical analysis of fresh waters.* Revised second printing. Blackwell Scientific Publications, Oxford and Edinburgh, IBP Hand book No. **8**, p. 166.

GORNALL A.G., BARDAVILLE C.J., DAVID M.M. (1949) Determination of serum proteins by means of the biuret reaction. *J. Biol. Chem.*, **177**, 751–766.

HADEN R.L. (1923) A modification of the Folin-Wu method for making protein-free blood filtrates. *J. Biol. Chem.*, **56**, 469–471

HARROW, B., BOREK E., MAZUR A., STONE G.C.H., WAGREICH H. (1958) *Laboratory Manual of Biochemistry.* IV. ed. Saunders, Philadelphia and London.

HASSID W.Z., ABRAHAM S. (1957) Chemical procedures for analysis of polysaccharides. In Colowick S. P., Kaplan N. O. (ed.), *Method in enzymology.* Academic Press, New York, vol. **3**, p. 34–54.

HAWK P.B., OSER B.L., SUMMERSON W.H. (1954) *Practical physiological chemistry.* Ed. 13, The Blakiston Co., New York, p. 1439.

HELANDER E. (1957) On quantitative muscle protein determination. *Acta physiol. Scand.*, **41**, suppl. 141, 1–99.

HESSE G., LINDNER R., MÜLLER R. (1971) Eine Mikromethode zur schnellen und spezifischen Seiren bestimmung des Proteingehaltes unzerstörter Mikroorganismen. *Z. allg. Mikrobiol.*, **11**, 585–594.

HEWITT B.R. (1958) Spectrophotometric determination of total carbohydrate. *Nature*, London, **182**, 246–247.

HOAR W.S. (1966) *General and comparative physiology.* Englewood Cliffs, New Jersey, Prentice Hall. XII, p. 815.

HOCH F.L., VALLEE B.L. (1953) Gravimetric estimation of proteins precipitated by TC acid. *Analyt. chem.*, **25**, 317–320.

HORNING E.C., KARMER A., SWEELEY G.C. (1964) Gas chromatography of lipids. In Holman R. T., (ed.), *Progress in the chemistry of fat and other lipids.* Pergamon Press, Oxford, vol. **7**, part 2, p. 167–246.

HRBACEK J. (1962) *Hydrobiologicke metody.* 2 vyd. Statni pedagog. nakl. Praha, p. 23, 130.

HUTCHINS R.F.N., MARTIN M.M., (1968) The lipids of the common house cricket, *Acheta domesticus* L. I. Lipid classes and fatty acid distribution. *Lipids*, **3**, 247–249.

ITZHAKI R.F., GILL D.M. (1965) A microbiuret method for estimating proteins. *Analyt. Biochem.*, **9**, 401–410.

IVANKO S. (1971) Changeability of Protein fractions and their amino acids composition during maturation of Bailey grains. *Biologia Pl.*, **13**, 155–164.

JACOBS S. (1965) The determination of nitrogen in biological materials. In Glick D., (ed.) *Methods of biochemical analysis*, Wiley, New York, **XIII**, p. 241.

JANAS J., PODKÓWKA W. (1972) Porównanie metod oznaczania włókna surowego w paszach. (Comparative studies on some methods of determining crude fibre in feeds). (*Engl. Russ.*, *Summ*). *Roczn. Nauk Roln.*, *Ser. B.* **94**, 149–159.

JENNINGS A.C., PUSZTAI A., SYNGE R.L.M., WATT W.B. (1968) Fractionation of plant material. III. Two schemes for chemical fractionation of fresh leaves having special applicability for isolation of the bolk protein. *J. Soc. Food Agric.*, **19**, 203–213.

JEŻEWSKA M.M., GORZKOWSKI B., HELLER J. (1963) Nitrogen compounds in the snail, *Helix pomatia*, excretion. *Acta Biochim. polon.*, **X**, 55–65.

JOHANNES R.E., SATOMI M. (1966) Composition and nutritive value of faecal pellets of a marine crustocean. *Limnol. Oceanog.*, **11**, 191–197.

KAPLAN A. (1969) The determination of urea ammonia urease. In *Methods of Biochemical Analysis*, Glick D., (ed.) *Inter. Sci. Publ.*, New York, vol. **17**, p. 311–324.

KARNAVAR G.K., NAIR K.S.S. (1969) Changes in body weight, fat, glycogen and protein during diapause of *Trogoderma granarium*. *J. Insect Physiol.*, **15**, 95–103.

KAUSHIK N.K., HYNES H.B.N. (1968) Experimental study on the role of autumn-shed leaves in aquatic environments. *J. Ecol.*, **56**, 229–243.

KAUSHIK N.K., HYNES H.B.N. (1971) The fate of the dead leaves that fall into streams. *Arch. Hydrobiol.*, **68**, 465–515.

KLECZKOWSKI K., JASIOROWSKI H., KUROWSKI H. (1961) Kolorymetryczna metoda oznaczania bialka w mleku. (A colorimetric method of milk protein estimation. *Roczn. Nauk Roln.*, Ser. B., **77**, 923–930. (England, Russian, Summ.).

KLEIBER M., (1961) *The fire of life: an introduction to animal energetics.* John Wiley a. Sons, Inc., New York, p. 454.

KOSENKO L.V., RASHBA E.JA., (1967) Vivchenija vmistu vuglevodiv u klitinah dejakih azotfiksujuchih sin'o—zelenih vodorostej. (A study of the carbohydrate content in the cells of some nitrogen-fixing blue-green algae). *Mikrobiol. Zh., Kiiv.*, No. **6**, 487–491. (Engl. summ.).

KRAUSE H.R. (1959) Beiträge zur kenntnis des Chitinabbaues im toten Zooplankton. *Arch. Hydrobiol.*, *Suppl. Bd.* **25** (IV, 1), p. 67–82.

LANG C.A. (1958) Simple microdetermination of Kjeldahl nitrogen in biological materials. *Analyt. Chem.*, **30**, 1962–1694.

LANG K., LEHNARTZ E. (1953–1964) (eds.) Hoppe-Seyler E. F., Thierfelder H., Handbuch der physiologisch-und pathologisch-chemischen Analyse. Springer, Berlin, Aufl. **10**, vol. 1–5.

LAYNE E. (1957) Spectrophotometric and turbidimetric methods for measuring proteins. In Colowick S. P., Kaplan N. O., (eds.), *Methods in enzymology.* Academic Press, New York, vol. **3**, p. 447–454.

LEE R.F., NEVENZEL J.C., PAFFENHÖFER G.A. (1970) Wax esters in marine copepods. *Science*, **167**, 1510–1511.

LENARTOWICZ E., ZALUSKA H., NIEMIERKO S. (1967) Carbohydrates in the wax moth during development. *Acta Biochim. pol.*, **14**, 267–276.

LOWRY O.H., FARR A.L., RANDALL R.J., ROSEBROUGH N.J. (1951) Protein measurement with the Folin phenol reagent. *J. biol. Chem.*, **193**, 265–275.

LUQUET P. (1970) Étude des divers composés azotés des salmonidés: Proteines, Azoter non protéique, acides nucleiques. Facteurs de variation. *Ann. Hydrobiol.*, **1**, 111–132.

MACFADYEN A. (1963) *Animal ecology: Aims and methods.* Sir Isaac Pitman a. Sons, London, p. 344.

MACGRATH R. (1972) Protein measurement by ninlydrin determination of amino acids released by alkaline hydrolysis. *Analyt. Biochem.*, **49**, 95–102.

MACIOLEK J.P. (1962) Limnological organic analyses by quantitative dichromate oxidation. *U.S. Dept. of the Interior. Fish and Wildlife service. Research Report* 60, Washington.

MACKERETH F.J.H. (1963) Some methods of water analysis for limnologists. *Freshwater Biol. Ass.* Scientific Publication No. 21.

MANOUKAS A.G. (1972) Total amino acids in hydrolysates of the olive fruit fly, *Dacus oleae*, grown in an artificial diet and in olive fruits. *J. Insect Physiol.*, **18**, 683–688.

MARINETTI G.V. (1967–1969) (ed.) *Lipid chromatographic analysis.* M. Dekker, New York, vol. **1–2**.

MASLOWSKI P., SKÓRKO R. (1966) Adaptacja kolorymetrycznej metody biuretowej do oznaczania bialek zbóż. (The adaptation of the biuret colorimetric method to cereal protein determination). *Rocz. Nauk Roln.*, Ser. A., **91**, 665–672. (Engl. summ.).

MERRILL A.T. (1924) A note on the relation pf pH to tungstic acid precipitation of protein. *J. Biol. Chem.*, **60**, 257–259.

MINDERMAN G., BIERLING J., (1968) The determination of cellulose and sugars in forest —litter. *Pedobiologia*, **8**, 536–542.

MORRIS D.L. (1948) Quantitative determination of carbohydrates with dreywood's anthrone reagent. *Science*, **107**, 254–255.

NEJEDLY K. (1953) Prubeh exkrece u hlemyzde zahradniho (*Helix pomatia*) (Excretion course in *H. pomatia*). *Rozpravy CSAV.*, **63**, 45–56.

NEWELL B.S. (1967) The determination of ammonia in sea water *J. mar. biol. Ass.* U.K., **47**, 271–280.

NIEMIERKO W., NIEMIERKO S., WLODAWER P. (1952) The extraction and fractionation of phosphorus compounds in animal tissues. Part. 1 *Acta Biol. Exp.*, Warsaw, **16**, 247–252.

Official method of analysis of the Association of Official Agricultural chemists. VII ed. AOAC Publ. Washington (1950), p. 910.

PAECH K., TRACEY M.V. (1955–1964) (eds.) *Moderne Methoden der Pflanzen Analyse.* Springer, Berlin, Bd. 1–7.

PEARSE J.S., GIESE A.O. (1966) The organic constitution of several benthonic invertebrates from McMurdo Sound, Antarctica. *Comp. Biochem. Physiol.*, **18**, 47–57.

PERKINS R.J., DAHLBERG M.D. (1971) Fat cycles and condition factors of Altamaha river Shads. *Ecology*, **52**, 359–362.

PETRUSIEWICZ K., MACFADYEN A. (1970) *Productivity of Terrestrial Animals: Principles and Methods.* IBP Handbook No. 13. Blackwells, Oxford.

198 *Chapter 5*

PIEZ K.A., SAROFF H.A. (1964) Chromatography of amino-acids and peptides. In Heftmann E., (ed.), *Chromatography*. Reinhold, New York, p. 347–377.

PIGMAN W. (1957), (ed) *The carbohydrates. Chemistry, biochemistry, physiology.* Academic Press, New York, p. 902.

POLLEY J.R. (1954) Colorimetric determination of nitrogen in biological materials. *Anal. Chem.*, **26**, 1523–1524.

POTTER V.R., ELVEHJEM C.A. (1936) A modified method for the study of tissue oxidations. *J. biol. Chem.*, **114**, 495–504.

POTTS T.J. (1963) Colorimetric determination of urea in feeds. *J. AOAC*, **46**, 303–306.

POURRIOT R. LEBORGNE L. (1970) Teneurs en protéines, lipides et glucides de zooplanctons d'eau Douce. *Ann. Hydrobiol.*, **1**, 171–178.

PROCHÁZKOVA L. (1964) Spectrophotometric determination of ammonia as rubazoic acid with bispirazolone reagent. *Anal. Chem.*, **36**, 865–871.

RAYMONT J.E.G., AUSTIN J., LINFORD E. (1964) Biochemical studies on marine zooplankton. I. The biochemical composition of *Neomysis integer. J. Cons. perm. int. Explor. Mer* **28**, 354–363.

RAYMONT J.E.G., AUSTIN J., LINFORD E. (1968) Biochemical studies on marine zooplankton. V. The composition of the major biochemical fractions in *Neomysis integer. J. mar. biol. Ass.* U.K. **48**, 735–760.

RAYMONT J.E.G., SRINIVASAGAM R.T., RAYMONT J.K.B. (1971) Biochemical studies on marine zooplankton. IX. The biochemical composition of *Euphausa superba. J. mar. biol. Ass.* U.K. **51**, 581–588.

RILEY J.P. (1953) The spectrophotometric determination of ammonia in natural waters with particular reference to sea-water. *Anal. Chim. Acta* **9**, 575–589.

ROE J. H., DAILEY R.E. (1966) Determination of glycogen with the anthrone reagent. *Anal. Biochem.*, **15**, 245–250.

ROUSER G., FLEISCHER S. (1967) Isolation. characterization and determination of Polar lipids of mitochondria. In Colowick S. P., Kaplan N. O., (eds.), *Methods in enzymology*, Academic Press, New York a. London. vol. X. 385–406.

RZYMOWSKA C., BERSTEINÓWNA I., GROCHOWSKA J. (1953) Zastosowanie metod mikrochemicznych do badania artykułów żywności, I. Mikrometoda i pólmikrometoda Kjeldahla (The use of microchemical methods in food research. I. Kjeldahl micro- and semi-mircomethod). *Roczn. państw. Zakl. Hig.*, **1**, 1–21. (Engl. summ.),

SEIFTER S., DAYTON S., NOVIC B., MUNTWYLER E. (1950) The estimation of glycogen with the anthrone reagent. *Archs Biochem.*, **25**, 191–200

SHNYNKOVA E.I., PIROZHENKO S.U. (1972) Kharakteristika polisakharidnogo kompleksa fitoplanktona w period cwetenija w wodema. (Engl. summ.), *Gidrobiol. Zh.*, **8**,(5). 36–41.

SHULMAN G.E. (1971) Variability of fat content in some fish species of Azov—Black sea basin (Russian) *Gidrobiol.*, **7**,(3) 86–91.

VAN SOEST P.J., (1963) Use of detergents in the analysis of fibrous feeds. II. A rapid method for the determination of fiber and lignin. *J. Ass. off. agric. Chem.*, **46**, 829–. 835.

SPERRY W.M. (1955) Lipid analysis. In Glick D., (ed.) *Methods of biochemical analysis*. Interscience Publ., New York, vol **2**, p. 83–11.

STAHL E. (1969) (ed.): *Thin—Layer chromatography. A laboratory handbook*. Ed. 2, Berlin, Springer Verl. XXIV p. 1041.

STEGEMANN H., LOESCHEKE V. (1962) Mikrobestimmung von Stichstoff als Indophenolblau nach Chloramin—T—Oxydation. *Hoppe Sei lers Z. Physiol. Chem.*, **329**, 241–248.

STEIN R.A., SLAWSON V., (1966) Column chromatography of lipids. In Holman R.T., (ed.), *Progress in the chemistry of fats and other lipids*. Pergamon Press, Oxford, vol. **8**, part 3, p. 375–420.

STERN J., SHAPIRO B. (1953) A rapid and simple method for the determination of esterified fatty acids and for total fatty acids in blood. *J. clin. Path.*, **6**, 158–160.

STRICKLAND J.D.H., PARSONS T.R., (1968) A practical handbook of seawater analysis. *Bull. Fish. Res. Bd Can.*, **167**, 1–311.

SZCZEPAŃSKI A. (1970) Primary production as a source of energy in aquotic biocenosis. *Pol. Arch. Hydrobiol.*, **17**, 39–101.

TETLOW J.A., WILSON A.L., (1964) An absorptiometric method for determining ammonia in boiler feed-water. *Analyst.*, **89**, 453–465.

TREVELYAN W.E., HARRISON J.S. (1952) Studies on yeast metabolism. I. Fractionation and microdetermination of cell carbohydrates. *J. Biochem.*, **50**, 298–303.

WALICKA E. (1972) Zastosowanie krajowych detergentów w metodzie van Soesta oznaczanie włókna i ligniny i ligniny. (Use of Polish detergents in the van Soest's method of fiber and lignin determination.) Roczn. Nauk Roln., Ser. B., **94**, 139–147. (Polish, Engl. Russ. summ.).

WALKER G. (1972) The biochemical composition of the cement of two barnacle species, *Balanus hameri* and *Balanus crenatus. J. mar. biol. Ass. U.K.*, **52**, 429–435.

WHISTLER R.L., SMART C.L., (1953) *Polysaccharide chemistry*. Academic Press, New York, p. 493.

WHISTLER R.L., WOLFROM M.L., (ed.), (1962) *Methods in carbohydrate chemistry*. Academic Press, vol. **1**, p. 22, 589.

WILLIAMS P.C. (1964) The colorimetric determination of total nitrogen in feeding stuffs. *Analyst.*, **89**, 276–281.

WINBERG G.G. (ed.), (1971) *Methods for the estimation of production of aquatic animals*. Academic Press, London, New York, p. 175.

WINZLER R.J. (1955) Determination of serum glycoproteins In Glick D. (ed.) *Methods of biochemical analysis*. Intersci. Publ., New York, London, vol. **2**, 279–311.

WISSING T.E., HASLER A.D. (1971) Intraseasonal change in caloric content of some freshwater invertebrates. *Ecology*, **52**, 371–373.

YEMM E.W., COCKING E.C. (1955) The determination of amino acids with ni hydrin *Analyst* , **80**, 209–213

ZALUSKA H (1959) Glycogen and chitin metabolism during development of the silkworm (*Bombyx mori* L.) *Acta biol. exper.*, Warsaw, **XIX**, 339–351.

ZIELIŃSKA Z.M. (1957) Studies on the biochemistry of the waxmoth (*Galleria mellonella*) 17. Nitrogen metabolism in the tissues and organs of the larvae. *Acta Biol. Exper.*, Warsaw, **XVII**, 2, 351–371.

6
Respirometry

6A CARTESIAN DIVERS MICRORESPIROMETRY FOR TERRESTRIAL ANIMALS

R. KLEKOWSKI

6A.1. Introduction

Cartesian divers serve for manometric gasometry. They were introduced by Linderstrøm-Lang (1937), theoretical backgrounds can be found in Linderstrøm-Lang (1943), descriptions and reviews in Frydenberg and Zeuthen (1960), Glick (1961), Holter (1943, 1961), Holter and Zeuthen (1966), Klekowski (1971), Løvlie and Zeuthen (1962), Zeuthen (1943, 1964). Cartesian divers have been used mainly to measure metabolism of protozoans and algae as well as that of cells and developing eggs of metazoans. Invented for objects investigated in fluid environment, they were seldom used for objects which breathe with atmospheric oxygen (e.g. Berthet, 1964,—Oribatei; Healey, 1966, 1967,—Collembola; Lints et al., 1967,—Drosophila eggs; Klekowski et al., 1967,—Tribolium eggs; Nielsen, 1961,—Enchytraeidae; Stępień and Klekowski in press—Acaridae). For ecological studies in bioenergetics, Cartesian divers are a good tool to determine O_2 consumption, that is—using respiratory quotient and calorific equivalent—cost of maintenance. They were used for measurements of energy budget of Oribatei (Berthet, 1964, 1967), Collembola (Healey, 1966, 1967), Cladocera (Klekowski and Ivanova, in prep.), Copepoda (Klekowski and Shushkina 1966a, b), Coleoptera (Klekowski et al., 1967), Acaridae (Klekowski and Stępień, in press).

For an ecologist dealing with bioenergetics, stoppered divers are most convenient of all Cartesian divers. This type of diver has been introduced by Zeuthen (basic description in Zeuthen, 1950). The construction and calibration of these divers are easy, they can be charged free-hand and without any complicated tools and instruments. A detailed description of divers of this type, their manufacturing, calibration and application for aquatic animals can be found in elaboration by Klekowski (1971). The majority of instructions included there are also applicable for modification of stoppered divers used for air-breathing animals; these divers will be described below.

6A.2. Principle of functioning.

The Cartesian diver is a constant volume, manometric gasometer. It is a container enclosing a gas bubble connected with surrounding flotation medium. Any change in pressure affects the gas bubble. Increasing pressure brings about diminution of gas bubble volume and the diver sinks. A decrease in pressure brings about an increase of gas bubble volume and the diver rises. For each particular diver there is a gas volume ('diver constant'), at which the diver floats at a definite level, i.e., equilibrium level. With diminishing amount of gas in the diver (for example O_2 consumed at respiration) one can compensate the changes in gas quantity to 'diver constant' and thus place the diver at equilibrium level by changing the pressure, controlled and measured manometrically. From manometer readings the change in amount of gas can be calculated.

6A.3. Equipment

6A.3.1 Water bath

(Fig. 6A.1) with submerged flotation vessels (1) holding divers (4, 5), with temperature control of $\pm$ 0·01°C, secured from vibrations. For the last reason, the stirrer (17) cannot be attached to the bath frame.

6A.3.2. The flotation vessels

(1) have coned-shaped lower ends. During measurements they are submerged in water bath, however they can be lifted over the water in order to place and remove the divers. Each flotation vessel (6–8 altogether) is connected with manifold (11), one end of which joins manometer (6), the inner diameter of the latter being 2–2·5 mm, its height 150–180 cm, millimetric scale behind behind both arms. The pressure is regulated and measured by coarse (7) and fine (8) adjustment screws. Brodie's solution (specific gravity = 1·033) is used as manometer fluid. The second arm of manometer is connected with an air bottle (12), 5–10 l, submerged in a water bath. The position of the diver in the flotation vessel is checked visually by means of a cathetometer.

The following tools are necessary: 1. Forceps of special make for placing the divers into flotation vessel and for their removal.

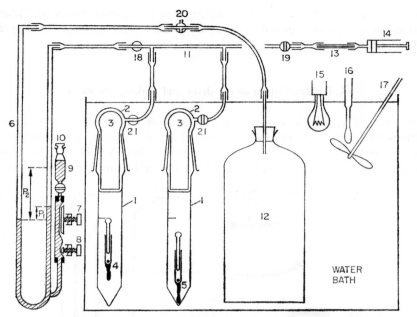

Fig. 6A.1. Schematic drawing of Cartesian diver assembly (scale distorted). 1—Flotation vessel; 2—Cap; 3—Space occupier; 4—Diver floating at the equilibrium mark; 5—Diver resting at the bottom between measurements; 6—Manometer; 7—Coarse; 8—Fine adjustment screws; 9—The reservoir with manometer fluid; 10—Reservoir stopper; 11—Manifold; 12—Air bottle; 13—Air brake; 14—Syringe (100 ml); 15—Healting bulb; 16—Thermoregulator; 17—Stirrer; 18–21—Taps.

2. Braking pipette (Fig. 6A.2). A capillary (1) of a constant inner diameter and hair-thin tip (air-brake) (4) is mounted in a glass jacket tube (2) by means of cement (3); the flow of liquid through the capillary is hindered by the 'air-brake' in hair-thin tip. The jacket tube is connected with a mouthpiece through rubber tubing. For a given pipette the ratio of length (mm) per 1 μl

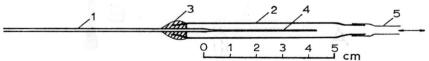

Fig. 6A.2. Braking pipette. 1—Capillary; 2—Jacket; 3—Cement; 4—Air brake; 5—Rubber tubing.

volume is known. 3. Micro-burner, made from an injection needle and screw clamp. 4. A diamond in holder or carborundum piece mounted in a pyrex glass rod, for capillary cutting.

6A.4. Description of the diver and making the diver.

The diver (Fig. 6A.3A) consists of a diver chamber (1) whose bottom part (2) can be spherical; in this part the studied object is placed. A gas bubble (3) reaches to the mark on the wall of the diver chamber. A flock of nylon fibres from a stocking prevents moving animal from getting into 0·1N solution of

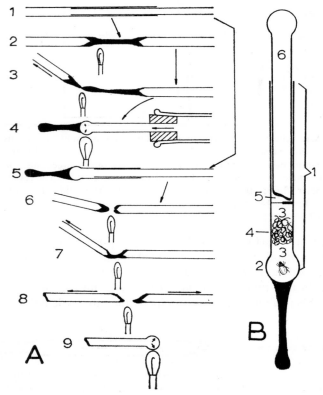

Fig. 6A.3. Making the diver; for explanation see text.

NaOH (5). Hollow glass stopper (6), inserted in the neck of the diver chamber, reduces the leakage of gas from (or to) the volume of the air bubble in the diver chamber. It is inserted so deep in the diver chamber that it comes close to the meniscus of 0·1N NaOH without making contact. The stopper is fixed in this position by gentle twisting until mechanical resistance is felt. The diver chamber has a solid tail at its lower end, stabilizing the diver in vertical position.

Capillaries for divers should be drawn from pyrex, rasotherm, thermisil, or other hard, borosilicate glass in a broad oxygen flame (for details consult Klekowski, 1971). The steps in making the diver (Fig. 6A.3.) are as follows: 1. Select two sections of tubing, one section fits inside the other. 2. Melt and thicken wall and close the capillary. 3. Shape the diver tail. 4. Insert the capillary into a hole in rubber stopper fitted into a glass jacket connected through a rubber tubing with a mouthpiece. Heat the diver chamber uniformly on all sides. Gently blow out the bulb. Cut off excess length. 5. Find suitable length of tubing for stopper. 6. Melt the end of the tube. 7. Close this end with another piece of glass. 8. At suitable point close the second end of the stopper. 9. Heat this end of stopper uniformly until expanding gas forms a bulb; this will act as a float. The technique for making and calibrating the braking pipette is the same as for other types of divers (Klekowski, 1971).

6A.5. Calibration of the diver

(Fig. 6A.4). 1. Insert stopper into the chamber and close firmly by turning the stopper gently. Make the mark about 0·5 mm, from the end of stopper, with glass paint on the diver wall, take out stopper and heat the mark in micro-flame with care not to melt the glass but fix the paint. It is advisable now to evaluate (or measure using braking pipette) if the air bubble volume up to the mark is suitable for expected experiments. 2. Submerge diver chamber in a vessel with NaCl solution (4 g per 1 l of solution) with a specific gravity identical with that of 0·1N NaOH. Insert braking pipette into diver chamber and suck off sufficient air so that the meniscus of liquid coincides with the mark. Take out pipette, insert and block stopper. 3. Try to float the diver in the same NaCl solution. 4. If the diver sinks remove glass from the stopper tail, if it rises, add glass. Repeat until the diver floats. 5. Suck up the whole air bubble into braking pipette. 6. Measure length of air bubble with mm

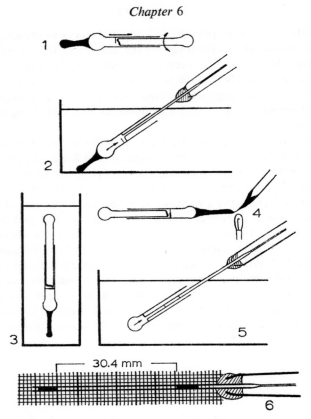

Fig. 6A.4. Calibration of diver; for explanation see text.

graph paper. Calculate volume of air bubble. Record the number of the calibrated diver and its air bubble volume, i.e. diver constant.

6A.6. Filling of the diver.

1. Place the object to be investigated in dry diver chamber. If it moves, place also a flock of nylon fibres. 2. Fill the diver neck with 0·1N NaOH up to the mark using braking pipette as described in point 2 of calibration technique (Fig. 6A.4). Insert and block the stopper.

6A.7. Measurements and calculations.

Close tap 18 (Fig. 6A.1). open taps 19 and 21, open tap 20 to atmosphere, close it after a few seconds. Lift up the flotation vessel along the supporting rod, open the flotation vessel; the flotation medium (0·1N NaOH) should be saturated with air free of CO_2. Put the diver into vessel, cover it and secure with rubber bands. Lower the vessel into water, close tap 21. Repeat for each flotation vessel.

Open tap 18 and 19 and set the manometer fluid half way up both arms, close tap 19. Open tap 21 and with screws 7 and 8 establish the diver at equilibrium level, i.e. top of the diver stopper coincides with the hair in the cathetometer. Record the time, hour and minute, and manometer fluid level in both arms (0·5 mm accuracy)*. Increase pressure by about 50 mm, close tap 21. Perform measurement for each diver and repeat at chosen time intervals (usually every 15 min). When the measurements are finished, necessarily close tap 18 (very important), open taps 21, 19, and 20. Lift up the flotation vessel, open it, take the diver out, replace the cover.

Remove the stopper under water surface. Rinse NaOH solution with water from the diver chamber neck (Fig. 6A.5). Suck out water from the neck using braking pipette, dry the remaining water up with a strip of filter paper. Take out the object, for this a piece of hair glued to a glass rod can be helpful.

Table 6A.1 presents data sheet (from Klekowski *et al.*, 1967). For each diver the following were recorded: time, manometric fluid level in the left and right arm, and the difference between them, i.e. equilibrium pressure. The latter values of the consecutive records for each diver are marked on milli-metre graph paper: Figure 6A.6 represents the data listed in Table 6A.1. Draw average straight line for each curve using a perspex rule with parallel lines and calculate the average equilibrium pressure difference, ΔP, per hour. The oxygen consumption per hour (QO_2) is calculated from formula:

$$QO_2 = \frac{V_g \cdot \Delta P}{P_o} \cdot \frac{273}{T}$$

---

*If the initial difference between the levels in the both arms of the manometer exceeds 20 cm, remove the diver and correct the amount of gas in the diver by adding or removing a minute air bubble, using braking pipette in a vessel with 0·1N NaOH.

where, V_g—is air bubble volume (diver constant), ΔP—the equilibrium pressure change in mm per hour, P_o—the normal pressure (10,000 mm Brodie sol.), T—temperature in Kelvin's scale.

Fig. 6A.5. Rinsing the fluid (water with NaOH solution or vice-versa).

6A.8. Cleaning and siliconing the divers.

Rinse the diver chamber and stopper in concentrated H_2SO_4 and successively in a number of fresh portions of distilled water; dry up at 105°C each diver in its labelled Petri dish. Fill the diver chamber with 5% v/v $(CH_3)_2$ $SiCl_2$ in CCl_4, suck it out with pipette, heat to 180°C for 1 hour. Before charging rinse many times with water and dry up in 105°C. This coating with silicone prevents creeping NaOH solution along the chamber wall within the air bubble.

Diver for CO_2—output measurement. Calibration, testing, filling, and measurements are very similar to those applied in divers for aquatic animals (Klekowski, 1971). The latter paper includes also broader discussion of other

Tribolium castaneum (Herbst). (Compiled from different runs of measurement).

Flotation vessels: 1, 2, 3 – eggs in different stages of development

― ″ ― : 4, 5 – larvae

No. diver	Vg	ΔP/h 273		ΔV O₂ μl·10/h
248	1	6.92	41.5 0.90	7.16
249	2	7.45	31.8 "	24.32
250	3	6.28	57.0 "	32.22
252	4	4.07	146.0 "	53.48
258	5	5.92	163.3 "	87.00
	6			
	7			
	8			

Table 6A.1. Data sheet for recording O₂ consumption measurements in developing eggs and newly hatched larvae of *Tribolium castaneum* (Herbst) compiled from different runs of measurement.

methodical and technical problems connected with Zeuthen's stoppered divers.

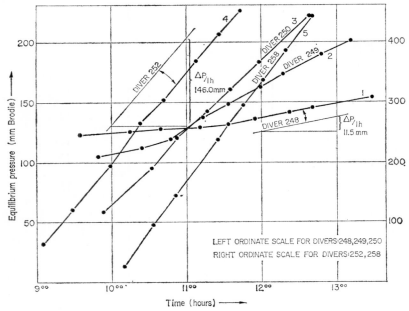

Fig. 6A.6. Respiration of an egg and newly hatched larva of *Tribolium castaneum* (Herbst).

References

BERTHET P. (1964) L'activité des Oribatides (Acari; Oribatei) d'une chénaie. *Mèm. Inst. r. Hist. natn. Belg.*, **152**, 1–152.

BERTHET P. (1967) The metabolic activity of oribatid mites (Acarina) in different forest floors. In Petrusewicz K. (ed.), *Secondary productivity of Terrestrial Ecosystems.* PWN, Warszawa–Kraków, **2**, 709–725.

FRYDENBERG O. & ZEUTHEN E. (1960) Oxygen uptake and carbon dioxide output related to the mitotic rhythm in the cleaving eggs of *Dendraster excentricus* and *Urechis caupo. C. r. Trav. Lab. Calsberg*, **31**, 423–455.

GLICK D. (1961) *Quantitative Chemical Techniques of Histo- and Cyto-chemistry.* J. Wiley, New York-London, 470 pp.

HEALEY I.N. (1966) The population metabolism of *Onychiurus procapatus* Gisin. In Graff O. & Satchell J. E. (eds). *Dynamics of Soil Communities.* Brunswick.

HEALEY I.N. (1967) The energy flow through a population of soil Collembola. In Petrusewicz K. (ed.) *Secondary Productivity of Terrestrial Ecosystems.* PWN, Warszawa-Kraków, **2**, 695–708.

HOLTER H. (1943) Technique of the Cartesian diver. *C. r. Trav. Lab. Calsberg, Sér. Chim.*, **24**, 399–478.

HOLTER H. (1961) The Cartesian diver. In Danielli J. F. (ed.), *General Cytochemical Methods.* Academic Press, New York–London, **2**, 93–129.

HOLTER H. & ZEUTHEN E. (1966) Manometric techniques for single cells. In Pollister A. W. (ed.) *Physical Techniques in Biological Research.* Academic Press, New York–London, **3 A**, 251–317.

KLEKOWSKI R.Z. (1971) Cartesian diver microrespirometry for aquatic animals. *Pol. Arch. Hydrobiol.*, **18**, 93–114.

KLEKOWSKI R.Z. & SHUSHKINA E.A. (1966a). The energetic balance of *Macrocyclops albidus* (Jur.) during the period of its development. (Russian). In *Ekologiya Vodnykh Organismov.* Izd. 'Nauka', Moskva, 125–136.

KLEKOWSKI R.Z. & SHUSHKINA E.A. (1966b) Ernährung, Atmung, Wachstum und Energie-Umformung in *Macrocyclops albidus* (Jurine). *Verh. int. Ver. Limnol.*, **16**, 399–418.

KLEKOWSKI R.Z., PRUS T. & ŻYROMSKA-RUDZKA H. (1967) Elements of energy budget of *Tribolium castaneum* (Hbst) in its developmental cycle. In Petrusewicz K. (ed.), *Secondary Productivity of Terrestrial Ecosystems.* PWN, Warszawa-Kraków, **2**, 859–879.

KLEKOWSKI R.Z. & IVANOVA M.B. (in prep.) Elements of energy budget of *Simocephalus vetulus.*

KLEKOWSKI R.Z. & STĘPIEŃ Z. (in press) Elements of energy budget of *Rhizoglyphus echinopus* (F. & R.).

LINDERSTRØM-LANG K. (1937) Principle of the Cartesian diver applied to gasometric technique *Nature, Lond.*, **140**, 108.

LINDERSTRØM-LANG K. (1943) On the theory of the Cartesian diver microrespirometer. *C. r. Trav. Lab. Carlsberg, Sér. chim.*, **24**, 333–398.

LINTS C.V., LINTS F.A. & ZEUTHEN E. (1967) Respiration in *Drosophila.* I.—Oxygen consumption during development of the egg in genotypes of *Drosophila melanogaster* with contribution to the gradient diver technique. *C. r. Trav. Lab. Carlsberg*, **36**, 35–66.

LØVLIE A. & ZEUTHEN E. (1962) The gradient diver—a recording instrument for gasometric micro-analysis. *C. r. Trav. Lab. Carlsberg*, **32**, 513–534.

NIELSEN C.O. (1961) Respiratory metabolism of some populations of Enchytraeid worms and free-living Nematodes. *Oikos*, **12**, 17–35.

ZEUTHEN E. (1943) A Cartesian diver respirometer with a gas volume of $0 \cdot 1$ μl. *C. r. Trav. Lab. Carlsberg, Sér. chim.*, **24**, 479–518.

ZEUTHEN E. (1950) Cartesian diver respirometer. *Biol. Bull.*, **98**, 139–143.

ZEUTHEN E. (1964) Microgasometric methods, Cartesian divers. In *Second. Int. Congr. Histochem. Cytochem.*, Frankfurt/Main, August 16–21, 1964. Springer–Verlag Berlin, 70–80.

Chapter 6

6B CONSTANT-PRESSURE VOLUMETRIC MICRORESPIROMETER FOR TERRESTRIAL INVERTEBRATES

R. KLEKOWSKI

6B.1. Introduction

The microrespirometer described below is a modification of the apparatus whose different types, based on the same principle, have been known for long (e.g: Winterstein, 1912; Krajnik, 1922; Drastich, 1934; Fenn, 1927; broader descriptions of this type of gasometer in: Dixon, 1952; Kleinzeller et al., 1954). This gasometer is devised to measure O_2 consumption and CO_2 production by animals of the size from that, for example, of an individual adult beetle of *Tribolium* sp. to adult Orthoptera or other insects, mature Gastropoda pulmonata etc. It has been used for bioenergetic studies on *Tribolium* (Klekowski et al., 1967), *Polydesmus* (Diplopoda) (Stachurska in press) as well as for studies on aquatic animals: larvae of Odonata (Fischer, 1967), *Asellus aquaticus* (Prus in press). The advantage of this respirometer is that its main glass parts can be easily made by a glassblower of an average skill and the remaining parts in any mechanical workshop.

6B.2. Principle and functioning

The respirometer presented here (Fig. 6B.1) is a volumetric gasometer of the constant-pressure system, i.e., the pressure within the system is being brought to the same level during subsequent measurements of one run, and the changes in the amount of gas are measured volumetrically.

The disappearance of the gas in the respiration chamber (I) due to respiring of the animal present there is compensated and measured by rising the level of the mercury (1) in a graduated (in μl) capillary (2). The respirometer is a 'zero apparatus', that is, the reading follows after a thorough compensation of the gas loss by the mercury has been made. The latter is indicated by the return of the 'index' drop (3) to point 'O' in the capillary connecting the respiration chamber with the compensation chamber (II). The latter is of the same volume as the former and has the same charge except for the organism

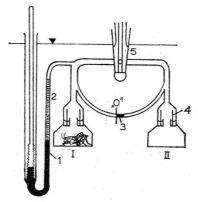

Fig.6B.1. The scheme of the system and its main parts. 1—Mercury, 2—Measuring capillary, 3—Index drop, 4—Filter paper saturated with solution of NaOH, 5—Tap. I—Respiration chamber II—Compensation chamber.

examined. Chamber II is to compensate possible changes of temperature around the respirometer. At measurement of O_2 consumption, the carbon dioxide produced by an organism is being absorbed by alkaline solution (NaOH) which moistures the filter paper (4). Tap (5) allows to open or to close the whole of the system from the air.

6B.3. Equipment

A water-bath with temperature controlled not less than with ca 0·03°C accuracy is supplied with a hanger or stands to position the respirometers. The water-bath should be deep enough as to immerse respirometers to the level marked in Fig. 6B.2 A. At least one of its walls (that one at which the respirometers hang) should be made of glass to allow readings of the respirometer scales.

6B.4. Description of the respirometer

The apparatus is made of pyrex, rasotherm, thermisil or other borosilicate glass. Respiration chamber (Fig. 6B.2) (I) and compensation chamber (II)

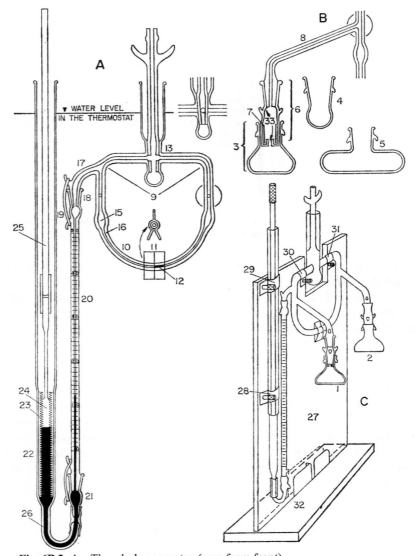

Fig. 6B.2. A—The whole apparatus (seen from front).
B—The arm with absorber, and 3 types of chambers.
C—Total view of the apparatus (perspective). For further explanations see text.

are identical, each consisting of 2 parts (Fig. 6.B2 B): the vessel (3, 4, 5) and the CO_2 absorber holding filter paper (7) soaked with 5% NaOH solution. The top aperture of the CO_2 absorber is connected by ground glass with the respirometer arm (8), which comes out of circular capillary (9), (Fig. 6B.2 A). Inner bore of the arms and circular capillary is of about 2·5 mm. The latter in its lower part changes into index capillary (10) with the inner bore of about 1·4 mm. The index capillary is coated with perspex double plate (11) on which there are two coinciding vertical marks. Inside the capillary, there is a drop of kerosene about 3–4 mm in length (distilled fraction at temp. 190–195°C, coloured with Congo red). The circular capillary has a tap (13) in the middle of the upper part. This tap allows the system to be opened to the air [when positioned as in Fig. 2 A, (13),], or to cut off the air with a simultaneous closing of the capillary light [when positioned as (14).]. In the index-circular capillary junction, there is an expanding which prevents the index drop to get into the circular capillary (15). NB, no narrowing of the capillary light is allowed in point (16), since this would have caused imprisoning of some of the index drop.

There is a lateral branching (17) of the circular capillary, stemming from it above the arm bearing the respiration chamber. This branching capillary is connected through a ground joint [secured with a rubber band (19)] with measuring capillary (20), graduated in μl or ml. The lower end of the measuring capillary is further connected through a ground joint (also secured with rubber band) with a short connective tube (21), ended with a bulge. This bulging in turn, is joined by a short tygon tube (26) with the outlet of mercury regulator. Several measuring capillaries of different bore and volume can be fitted to one apparatus. In our practice, we have employed capillaries with the capacity of 50 μl and 0·7 ml. These capillaries have the same distance between the ground joints and therefore are interchangeable. The mercury level regulator (22) is of a shape of glass tube, whose lower section has somewhat narrower bore than the upper one. The upper end of the tube is above the water level in the water bath. Into the lower, narrower section of the tube, a perspex tube (23) is glued in with DeKothinsky cement or sealing wax, female thread inside which moves threaded piston made of perspex or aluminium greased with vaseline + caoutchouc. The upper end of the piston is attached through a section of rubber tubing with an extensing rod (25). By turning the rod, the level of mercury (26) in the measuring capillary can be altered. Both the length of the perspex tube and bore of the piston should have such dimensions as to cause the mercury to move within the measuring capillary of the largest capacity along its entire graduated section when the piston moves from its upper position to the lowest one. (NB, at the upper position about 2 cm of the thread section of the piston should be screwed in the perspex tube!). The piston diameter should not be too large since that will cause difficulty in reading the small-bore capillary, 5 mm seems to be most suitable.

The apparatus is attached to a perspex plate (27), 5 mm thick with brass holders (28, 29, 30, 31) lined with rubber. The lower edge of the perspex plate is mounted in brass holders attached to heavy lead plate (32).

6B.5. Calibration of measuring capillary

The measuring capillary (20) *must* have constant bore along its whole length. If capillaries of 'verydia–type' are not available, one should select a capillary in which a column of mercury (say 20–30 mm long) has the same length (as measured with mm graph paper) in their different sections. After having made ground joints on both ends of the capillary, the latter is calibrated by means of measuring (with 0·1 mm accuracy) the length of mercury column in the capillary, pouring out the mercury, weighing it (with an accuracy of 0·5 mg), and calculating the ratio: mm per 1 μl; NB—when finding specific gravity of mercury from tables, the temperature factor should be taken into account.

The scale is drawn with glass paint, hardened later at temperature of about 500–600°C depending on the type of paint used.

6B.6. Changing of the measuring capillary

This change can be done as follows:

1. Turn out piston (24) and tilt respirometer to make mercury leave capillary.
2. Loosen holders (28, 29). 3. Remove rubber rings and disconnect ground joints. 4. Move the mercury level regulator (22). 5. Take out measuring capillary, replace it by another. 6. Repeat manipulations in points 1–4 in opposite sequence.

6B.7. Filling the respirometer

Place respirometer in the stand (as in Fig. 6B.2 C); open tap (13) so as to connect the system with atmosphere. Place cylinders of Whatman's filter paper (starchless) of equal area in both CO_2 absorbers and grease their edges (33) with vaseline in order to prevent creeping of NaOH. Grease slightly the ground joints of absorbers with vaseline-caoutchouc mixture, soak filter paper with a constant amount (usually 0·05–0·1 ml piece) of 25% NaOH. Make sure that the excess of solution does not pour out of the lower fold of absorber, then attach the absorbers to the respirometer arms and secure with rubber rings. Place the animal (animals) in the respiration chamber, then attach both chambers, respiratory and compensatory, to the ground joints of appropriate absorbers greased earlier with vaseline, and secure them with rubber rings.

The respirometer is then carried (movement is allowed only in vertical position) to the water-bath hung on the suspensor or placed in the stand on the water-bath bottom. In both cases, the water level should reach the line drawn in Fig. 6B.2 A. By turning the piston (24) adjust the level of mercury just above 'O' in the measuring capillary. Wait (about 15 min) until the temperature of the respirometer attains that of the water-bath. Close tap (13). Enter the data sheet with general information (description of the animal, date, code of a given experiment, etc), register time when animal had been introduced to the respiration chamber, and time of closing tap (13), temperature of water-bath, and barometric pressure (mm Hg).

6B.8. Measurements of O_2 consumption

By gentle turning of the piston position index drop (12) so that one of its menisci (always the same one!) coincides with 'O' mark. Record in data sheet (fig. 6B.3): hour and minute as well as the level of mercury in the measuring capillary (estimated to one tenth of the scale division). Repeat readings at suitable time intervals, usually every 10–15 min, each time starting with bringing the index drop to 'O' position. The intervals should be short enough to prevent the index drop from moving out of the index capillary. However, if this has happened, one should open the system to the air using tap (13), wait until the index drop flows back to the centre of the capillary,

Grillus domesticus

Start-hour	12^{52}
Date	12. III. 1968
Temp.	29°C
Time closing the system	13^{08}
Bar.	743,0
Factor for converting gas volume in STP-conditions	0,862
Notes:	

1		2		3		4		5		6		7		8		9		10	
hour	µl	hour	µl	hour	µl	hour	µl	hour	µl	hour	µl	hour	µl	hour	µl	hour	µl	hour	µl
13^{10}	53	13^{10}	63	13^{10}	42														
13^{20}	65	13^{20}	85	13^{20}	90														
13^{30}	75	13^{30}	105	13^{30}	130														
13^{40}	85	13^{40}	122	13^{40}	174														
13^{50}	95	13^{50}	138	13^{50}	220														
14^{00}	105	14^{00}	152	14^{00}	265														
14^{10}	115	14^{10}	169	14^{10}	312														
14^{20}	125	14^{20}	184	14^{20}	355														
14^{30}	135	14^{30}	203	14^{30}	400														
14^{40}	145	14^{40}	225																
14^{50}	156	14^{50}	242																
15^{00}	166	15^{00}	254																

	1	2	3	4	5	6	7	8	9	10
QO_2 µl/hr; as read from graph	61,0	102,0	269,0							
QO_2 µl/hr; corrected for STP (0°C, 760 mm Hg)	52,6	87,9	231,9							
Fresh weight of animal(s)	0,050g	0,135g	0,200g							
Other data about animals (e.g. behaviour in resp. chamber etc)										

Fig. 6B.3. Data sheet for measurements of oxygen consumption by 3 specimens of *Grillus domesticus*.

and start measurement again, repeating all initial manipulations. If it is still necessary to continue the measurements although the mercury column has reached the top of the measuring capillary scale, one should open tap (13), turn the piston upward lowering thus the mercury level just above 'O' point of the scale, close tap (13), and go on with measurements. When closing tap, it should be checked whether there was no significant change in barometric pressure.

After completion of the measurements, which usually last 3–4 hours, open tap (13), remove respirometer from the water-bath and disconnect both chambers. After removing absorber (6), and wiping off the excess grease from ground joint, take out the animal and weigh it. Having discarded the filter paper with NaOH, rinse the absorber with water, remove grease from ground surfaces and absorbers with benzene and ethylene chloride mixed in equal volumetric proportions, then wash them in concentrated H_2SO_4, finally rinse many times with water and dry.

6B.9. Calculations

Results of measurements, for example, those of respiration of *Grillus domesticus*, presented in Fig. 6B.3, should be drawn on mm paper (Fig. 6B.4); usually on abscissa 1 min = 1 mm on ordinate the scale depends on intensity of respiration of the object studied, thus from the dimensions of measuring capillary which was in use. At the beginning of the experiment, changes in the amount of gases present in the respiration chamber usually differ from those obtained later. This often is due to incompleted gaseous equilibrium between the air and NaOH solution as well as to a higher uptake of CO_2 by the animal directly after it has been placed in the respiration chamber. This period should be excluded from further calculations. For each curve pertaining to a given respirometer, one draws 'average' straight line using a perspex rule with parallel lines, and calculates O_2 consumption per hour.

Thus obtained result needs to be reduced to Standard Temperature Pressure conditions (*STP*), i.e., 0°C and 760 mm Hg. according to the formula:

$$QO_{2\,(STP)} = QO_{2\,(TP)} \cdot \frac{P - Pw}{760} \cdot \frac{273}{T}$$

where,

$QO_{2(STP)}$ —the volume of oxygen consumed per hour at 0°C and 760 mm Hg, $QO_{2(TP)}$ —the volume of oxygen consumed per hour at T°C and P mm

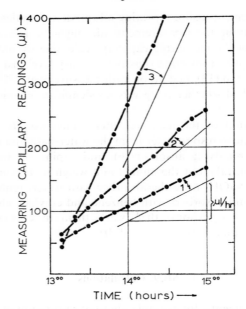

Fig. 6B.4. Diagram of results and calculation of O_2 consumption acc. to the data in Fig. 6B.3.

Hg, T —the temperature of the bath (in Kelvin's scale), P —the atmospheric pressure in mm Hg read when closing the respirometer, Pw—the pressure of saturated water vapour of liquid phase in respirometer chambers (e.g. of 25% solution of NaOH) in mm Hg at $T°C$.

Table 6B.1 below shows the pressure of saturated water vapour of water and of 25% solution NaOH at different temperatures

Table 6B.1.

Temp. °C	Water vapour pressure mm Hg, (Pw)		Temp. °C	Water vapour pressure mm Hg, (Pw)	
	Water	25% NaOH		Water	25% NaOH
0	4·58	2·9	20	17·54	10·9
5	6·54	4·1	25	23·76	14·9
10	9·21	5·8	30	31·82	19·9
15	12·79	8·0	35	42·18	26·4

STP recalculation can be also done by multiplying $QO_{2(TP)}$ in μl read from the graph (Fig. 6B.4) by the factor read from nomogram (Fig. 6B.5):

$$QO_{2(STP)} = QO_{2(TP)} \cdot C$$

(N.B. the nomogram has been calculated for Pw of 25% NaOH).

The 25% solution of NaOH produces relative humidity of about 62·5%. This is an appropriate humidity for many terrestrial insects, however it can be insufficient e.g. for slugs. In the latter case one can use lower concentrations of NaOH since the oxygen demand and thus CO_2 output in slugs are relatively lower. It will result in higher relative humidity of the air. In this case it is necessary to use appropriate Pw for a given NaOH concentration in the formula for $QO_{2(STP)}$ calculation. Nevertheless the differences in Pw depending on NaOH concentrations are rather small (cf. Table 6B.1), so they can be in many cases either neglected or roughly estimated only.

6B.10. Measurement of CO_2 output

The principle of this measurement is as follows: for the same animal one measures change in gas volume in respirometer at 3 consecutive runs; during the first and third run, CO_2 evolved is absorbed by 25% OH, in the second run, CO_2 is not absorbed. Only these measurements whose results in the 1st and 3rd runs are sufficiently similar, can be taken into account in calculating CO_2 output.

Sequence of manipulations:

I. Prepare 2 chambers and 4 absorbers with cylinders of filter paper for each respirometer. Place the animal in the respiration chamber, add to one pair of absorbers, say, 0·1 ml 25% NaOH and carry on measurements of O_2 consumption in the manner described above. Plot each subsequent result on the graph, to the moment when the course of curve suggests stabilization of the respirometric results.

II. Remove the respirometer from the water-bath, disconnect both chambers and absorbers. In place of the latter, attach the other pair with 30% solution of $CaCl_2$ whose hygrostatic properties are similar as those of NaOH (i.e. R.H. = 62·5%), but it does not absorb CO_2. The quantity of $CaCl_2$ solution should be small (e.g., for the vessels presented in Fig. 6B.2 B of the order of 0·05 ml), in order to minimize CO_2 retention. Immerse the respirometer in

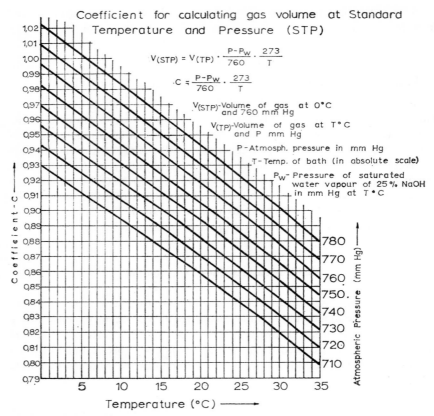

Fig. 6B.5. Coefficient for calculating gas volume at Standard Temperature and Pressure (*STP*) (Nomogram)—

$$V_{(STP)} = V_{(TP)} \cdot \frac{P - Pw}{760} \cdot \frac{273}{T}$$

$$C = \frac{P - Pw}{760} \cdot \frac{273}{T}$$

$V_{(STP)}$ —volume of gas at 0°C and P mm Hg.

$V_{(TP)}$ —volume of gas at T°C and P mm Hg.

P —atmospheric pressure in mm Hg.

T —temp. of bath (in Kelvin's scale).

Pw —pressure of saturated water vapour of 25% **NaOH in mm Hg at** T°C.

the water-bath and carry on measurements, recording their results and drawing the graph.

III. Remove the respirometer from the water-bath again, disassemble it and, instead of the absorbers with $CaCl_2$, use the first pair with NaOH, but with new filter paper and fresh alkaline solution. Carry on a series of measurements and draw the graph.

The manner of computing CO_2 output depends on the ratio of the gas phase to the liquid phase in the respiration chamber, and thus from the magnitude of CO_2 retention in the liquid. If, for example, the liquid phase ($CaCl_2$ solution) $\approx$ 0·05 ml, and the gas phase (the air) $\approx$ 10 ml, then the retention of CO_2 in the liquid phase does not exceed 0·5% of the total amount of CO_2 evolved and it can be omitted in calculations; for details on calculation of correction term for CO_2 retention in the liquid phase—consult Klekowski 1971.

The amount of CO_2 produced is obtained as follows*:

$$QCO_{2(TP)} = Q_{II} - Q_{I\text{-}III}$$

where:

$QCO_{2(TP)}$—the volume of CO_2 produced at $T°C$ and P mm Hg, $Q_{I\text{-}III}$—the decrease in the gas volume per hour in measurement runs I and III (average), i.e. the volume of oxygen consumed per hour at $T°C$ and P mm Hg, Q_{II}—the difference in gas volume per hour in measurement run II, i.e. without CO_2 absorption.

Thus obtained $QCO_{2(TP)}$ should be recalculated as $QCO_{2(STP)}$ acc. to formula given for $QO_{2(STP)}$.

From thus obtained rate of CO_2 output the respiratory quotient can be found:

$$RQ = \frac{QCO_2}{QO_2}$$

Applying RQ for bioenergetic studies, the oxy-calorific coefficient can be found and the amount of oxygen consumed by the animal can be converted into energy units (calories), estimating thus the so-called 'cost of maintenance'.

---

*Disappearing of gas is marked with '—' and production with '+', therefore, QO_2 values are negative, and QCO_2 values—positive.

6B.11. Remarks

For investigations of ecological bioenergetics in the field, it is of a great significance what type of the metabolic rate has been measured in the respirometer. For such evaluation, observations on the behaviour of the animal in the respiration chamber are very important, since they will permit to compare it with the corresponding behaviour of this animal under natural conditions. In general can be suggested that the data obtained by respirometric method give values between those of Basal Metabolic Rate (or more precisely— Resting Metabolic Rate) and Routine Metabolic Rate (or at 24 hrs measurements, Average Daily Metabolic Rate).

Thus obtained results are still more underestimated when compared with Active Metabolic Rate; the latter has however no major meaning for ecological energetics.

By and large, these results are more or less underestimated as far as their usefulness for calculation of energy flow by natural populations is concerned. There are some exceptions, however when, e.g. in *Tribolium*, the animals in the respirometer are under conditions almost identical with those in nature, or at least in large cultures. Then the results should not evoke any doubt as to their validity in estimation of energy flow by a population.

References

DIXON M. (1952) *Manometric methods, as applied to the measurement of cell respiration and other processes.* Ed. 3, Cambridge Univ. Press Cambridge, 165 pp.

DRASTICH L. (1934) Microrespirometer v nove uprave. *Biol. Listy*, **10**, 1–20.

FENN W.O. (1927) The gas exchange of nerve during stimulation. *Amer. J. Physiol.*, **80**, 327–346.

FISCHER Z. (1967) The energy budget of *Lestes sponsa* during its larval development. IBP Meeting on Methods of Assessment Secondary Production in Freshwaters. Prague. 9 pp. (mimeographed copy).

KLEINZELLER A. & MÁLEK J. & VRBA R. (1954) *Manometrické metody a jejich použiti v biologii a biochimii.* Statni Zdravotnické Naklad., Praha, 391 pp.

KLEKOWSKI R.Z. (1971) Cartesian diver microrespirometry for aquatic animals. *Pol. Arch. Hydrobiol.*, **18**, 93–114.

KLEKOWSKI R.Z., PRUS T. & ZYROMSKA-RUDZKA H. (1967) Elements of energy budget of *Tribolium castaneum* (Hbst) in its development cycle. In Petrusewicz K. (ed.) *Secondary Productivity of Terrestrial Ecosystems.* PWN, Warszawa–Kraków, **2**, 859–879.

KRAJNIK B. (1922) Über eine Modifikation des Mikrorespirations—apparates. *Biochem. Z.* **13–**, 286–293.

PRUS T. (in press) Energy requirement, expenditure, and transformation efficiency during development of *Asellus aquaticus* L. (Crustacea, Isopoda). *Pol. Arch. Hydrobiol.*, **19,** 97–112.

STACHURSKA T. (in press) Survival and metabolism of *Polydesmus complanatus* (L). (Diplopoda) in conditions of heavy infection with gregarines.

WINTERSTEIN H. (1912) Ein Apparat zur Mikroblutgasanalyse und Mikrorespirometrie. *Biochem. Z.* **46,** 440–449.

7
Feeding and Nutrition

7A REVIEW OF METHODS FOR IDENTIFICATION OF FOOD AND FOR MEASUREMENT OF CONSUMPTION AND ASSIMILATION RATES

R. Z. KLEKOWSKI AND A. DUNCAN

7A.1. Introduction

It is a major problem to identify what food is being eaten by a particular species in its habitat and an attempt is made below to review the various methods used by previous authors for specific organisms in the hope that some may be applicable for the reader's situation. Most of these methods can be used only to identify the food species but some can be adapted to obtain quantitative estimates of food consumption. Otherwise, there are only a limited number of techniques available for the measurement of feeding, all of them not easy, time-consuming and requiring biological skill in handling animals. It is true to say that the least developed part of the energy budget is that relating to feeding, although it is one of the most direct 'inter-links' between the trophic components in an ecosystem. It was therefore thought worth-while to attempt to analyse in Table 7A.2 (p. 239) characteristics and conditions of those quantitative methods available.

7A.2. The identification of food actually consumed

It seems that most animals do not eat haphazardly what is available but actively select their food species or food substances, with the exception, perhaps, of deposit-feeders, although Darwin (1898) showed that the detritus-feeder earthworms have distinct preferences for certain kinds of deciduous leaves and with the exception perhaps of filter-feeders, although the filtering mechanism usually selects for size.

Usually several different kinds of food are eaten especially during periods of active body growth or growth of reproductive tissue or if at one season a particularly nutritious food is available (as in harvester ants which feed on insects when available but store seeds during the summer and autumn for

227

winter food). Some species have a very restricted diet, such as honey-comb wax (*Galleria* spp.), keratin (dermestids; Mallophaga clothes moth), wood with cellulose and lignin (termites; anobiid and cerambycid beetles), xylem and phloem sap of plants (heteropteran species), mammalian blood (mosquitoes; fleas; bugs). It is interesting that those plant or blood-sucking insects whose diet is exclusively sap or blood throughout their life cycle all possess symbiotic micro-organisms whose function may be to supplement these diets with the essential accessory food substances mentioned earlier; in other animals, such as termites and ruminant mammals, it is the symbiotic intestinal protozoa which possess the necessary cellulases to breakdown the wood cellulose. Other species are monophagous, often terrestrial species restricted to one plant species, but nevertheless still with a capacity to discriminate between different parts of plants (Crossley, 1963; Reichle and Crossley, 1967; Hiratsuka, 1920; Moss, 1967).

7A.2.a. Direct observation

One of the best ways of determining both qualitatively and quantitatively the food eaten by a species in a given time is by direct observation in the field. This is not possible in many species but where possible the results obtained are realistic and unbiased by the technique employed. Holišova *et al.*, (1962) was able to determine not only which plants were preferred but also which parts of plants were eaten by water voles (*Arvicola terrestris* L.) by direct observation. Petal (1967) was able to observe in colonies of the meadow ant (*Myrmica laevinodis* Nyl) not only the species and developmental stages of the food organisms in the mandibles of the forager workers but also the amount of food carried into the nest both in the crop and in the mandibles of the forager workers. The food of nestling birds may also be studies by direct observation and some birds will readily re-gurgitate what they have eaten or a ring may be employed to prevent swallowing.

The food consumption of arboreal lepidoteran larvae may also be studied both qualitatively and quantitatively in the field.

7A.2.b. Laboratory observation and experiments

Observation in the laboratory may also be made upon the nature of an animal's food but animals cultured in or brought into artificial conditions alter their feeding behaviour and may even feed on un-natural food; useful observations on the feeding mechanisms (e.g. method of food detection, size limits of capturable prey etc) can be made in the laboratory.

Food preference experiments such as are described by Pinowski for granivorous birds or by Drożdż for small rodents in the present handbook (pp. 334–337) may be carried out in laboratory conditions on animals acclimated to laboratory conditions. Healey (1965) detected strong preferences for certain named species of soil fungi in soil collembola and Wallwork (1958) examined a wide variety of possible food choices presented to oribatid mites. Semi-field food preference experiments of the kind carried out by Phillipson (1960a, b) on the carnivorous phalangid (*Mitopus mori*), on the detritus feeding wood-louse (*Oniscus asellus*) or on other carnivorous phalangids are described earlier in the handbook.

By recording over 24 hours the activity of predators and prey in a container placed in their original habitat, Phillipson was able to obtain both qualitative and quantitative results on food consumption.

7A.2.c. Analyses of gut contents and faeces

Another frequently adopted method is to identify the species eaten from their remains either in various parts of the alimentary canal or in the egested faeces. This method has been used for many different kinds of animals, for bird gizzards (Hartley, 1948; Kennedy, 1950; Betts, 1954, 1955; Steward, 1967), mammalian stomachs (Drożdż, 1967, 1968; Holišova *et al.*, 1962; Williams, 1962), insect guts (Fischer, 1966, 1967) and others (Reynoldson and Young, 1963). Such methods may be used for insects and rodent or insectivore mammals which chew their food into small pieces and the remains of different types of plant and animals may be identified under the microscope, especially if some sort of standard set of 'remains' is built up (Hanna, 1957; Mulkern and Anderson, 1959; Holišova *et al.*, 1962; Williams, 1955). The various plant remains (epidermal cells, tracheal tissue, trichias etc.) may be compared with a standard atlas (Worytkiewicz, 1968, in press) or identified directly (striated muscle, chitinous fragments, annelid chaetae, mammalian bones etc.). Where the remains consist of countable and identifiable items, such as chironomid head capsules, beetle elytra, mammalian pelvic girdles or other identifiable bones, fish otoliths etc., this method may provide approximate quantitative results, especially if the mean weight of the eaten individuals may be assumed from known mean weights of individuals from the feeding area. Thus, for some time, ornithologists have used the concept of 'average rat unit' in their investigations on the food of predatory birds from gizzard and stomach analyses. With material such as chironomid head capsules, the remains may even be analysable into size classes comparable with field data and so provide more accurate estimates of the weight of food eaten. However, this method of quantitatively estimating the food consumed is only very approximate because it is based on two major assumptions, namely, that different foods pass through the alimentary canal at the same rate, which is almost certainly not true (see Kostelecka-Myrcha and Myrcha, 1964; and p. 99f of handbook) and in any case requires testing; the second assumption is that soft-bodied organisms which leave no indigestible remains are not eaten. Some techniques for testing these two assumptions are presented later in this section.

That certain animals are fluid feeders is readily observable in the field. In plant-sucking insects, it is necessary to determine the actual tissue, whether phloem, xylem or mesophyll tissue, from which the nutrient is obtained since these will differ greatly in their energetic content. Wiegert (1964) determined that *Philaenus spumarius* fed on xylem sap which in tomato plants consisted of amino acids (98%) with a calorific value of between 2·93–9·54 calories per gram; since the calorific value of the egested material which contained no amino acids was also determinable, it was possible to determine the rate of food consumption from the difference in the calorific value of these and the rate of fluid intake. In fact, the fluid intake was assumed to be equal to the rate of fluid excretion which was readily measurable in nymphs but not in adult spittle-bugs. Andrzejewska (1967) working on *Cicadella viridis*, (Homoptera), attempted to measure consumption rates in this plant-sucker by comparing the rate of uptake of a mineral medium by grass plants with and without a known number of sucking insects.

7A.2.d. Coefficients of 'food selection' or 'Electivity indices'

The anlyses of stomach or gut contents forms the basis for two coefficients of 'food selection' suggested by two Russian biologists, A. Šorygin (1939) and V. S. Ivlev (1955). Both coefficients attempt to bring together the two quite un-related phenomena which are involved in what food is taken from the environment by an animal, firstly, which food species the animal prefers to eat when all species are equally available and secondly, the availability of the food species in the environment. Both these phenomena are very complex and neither coefficient is very successful in relating them together although

Ivlev's is useful as some kind of summing index; moreover, Ivlev's concept of electivity tries to indicate the compromise made by organisms faced with the discrepancy between their food preference and what food is available.

Food preference was estimated by both Šorygin and Ivlev from the percentage number of 'individuals' belonging to one food species in the gut; it is also possible to give food preference in terms of biomass or calories by multiplying the numbers of 'individuals' by either the mean field weight or calorific content (Fischer, 1966, 1967). Another approach would be to estimate food preferences from selections in 'choice experiments' such as described by Phillipson (1960a, b), Pinowski (p. 334) Drożdż (p. 336). The idea of 'availability' of food in the environment is also complex since it may be related not only to the abundance or scarcity of the food species (as detected by the biologists' collecting technique rather than the animal's own sense organs!) but also to the consumer's morphological and physiological 'limitations', for example, whether the mouth is big enough, whether the teeth are strong enough or of a suitable shape, whether it can run fast enough, whether it has suitable range of digestive enzymes etc. Moreover, the species that occur in the gut are those available (i.e. both present and detectable by the animal's sense organs) within the species' feeding area, an area delimited by its locomotory ability and normal behaviour, so that the unit area or volume used to define 'p' (the relative abundance of a food species in the field) should bear some realistic relationship to this.

Both coefficients of 'food selection' are therefore based on some estimate of the relative abundance of a food species in the gut (r) and some estimate of its relative abundance in the field (p) ,both estimates being in percentages based upon numerical data. Šorygin's coefficient is given by $\dfrac{r}{p}$ and Ivlev's by $\dfrac{r-p}{r+p}$. Figures 7A.1 and 7A.2 present a series of different hypothetical conditions of 'r' and 'p' and clearly demonstrates that Ivlev's coefficient discriminates more sensitively between different levels of 'r' and 'p' than Šorygin's, which adequately distinguishes only conditions of extreme selection (i.e. when 'r' is very high and 'p' extremely low). Moreover, it is very convenient that Ivlev's coefficient ranges from $+1$ to -1 with a zero value when the relative abundance of a food species in the gut and in the field is the same; positive values of the coefficient suggest some preference for this species as a food and the nearer the value is to 1, the greater the preference. Negative values of the coefficient suggest rejection of the species as a food. However,

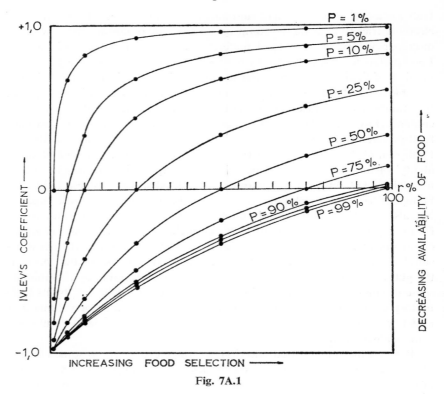

Fig. 7A.1

convenient as Ivlev's coefficient is, there seems no biological justification for the basic assumption that a 50% occurrence of a species in the gut is the same as a 50% occurrence of that species in the field. Figure 7A.3 is a nomogram for determining Ivlev's coefficient.

Examples of the use of both coefficients have been taken from the work of Z. Fischer on food selection in aquatic animals. In her 1966 and 1967 papers, Fischer used Šorygin's coefficient to indicate the preference of larval *Lestes sponsa* (Odonata) to species belonging either to the weed fauna or to the plankton of astatic ponds. In pond Zoldanka on 20.VIII (Table 1, 1967), $p = 2·4\%$ and $r = 36·0\%$ for the percentage number of species composing the weed fauna, whereas $p = 97·6\%$ and $r = 64·0\%$ for planktonic species. From this data, Šorygin's coefficients are 15 and 0·66 for the weed fauna and

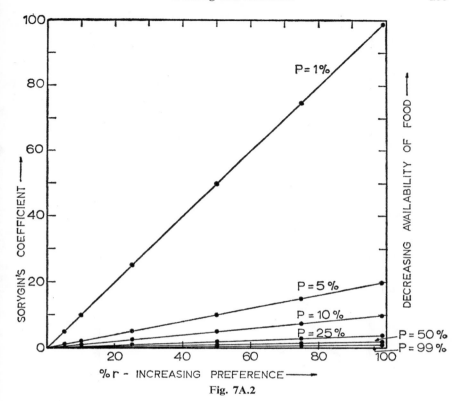

Fig. 7A.2

plankton respectively and are plotted in Figs. 7A.1 and 7A.2. Ivlev's co-efficients, calculated by us for the same data, are $+0.89$ and -0.21 respectively and, when plotted on their appropriate p-lines (i.e. $p =$ about 1 for the weed fauna and $p =$ about 100 for the plankton), show that *Lestes sponsa* larvae have a high preference for species from the weed fauna on that sampling date, whereas nothing useful biologically can be said about the negative value of Ivlev's coefficient for plankton because when $p =$ about 100, this coefficient is negative for all values of 'r'.

In another work (1968), Fischer used Ivlev's coefficient to indicate the preference of a group of ten young grass carp (*Ctenopharyngodon idella* Val.) for species of plants when presented in equal and excessive quantities; ten species of plants were tested, presented simultaneously in twos and threes

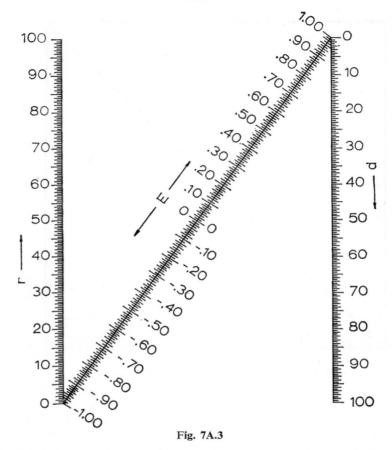

Fig. 7A.3

and the weight eaten was obtained by weighing the quantities remaining after 24 hours. As Table 7A.1 shows, when pairs of plants were presented in equal quantities and so $p = 50\%$ (e.g. for *Typha-Lemna*, *Lemna-Juncus*, *Typha-Juncus*), the weight of each species eaten depended on the choice available, so that different values of 'r' and of the coefficient were obtained; plotted in Figs. 7A.1 and 7A.2, these values clearly indicate the preference of grass carp for *Lemna* and *Juncus* compared with *Typha*. This preference may already be seen in the original data (Table 1, Fischer, 1968) without the help of a coefficient but these results give us some confidence in the possibility of

Table 7A.1. Examples of Ivlev's coefficient of 'food selection' obtained from 'choice experiments' in grass carp (*Ctenopharyngodon idella* Val.) (Fischer, 1968)

	Combinations of plant species offered					
	Typha–Lemna		*Lemna–Junctus*		*Typha–Junctus*	
'*p*'	50%	50%	50%	50%	50%	50%
'*r*'	0·7%	99·3%	44%	56%	34%	66%
I	−0·96%	+0·30%	−0·06%	+0·06%	−0·19%	+0·14%

using Ivlev's coefficient to untangle the complexities of food preferences under field conditions, provided that the behavioural feeding biology of the species being studied is sufficiently well known.

7A.2.e. The precipitin test

Techniques other than analysis of stomach contents have been used to identify the food of predatory and other animals. One of these involves using serological techniques in which the specific proteins of the prey are identified by their reactions in a precipitin test with the serum of a sensitised mammal, usually a rabbit; another method involves the use of prey animals labelled with a radio-active isotope which, if eaten, passes on its radioactivity to a predator.

Serological techniques have been used for some time to identify prey eaten by predators, especially with insect predators (Brooke and Proske 1946; West, 1950; Hall *et al.*, 1953; Downe and West, 1954; Fox and Mac-lellan, 1956; Dempster *et al.*, 1959; Dempster, 1960; Loughton *et al.*, 1963; Davies, 1969) or to determine the number of predators feeding on a prey species (Dempster, 1960; Young *et al.*, 1964) were able to demonstrate by means of the precipitin test that *Asellus* spp, were eaten by the predatory triclad *Dendrocoelum lacteum*, although no chitinous remains were detected in gut smears. For these purposes, the technique is useful; however, Southwood (1966) reviews in detail some of the difficulties of the method and gives practical details about the precipitin test itself.

Basically, antigen is prepared from prey with empty guts, crushed in a pestle and mortar with some saline solution. The clear supernatant is decanted, sterilised and, after freeze-drying, the antigen can be stored. Antiserum is produced by injecting the reconstituted antigen into a rabbit and taking samples of its blood at the point of maximal sensitivity by testing regularly against an extract of the prey; this takes about 10–12 days. The serum may then be separated off from the blood samples and stored, after freeze-drying, in a deep freeze for later use. The collection of predator samples may be made earlier and stored for some time by smearing the whole animal if it is small or if large by smearing the gut onto filter paper and drying. However, tests may also be made directly on fresh extracts of the whole predator or its gut in saline and placing equivalent volumes of first the antiserum and then the clear supernatant of the extract, underlying the former, in a serological test tube. A positive result is given by the formation of a visible ring of precipitin at the interface between the two liquids. The usefulness of the serological test for detecting the prey eaten by a predator is limited by the rate of digestion and so the length of time the prey is detectable. The work of Dempster (1960) and Loughton *et al.* (1963) suggests that lepidopteran and coleopteran eggs were detectable for 18–40 hours and large larvae for 5 days after feeding. The method is less useful for quantitative studies on food consumption unless it is known that only one food individual is normally eaten per period of detectability.

7A.2.f. Food markers

The serological test for food identification is based on the presence of proteins specific to a particular prey-species in the gut of a predator, that is, these proteins acted as very efficient markers characteristic of a species. Other such 'natural markers' also exist in the highly coloured wing elytra of beetles or in the body colours of various litter arthropods (green heteropteran *Calocaris;* black collembolan *Tomoceros* or red cercopid nymphs, Phillipson, 1960a, b) which are readily detectible in the gut or in the faeces. There are probably other naturally coloured 'markers' available in the food of various animals.

Where such natural markers do not exist, it is possible to use various vital dyes such as neutral red (South, 1965), rhodamine B (Reeves *et al.*, 1948), thiorescin (Sura-Bura and Gregean, 1956), fluorescin (Zaidenov, 1960) and trypan blue (Knight and Southon, 1963) which may be detected visually in the gut contents or in the faeces. Quoting from Southwood (1966): 'Blood-sucking Diptera have been marked by allowing them to feed in a cow which had had 200 ml of an aqueous solution, containing 4 g of trypan blue, intravenously administered over 20 minutes. The dyestuff can be detected by a paper chromatographic technique in which the gut contents of the fly is mixed with 0·1 N sodium hydroxide solution and applied to a narrow strip of Whatman no. 1 filter paper, which is developed in 0·1 N sodium hydroxide solution; the trypan blue remains at the origin, other marks due to the gut contents move away (Knight and Southon 1963).'

Another method of marking animals which is identical in principle to that of marking by vital dyes is by radioactive isotopes in which the isotope may be used either as a label outside the animal or it may be fed to the animal and subsequently incorporated into its tissues. The latter method of internal marking is preferable, using isotopes which are readily incorporated into an animal's tissues, because the marking process involves minimal manipulation and the mark is invisible to the consumer. Moreover, it is not completely discarded on moulting. Some preliminary laboratory tests can show how quickly the radioactive isotope atoms are excreted and it is convenient to choose an isotope whose half life is longer than that of the animal's life cycle. Detection in the field may be made by means of a Geiger-Muller tube and in the laboratory by autoradiography. The latter method involves killing the animals but is very convenient in that only very low levels of radioactivity need to be applied. A good description of this technique can be found in Southwood (1966, pp 69–70) or in original papers, Gillies (1958) or Lewis and Waloff (1964).

7A.3. Quantitative determination of consumption and assimilation rates

Far fewer methods are available for the measurement of food consumption and assimilation rates. The following methods are reviewed here:

a. gravimetric methods:

I. by obtaining the weight of food eaten in some direct or indirect manner as well as of the faeces egested;

II by summing the production, respiration and faeces or rejecta so that, $A = P + R + U$ and $C = P + R + U + F$ (or FU);
Both P (growth + exuviae + other products) and F are determined gravimetrically or calorimetrically;

b. marker methods:

I. using radioactive isotopes; by labelling the food and following either the passage of the marked food to the animal or the decrease in concentration of marked food in the medium;

II. using non-assimilable markers; by using some natural marker such as silica already present in the food or adding an artificial inert one, such as powdered platinum. Knowing the calorific equivalent of a unit of marker in the food and in the faeces as well as the amount of marker in the faeces, both consumption and assimilation may be determined.

Table 7A.2 consists of a pictorial summary of most of the methods which are available for quantifying consumption and assimilation, using these gravimetric and marker techniques. Where examples could be found of authors employing specific methods, these works are cited, otherwise a technique is described from first principles without knowing whether it will work or not in the hope that it may be applicable to particular species. However, when one considers the nature of animal food and the different methods of feeding that exist, it is clear that only one or two of the methods described are appropriate to any particular circumstance. (Table 7A.3)

7A.3.a Gravimetric techniques

The most frequently applied method of measuring both consumption and assimilation rates is the gravimetric technique where food is weighed before and after the animal had fed, so that, for a 24-hour period or for the period of the feeding cycle and in terms of dry weight (or calories),

$$\text{food offered}-\text{food left} = \text{food consumed } (C),$$

and by weighing the faeces or rejecta egested, then food assimilated $(A) =$ food consumed $(C) -$ faeces egested (F) or rejecta egested (FU). The steps involved to produce these two equations are very clearly expressed in Varley (1967, p. 451). These equations are adequate for situations where the gut is known to be empty initially (as may occur in predators) so that the faeces collected come from a known weight of ingested food or where feeding is continuous and the food passes through the gut at a constant rate so that the ratio of faeces produced to food ingested is known to be constant or, as discussed earlier, when the period employed is that of the species' characteristic feeding cycle.

  Phillipson (1960a, b) has proposed a very successful technique for solving this difficulty of obtaining the exact weight of faeces produced from a known weight of food consumed irrespective of digestion rate, by feeding the phalangid *Mitopus mori* on naturally coloured food species and collecting

TABLE 7A.2 SOME METHODS FOR QUANTITATIVE ESTIMATION OF CONSUMPTION AND ASSIMILATION (IN CAL/IND/hr)

VARIANT OF METHOD, ESSENTIAL CONDITIONS	SEPARABILITY OF FAECES FROM FOOD	GRAVIMETRIC METHOD – WEIGHT W	GRAVIMETRIC METHOD – NUMBERS N	MARKER METHOD + RADIOACTIVE ISOTOPES R	MARKER METHOD – NON-RADIOACTIVE M

1 MEASUREMENT OF FOOD EATEN (fe)

NATURE OF FOOD IS SUCH, THAT AREA OR VOLUME EATEN IS MEASURABLE OR NUMBER COUNTABLE, AND CONVERTIBLE TO CALORIES, GUT OF CONSUMER NEED NOT BE EMPTY

- **a** FAECES NOT MEASURABLE BY EXPERIMENTOR; $t <= > t_d$, REFECATION DOES NOT OCCUR
- **b** FAECES SEPARABLE FROM FOOD AND MEASURABLE BY EXPERIMENTOR; $t <= > t_d$[3]

W 1a / W 1b

N 1a: $C = \frac{N_{fe} + C_{gf}}{t+N}$ OR $\frac{N_{fe} + C_{gf}}{t+N}$

N 1b: $C = \frac{N_{fe} + C_{gf}}{t+N}$ OR $A = C \frac{F + C_{gf}}{t+N}$

$A = P + R + U$

R 1a / R 1b

M 1a / M 1b

2 MEASUREMENT OF REDUCTION IN FOOD ($f - f_n$)

GUT OF CONSUMER NEED NOT BE EMPTY

- **a** FAECES INSEPARABLY MIXED WITH FOOD, THEREFORE $t < t_d$
- **b** FAECES SEPARABLE FROM FOOD BY EXPERIMENTOR AND REFECATION DOES NOT OCCUR, THEREFORE $t > t_d$[3]

W 2a / W 2b: $C = \frac{f - f_n}{t+N} - C_{gf}$

$A = \frac{f - f_n}{t+N} - C_{gf} - \frac{F + C_{gf}}{t+N}$

N 2a: $C = \frac{(N_f - N_{fh}) + W_x \cdot C_{nf}}{t+N}$ OR $\frac{(N_f - N_{fh})}{t+N} + W_x \cdot C_{gf}$

N 2b

R 2a / R 2b

$A = P + R + U$

M 2a / M 2b

3 MEASUREMENT OF INCREASE IN BODY WEIGHT DUE TO FOOD CONSUMPTION ($W_1 - W_0$)

- **a** FAECES INSEPARABLY MIXED WITH FOOD, $t < t_d$, GUT MUST BE EMPTY AT THE TIME t_0
- **b** FAECES SEPARABLE FROM FOOD BY EXPERIMENTOR AND REFECATION DOES NOT OCCUR; $t <= > t_d$[3]

W 3a: $C = \frac{(W_1 - W_0) + C_{gf} \cdot L}{t+N}$

W 3b: $C = \frac{[W_1 - W_0] - C_{gf}}{t+N} + \frac{F \cdot C_{gf} + L}{t+N}$

$A = P + R + U$

N 3a / N 3b

R 3a / R 3b

$A = C$

(See M4)

M 3

4 MEASUREMENT OF TOTAL AMOUNT OF INERT NON ASSIMILABLE FOOD – MARKER IN FAECES, KNOWING THE CALORIFIC EQUIVALENT OF MARKER IN FOOD CONSUMED

- **a** FAECES INSEPARABLY MIXED WITH FOOD, $t < t_d$
- **b** FAECES SEPARABLE FROM FOOD BY EXPERIMENTOR AND REFECATION DOES NOT OCCUR; $t <= > t_d$[3]

W 4a: $C = \frac{M + C_{mf}}{t+N}$

W 4b

$\frac{A}{C} + 100 = \frac{f' - F'}{(1-F')f'} + 100$ OR $\frac{C_{mf} - C_{mf}'}{C_{mf}} + 100$

N 4a / N 4b

R 4a / R 4b

M 4a: $C = \frac{M + C_{mf}}{t+N}$

M 4b: $C = \frac{M + C_{mf}}{t+N}$

$A = \frac{M + C_{mf}}{t+N} - \frac{F + C_{gf}}{t+N}$

1) SUCH A CONDITION IS PERMISSIBLE ONLY FOR THOSE CONSUMERS – SPECIES WHERE AN EMPTY GUT DOES NOT AFFECT C (IF SUCH ANIMALS EXIST!)
2) METHOD FOR AQUATIC ANIMALS MAINLY.
3) WHERE t CAN BE $> t_d$, IT IS ADVISABLE FOR t TO BE RELATED OR EQUAL TO t_C, FOR MEASUREMENT OF C, OR t_F, FOR MEASUREMENT OF F.

Symbols and donations

a. *Symbols used to describe the measured parameter*
N number of consumers
C food consumption
F faeces
U urinary waste
FU 'rejecta' where faecal and urinary waste are inseparable (*not* including secretions).
L summed losses or gains during the experimental period (e.g. respiratory CO_2, urinary waste, sweat, exuviae, sexual products water drunk).
TA total area being investigated where feeding occurs
TV total volume being investigated
f food offered
f_n food not eaten
f_e food eaten

b. *Denotations bearing known units*
t_d time of food retention in digestive tract; $t_{d \cdot min}$–time of food retention in digestive tract up to appearance of first faeces.
t_c characteristic period of feeding cycle of species
t_F characteristic period of defaecation cycle of species
t experimental time of measured feeding irrespective of feeding cycle

Gravimetric method (weight grams)	Gravimetric method (numbers)	Marker method (radioactive)	Marker method (non-radioactive)
f wt. of food	N_f nos. of food	R_f cpm of food	M_f mg. marker in food
F wt. of faeces	N_F	R_F cpm of faeces	
w_f wt. per indiv.	n_f nos. food items per area	r_f cpm per unit of food	M_F mg. marker in faeces
			m_f mg. marker per unit of food
			m_F mg. marker per unit of faeces

Calorific equivalents

C_{gf} cal/mg. food	C_{nf} cal/indiv. food	C_{rf} cal/cpm. food	C_{mf} cal/mg. marker in food
C_{gF} cal/mg. faeces	C_{nF}	C_{rF} cal/cpm. faeces	C_{mF} cal/mg. marker in faeces

Basic assumptions associated with the radioactive isotopic methods
That the calorific equivalent C_r (cal/cpm) is known and is the same for all measured parameters; that is, for *food* (checkable), for *assimilated food* (difficult to measure), for *faeces* (checkable), for *urinary waste* (difficult to check) and in respired CO_2 (unlikely to be constant). The C_r will differ for these parameters because of 1. the different degree of isotopic labelling of fats, proteins and carbohydrates; 2. the different percentage assimilation of these food substances; 3. the different R.Q.'s for different food substances.

Table 7A.3. Methods for measurement of feeding available for different feeding types

Nature of food	Feeding type	
Small particles	Aquatic filter feeders	N2, R2, R3
	Aquatic scrapers	W1, N1, W4
Large masses	Deposit feeders	M4
	Detritus feeders, scavengers	?M4.—where the marker can be dusted on the dead food such as 'dusted' hay for feeding cows.
	Carnivores	W1, W2, W3, W4, N1, N2, N3,
	Herbivores on macrophytes	R2, R3
	Herbivores on small algae	N2
Fluids	Suckers of plant fluids	—
	Suckers of animal fluids	—
	Blood Suckers	—
Dissolved substances	Parasites	R3 (Kleiber, 1951; Luick, Lofgreen 1957)

all the coloured faeces produced. The percentage assimilation of the food consumed, obtained by this 'direct' method in laboratory experiments for various developmental stages of the species, served as standards for checking his 'indirect' method of assessing assimilation using other non-coloured food species.

In his 1960a paper, using adult *Mitopus*, Phillipson applied the following equations, for a 24-hour period and in terms of dry weights:

a. food eaten = food — food not eaten;

b.* non-assimilated food = gain in weight of *Mitopus* after 24 hours + faeces;

c. food assimilated = food eaten — non-assimilated food.

The average percentage assimilation of food consumed by the direct method using coloured food was 49·4% (range 25·0–89·2%) for adult males and 41·7% (range 38·0–98·0%) for females which compared very well with the results obtained by the indirect method using non-coloured food, where,

---

*In his 1960a paper, Phillipson actually gives the following equation: b. non-assimilated food = gain in weight of *Mitopus* after 24h + food not eaten + faeces.

in all but two cases, the percentage of ingested food assimilated was between 40–50%, with an average of 47%.

In his 1960b paper, Phillipson applied a similar direct method for measuring assimilation efficiency in young stages of *Mitopus mori* (F.), using coloured food such as black or green collembolans or red cercopid nymphs. In this method, the animals were starved in the laboratory for 24 hours after capture during which the colour of their faeces was noted. They were then given food that produced faeces of a different colour. For an experimental period of 24 hours, food that produced faeces of the initial colour was given and measurements made of the live weight of food, weight of food remains, weight of appropriately coloured faeces produced (indirectly from volume of faecal pellets) and the live weight of the animal; all the weights were expressed as dry weights except that of the animal's live weight. Since there was no large increase in the weight of the animal during the experimental period, the following equations were applied to these measurements:

1. weight of food eaten = weight of food − weight of food not eaten;

2. weight of non-assimilated food = weight of faeces;

3.* weight of food assimilated = weight of food − (weight of faeces − weight of food not eaten)

whereas, quoting the column numbers in his Table 5 which presents his results, 'food assimilated' (column 6) was calculated from column 2 ('food') minus column 5 ('non-assimilated food') instead of column 7 ('food eaten') minus column 5, according to equation c. We suggest that by ommitting the item 'food not eaten' from equation b. and column 5 in Table 5, the results and equations will be in agreement.

The percentage assimilation of food consumed decreased from 74% for instar II to 59%, 55%, 55% and 44% for instars III, IV, V and VI, and compared well with the mean adult assimilation efficiency of 47% obtained from the 1960a paper.

The main difficulty in this method is that changes in the water content of the food during the experiment cause either losses (e.g. from cut leaves) or gains (e.g. in flour) in its weight not associated with feeding. This can be solved by using constant dry weights where possible (e.g. for food not eaten, or for faeces) and applying a pre-determined ratio of fresh weight to dry weight for the food offered. Unfortunately, such a ratio may be very variable for

---

*In his original paper (Phillipson, 1960b), equation 3 is given as 'weight of food assimilated = weight of food − weight of non-assimilated food'.

fresh plant material. Soo-Hoo and Fraenkel (1966) offered their *Prodenia* larvae only half of any leaf and used the other half to determine the fresh to dry weight ratio. Klekowski and Stępień (in press), on the other hand, working with flour mites (Acarina), carried out all their weighings under standard conditions of partial drying (that is, vaccuum dried to constant weight at 50°C) so that the rye germ offered as food returned to its original edible, well-hydrated condition as the experiment proceeded in a thermo-hygrostat. Alternatively, controls may be employed for each experiment to determine the degree of loss or gain in weight of food offered.

In order to convert dry weights of food consumed to calories, the calorific value per gram dry weight must be determined for the food offered and faeces egested (see p. 158); should the animal select the more nutritious parts of food which may be difficult to separate, then it is also useful to know the energy content of the food left. Sometimes, the difference in the energy content of the consumed and egested parts of food are more striking than the difference in weight and moreover, in some cases, the calorific content of the egested faeces is the higher (e.g. in the wax moth larvae). Although conversions from dry weight to calories may be made using calorific values as calories per gram dry weight with ash, it is better to determine the percentage ash content of all the material and to quote values as calories per gram dry ashless weight (see Table 5A.1). Several summaries of calorific data on wild plants and animals have appeared in recent years (Slobodkin and Richman, 1961; Golley, 1960; Straškraba, 1968; Prus, 1970). The most comprehensive is that of Cummins and Wuycheck, 1971, who have presented a list of values for organisms, arranged by trophic level, habitat and taxonomic category, and give a good bibliography. The other authors discuss the variability of calorific values.

Many workers have adopted the gravimetric technique for estimating C and/or A rates and usually in association with one or other of the methods for food identification described earlier. Among many authors are Hiratsuka (1920), a very early and complete work on silk worms (Lepidoptera); Smith (1959) and Gyllenberg (1969) on grasshoppers (Orthoptera); Phillipson (1960a, b) on phalangids (Opiliones) and Phillipson and Watson (1965) on wood-lice (Isopoda); Nakamura (1965) on scarabeid beetles; Soo-Hoo and Fraenkel (1966b) on lepidopteran larvae; Petal (1967) on ants (Hymenoptera). Hubbell *et al.*, (1965) compare the gravimetric technique for estimating C and A rates in an isopod with the radiotracer and calorimetric methods.

Most workers using gravimetric techniques apply the results obtained under laboratory conditions for different stages in the life cycle of a species to field populations. Sometimes, with certain species, field measurements of consumption rates are possible. For example, the area of leaves of grass blades eaten per unit time by lepidopteran larvae or grasshoppers may be measured and a similar area cut out from a fresh leaf, dried and weighed (method W1 in table 7A.2; Gyllenberg (1969), Bray (1961). Or, the difference in the weight of worker ants gives an estimate of the weight of food in the crop from which together with the number of ants leaving and entering their nest a value for the food consumption rate for the whole nest may be obtained (Petal, 1967).

Determination of faecal production under field conditions, however, is rather more difficult although Phillipson (1960a) describes a method of estimating the mean rate of faecal production per animal per 24 hours for litter animals under semi-natural conditions and suggests that these results together with assimilation efficiencies obtained by the methods described earlier provide an assessment of consumption rates under semi-natural conditions. Animals were captured at different times during a 24-hour cycle (5.00, 11.00, 17.00 and 23.00) in order to encompass any diurnal variation in feeding activity and were kept outside for 24 hours in petri dishes containing a variety of potential prey species. The faecal pellets produced during eight three-hour periods were counted in each dish and a mean number per 24 hours obtained; from the calculated volume of the faecal pellets, the mean weight per 24 hours was estimated. Knowing the percentage assimilation of food consumed by the species, an estimate of the food consumption under semi-natural conditions could be calculated. Later, more sophisticated means were used, by filming the feeding activity in a tray containing a known number of predators and prey organisms and placed outside for 24 hours.

Some authors (Smalley, 1960; Klekowski *et al.*, 1967; Prus, 1971) have not measured consumption or assimilation rates as such but determined the production per 24 hours by gravimetric methods which summed with respiration over 24 hours give an estimate of assimilation. If the faecal egestion per 24 hours is also measurable (as in Smalley's grasshoppers or in Prus' *Asellus*) an estimate of consumption can also be obtained. For this method of summing P and R, all forms of production should be measured, that is, body growth plus moulted material plus released sexual and other products during a 24-hour period and their calorific value determined; the study on *Tribolium* by Klekowski, Prus and Zyromska-Rudzka (1967) is a convenient example. The oxygen consumption per 24 hours can be converted into calories by the oxycalorific coefficient (see pp. 223, 266) or by some other calorific equivalents when the nature of the metabolised substrate is known. Prus (1971) working on *Asellus aquaticus* measured simultaneously C, F, P and R rates and was able to compare C estimated gravimetrically from the difference between food offered and food not eaten and C calculated from summed $P + R + F$. The latter method produced values of C which were 30% lower than those from the former method.

Winberg (1956) developed the idea of using a 'balanced equation' to estimate the food consumption or 'ration' for fish under natural conditions. He proposed that assimilation of food energy be estimated from the sum of the energy equivalent of their growth in nature and some multiple (twice for fish) of the routine metabolism as determined in the laboratory. Then, assuming that for fish assimilated food energy is about 80% of the food consumed and knowing approximately the energy losses through nitrogenous waste (according to Winberg about 3% of the consumed energy), he suggests that the food consumption (C) under natural conditions can be estimated from the following equation: $C = 1{\cdot}25 (2R + P)$. It is necessary to check, as did Mann (1965), these basic assumptions before applying such an equation to natural populations. Moreover, for a terrestrial species, new assumptions based upon laboratory experimentation would need to be made in order to find a more or less realistic balanced equation. Particularly necessary is study of the relationship between metabolic rates measured under laboratory conditions and field metabolism; this question is considered in a later section (pp. 109–114).

7A.3.b. Marker Methods

I. Radioactive isotopes as markers.

Radioactive isotopes of elements are unstable and disintegrate into other non-active isotopes, emitting a radiation at a rate characteristic for each isotope which is expressed by its half life, that is the time taken to disintegrate 50% of the radioactive substance. For most ecological studies, only those isotopes with beta and gamma radiations are generally employed; in assimilation studies, only those isotopes which are readily incorporated into the tissues can be used, such as carbon (C^{14}), phosphorus (P^{32}), calcium (Ca^{45}) or sulphur (S^{35}). P^{32} is a convenient isotope with which to work and has often been employed because its radiation is 'hard' enough to be readily detected and measured even in whole not homogenised small animals (see pp. 257–260). Moreover, its very short half life (14·2 days) ensures that any contaminated laboratory or environment is soon clear of radiation. However, it cannot be applied to long term experiments such as are involved, for example, in the production of labelled flour for feeding experiments by growing wheat plants in a radioactive medium. For such work, C^{14} is more convenient with its half life of 5760 years, particularly as it can be incorporated into plants either via gaseous $C^{14}O_2$ or via active bicarbonate.

C^{14} is a useful label because it may be inserted into the actual molecular structure of fats, proteins and carbohydrates and as such acts as an excellent 'marker' or 'indicator'. However, there are problems associated with its use in feeding and assimilation studies as the degree of labelling may differ in different food molecules and moreover, in any one animal, different food substances are often assimilated to a different extent. Thus, even if it were possible to label organisms offered as food with equal 'burdens' of radioactivity, the calorific equivalent per cpm (count per minute) of the assimilated food components is more than likely to be different. It is therefore necessary to be cautious in interpreting results obtained from radioisotopes in feeding experiments. This source of error is likely to appear in any method involving the assimilation of the C^{14} label.

What is important in choosing the isotope to employ is the relationship between its 'biological half life', that is, the time required to eliminate 50% of an ingested dose of radio-isotope, and the duration of the ingestion-digestion-egestion cycle. Reviews of methods of applying radio-isotopes to nutritional studies can be found in Sorokin (1966, 1968) Southwood (1966, chap. 14).

II. Markers other than radioactive isotopes.

The methods of an experimental design described earlier for estimating consumption and/or assimilation rates may also be used with the help of markers other than weight (gravimetric techniques) or radioactive isotopes. The most useful markers, however, need to possess certain properties, namely, that they can be readily measured quantitatively in the food or faeces, that they are neither toxic nor affect the animal's feeding behaviour and that either they are completely inert and not assimilated in the gut or that they are excreted at the same rate as ingested.

It is preferable that such markers should be natural ones already present in the normal food; with such natural markers estimates of ingestion and assimilation of field animals may be undertaken. Thus, Moss (1967) utilised the percentage concentration of magnesium in the heather plants eaten by the red grouse as well as in the bird droppings, both collected from birds caged singly on the heather moor. A laboratory test using artificial food confirmed the assumption that the magnesium was excreted at the same rate as it was eaten and the following formula was applied to estimate metabolisable energy:

metabolisable energy = energy in food in Kcal/g — energy in droppings

in Kcal/g. $\dfrac{\% \text{Mg in food}}{\% \text{Mg in droppings}}$ that is, the ratio of the proportions of magnesium in the food and droppings was used as an estimate of the percentage assimilation of food consumed.

Conover (1966) assumed that only the organic matter and not the inorganic material of food is affected by the digestive process in the gut of animals and measured the proportion of organic matter in the food being eaten (F, using Conover's symbols) and in the faeces egested (E). He applied these two parameters to the following formula

% Utilisation of food (U)
(assimilation of food consumed)

$$U = \frac{F' - E'}{(1 - E') \cdot F'} \cdot 100$$

where $F' = \dfrac{\text{ash free dry weight in food}}{\text{total dry weight in food}}$

and $E' = \dfrac{\text{ash free dry weight in faeces}}{\text{total dry weight in faeces}}$

Conover applied this method to the food and faeces of zooplanktonic animals collected in the field but clearly it is generally applicable where representative samples of the food eaten and the faeces produced may be collected either in the field or laboratory. Moreover, it is convenient that these samples need not be quantitative if only the assimilation efficiency of the food consumed is required. However, where the faeces can be collected quantitatively in the field, a field consumption rate may be obtained together with the possibility of identification of the food species eaten from faecal analysis.

More frequently, artificial indicators have been added to the food offered under experimental conditions and then,

$$\%\text{assimilation of food consumed} = (1 - \frac{\text{conc. indicator in food per unit wt.)}}{\text{conc. indicator in faeces per unit wt.}}$$

Chromic acid, cellulose, lignin, chromogens and silica have been used as indicators in this way in studies on vertebrate feeding (Corbett *et al.*, 1960; Golley, 1967; van Dyne *et al.*, 1964).

McGinnis and Kasting (1964a, b) describe how chromic oxide paper may be used in insects as an indicator to measure assimilation from finely divided food, provided that it is well mixed. The main problems with such artificial indicators are that some may be digested or altered in some way during digestion and so are difficult to determine quantitatively. Moreover, there is always uncertainty about how the presence of such an artificial additive affects feeding.

All these methods of determining the percentage assimilation of food consumed assume that the calorific values of food and faeces are the same, an assumption which is unlikely to be true. The greater the difference between these two calorific values, the more erroneous will be the assessment of assimilation efficiency.

Ivlev (1939b) used black powdered platinum as a marker in order to estimate the consumption of the deposit-feeding oligochaete, *Tubifex tubifex*. He added the platinum to the mud in which the worm was living and the worms together with mud-platinum mixture were kept in a container suspended in water so that the faeces were collected in a dish below. The experiment lasted 24 days. He determined the calorific equivalent of one mg of platinum in the food-mud (cal/mg. platinum) and the total weight of platinum in the faeces collected over the 24 days. Then, assuming that the inert platinum powder was not assimilated or changed chemically in the gut, he calculated the consumption for 24 days as $C =$ (cals/mg. platinum in food). total mg. platinum in total faeces . He was not particularly interested in the percentage assimilation of food consumed but did determine it from the total weight of faeces produced during 24 days multiplied by the calorific value of one mg of

faecal material.; then $\frac{A}{C}\% = \frac{C - F}{C} \cdot 100$. Platinum seems to be such an

excellent artificial marker compared with the others mentioned above that it is the one quoted in our table of methods under method M4.

References

See Chapter 2

Formulae for methods presented in Table 7A.2.

Table 7A.2 consists of table of methods arranged under 4 columns (two gravimetric methods plus two marker methods), each column further subdivided into 4 treatments (i.e. measurement of f_e, $(f - f_n)$, $(w_t - w_o)$ and an inert marker in faeces).

W1b: $\quad C = \dfrac{f_e \cdot C_{gf}}{t \cdot N} \quad ; \quad A = \dfrac{f_e \cdot C_{gf}}{t \cdot N} - \dfrac{F \cdot C_{gF}}{t \cdot N}$

W2b: $\quad C = \dfrac{f - f_n}{t \cdot N} \quad ; \quad A = \dfrac{f - f_n}{t \cdot N} - \dfrac{F \cdot C_{gF}}{t \cdot N}$

W3a: $\quad C = \dfrac{(w_t - w_o) \cdot C_{gf} \pm L}{t \cdot N} \quad ; \quad A = P + R + U$

W3b: $\quad C = \dfrac{(w_t - w_o) \cdot C_{gf} + F' \cdot C_{gF} \pm L}{t \cdot N} \quad ; \quad A = C - \dfrac{(F' + F'') \cdot C_{gF}}{t \cdot N}$

W4b: $\quad C = \dfrac{M_f \cdot C_{mf}}{t \cdot N} \quad ; \quad A = C - \dfrac{F \cdot C_{gF}}{t \cdot N}$ or Conover's method

N1a: $\quad C = \dfrac{N_{fe} \cdot C_{nf}}{t \cdot N} = \dfrac{N_{fe} \cdot w_f \cdot C_{gf}}{t \cdot N} \quad ; \quad A = P + R + U$

N1b: $\quad C = \dfrac{N_{fe} \cdot C_{nf}}{t \cdot N} = \dfrac{N_{fe} \cdot w_f \cdot C_{gf}}{t \cdot N} \quad ; \quad A = C - \dfrac{F \cdot C_{gF}}{t \cdot N}$

N2a: $\quad C = \dfrac{(N_f - N_{fn}) \cdot C_{nf}}{t \cdot N} = \dfrac{(N_f - N_{fn}) \cdot w_f \cdot C_{gf}}{t \cdot N}$

$$= \frac{\text{total area} \cdot (n_f - n_{fn}) \cdot w_f \cdot C_{gf}}{t \cdot N}$$

$$A = P + R + U$$

N2b C = as in method N2a ; $A = C - \dfrac{F \cdot C_{gF}}{t \cdot N}$

R2a: $C = \dfrac{TA\,V\,(r_f - r_{fn}) \cdot C_{rf}}{t \cdot N}$; $A = P + R + U$

R2b: $C = \dfrac{TA\,V\,(r_f - r_{fn}) \cdot C_{rf}}{t \cdot N}$; $A = C - \dfrac{(R_{Fa} + R_{Fb}) \cdot C_{rF}}{t \cdot N}$

R3a: $C = \dfrac{(R_{ft} - R_{fo}) \cdot C_{rf}}{t \cdot N} \pm \dfrac{R_L \cdot C_{rL}}{t \cdot N}$; $A = P + R + U$

R3b: $C = \dfrac{(R_{ft} - R_{fo}) \cdot C_{rf} + (R'_{Fa} + R'_{Fb}) \cdot C_{rF} \pm R'_L \cdot C_{rL}}{t \cdot N}$

$$A = C - \frac{(R'_{Fa} + R'_{Fb} + R''_{Fa} + R''_{Fb}) \cdot C_{rF} + (R'_L + R''_L) \cdot C_{rL}}{t \cdot N}$$

M4a $C = \dfrac{M_f \cdot C_{mf}}{t \cdot N}$; $A = P + R + U$

M4b $C = \dfrac{M_F \cdot C_{mf}}{t_F \cdot N}$; $A = C - \dfrac{F \cdot C_{gF}}{t_F \cdot N}$

W1b: Supply some food (e.g. a leaf) for time t; measure the area eaten, cut out a similar area from another leaf and weigh it. Determine the dry to fresh weight ratio and convert fresh weight of area eaten to dry weight (f_e). Determine calorific value of food (C_{gf}). Thus

$C = \dfrac{f_e \cdot C_{gf}}{t \cdot N}$ cal. ind. hr. Collect faeces (F) after time t, dry and

weigh. Determine the calorific value of faeces (C_{gF}). Thus,

$A = \dfrac{f_e \cdot C_{gf}}{t \cdot N} - \dfrac{F \cdot C_{gF}}{t \cdot N}$ cal. ind. hr. Gyllenberg (1969); Soo-Hoo and

Fraenkel (1966b).

W2b: Supply known weight of food (f-as fresh weight); collect uneaten food after time t, dry and weigh (f_n). Determine the dry to fresh weight ratio of food and convert f to dry weight. Determine the calorific value of food (C_{gf}).

Thus, $C = \dfrac{f - f_n \cdot C_{gf}}{t \cdot N}$ cal. ind. hr.

To measure A gravimetrically, it is necessary to know the weight of faeces (F) coming from a known weight of food eaten (f_e), using some coloured food marker (as in Phillipson 1960a) or to determine and use the time t_F (see key of symbols). Collect faeces (F), dry and weigh. Determine the calorific value of faeces (C_{gF}).

Thus, $A = \dfrac{(f - f_n) \cdot C_{gf}}{t \cdot N} - \dfrac{F \cdot C_{gF}}{t \cdot N}$ cal. ind. hr.

Phillipson (1960a, b); Varley (1967); Klekowski and Stepien (in press)

W3a: Determine t_d (see key to symbols). Starve animal until gut empty, weigh animal (w_o). Feed animal until time t and weigh (w_t). Determine the dry to fresh weight ratio for food and convert ($w_t - w_o$) to dry weight. Determine the calorific value of food (C_{gf}). Estimate L and convert to calories unless small enough to ignore.

Thus, $C = \dfrac{(w_t - w_o) \cdot C_{gf} \pm L}{t \cdot N}$ cal. ind. hr.

Because t is shorter than t_d, F cannot be collected and A can only be determined from A = P + R + U.

W3b: Determine t_d. Weigh the animal (w_o) irrespective if gut full or empty. Feed the animal until time t and weigh (w_t). Determine the dry to fresh weight ratio of food and convert ($w_t - w_o$) to dry weight. Collect any faeces produced during time t, dry and weigh (F). Determine the calorific values of food (C_{gf}) and (C_{gF}). Estimate L and convert to calories, unless small enough to ignore.

Thus, $C = \dfrac{(w_t - w_o) \cdot C_{gf} + F' \cdot C_{gF} \pm L}{t \cdot N.}$ cal. ind. hr.

Then, stop feeding the animal after time t but continue to collect all the faeces produced after time t; dry and weigh (F' ').

Thus, $A = \dfrac{(w_t - w_o) \cdot C_{gf} + F' \cdot C_{gF} \pm L}{t \cdot N} - \dfrac{(F' + F' ') \cdot C_{gF}}{t \cdot N}$ cal. ind. hr.

Petal (1967)

W4b: Determine t_d. Determine the calorific equivalent of one mg of ash content in the food (C_{mf})—cal/mg.ash). Feed the animal on food for time t_F (see key of symbols); collect all the faeces produced during time t_d and determine the total ash content (M_F mg ash) in these faeces.

Thus, $C = \dfrac{M_F \cdot C_{mf}}{t \cdot N}$ cal.ind.hr.

Dry and weigh some more faeces produced during time t_d (F). Determine the calorific value of faeces (C_{gF}).

Thus, $A = \dfrac{M_F \cdot C_{mf}}{t \cdot N} - \dfrac{F \cdot C_{gF}}{t \cdot N}$ cal. ind. hr.

An alternative method which assesses assimilation efficiency is that suggested by Conover (1966). Determine the ash-free dry weight: total dry weight ratios for ingested food (f') and for faeces (F'). The percentage assimilation of consumed food is given by:

$$\frac{A \cdot 100}{C} = \frac{f' - F'}{(1 - F') \cdot f'} \cdot 100$$

This method is basically similar to the method described in M4 for assessing assimilation efficiency and carries the same erroneous assumption that the calorific values of food (C_{gf}) and faeces (C_{gF}) are the same. It is probably better to determine the calorific equivalents of food (C_{mf}, cal/mg ash) and faeces (C_{mF}, cal/mg ash) and apply to the following:

$$\frac{A \cdot 100}{C} = \frac{C_{mf} - C_{mF}}{C_{mf}} \cdot 100$$

The method described above involves using ash as a food marker and assumes that ash is not assimilated in the gut; this is certainly unlikely to be true since animals must have a supply of various ions, especially Na, K, Ca and others. However, the proportion of assimilable ash is probably rather small and the above method may be used to assess approximately C.

N1a: Supply a number of food items (individuals or made-up food pellets) of uniform weight (w_f) or uniform calorific content (C_{nf}). By direct observation, count the number of items eaten (N_{fe}) during time t.

Thus, $C = \dfrac{N_{fe}\cdot C_{nf}}{t\cdot N}$ OR $\dfrac{N_{fe}\cdot w_f\cdot C_{gf}}{t\cdot N}$ cal. ind. hr.

Because t is shorter than t_d, F cannot be collected and A can only be determined from $A = P + R + U$.

N1b: Supply a number of food items (individuals or made-up food pellets) of uniform weight (w_f) or uniform calorific content (C_{nf}). By direct observation, count the number of items eaten (N_{fe}) during time t_d.

Thus, $C = \dfrac{N_{fe}\cdot C_{nf}}{t\cdot N}$ OR $\dfrac{N_{fe}\cdot w_f\cdot C_{gf}}{t\cdot N}$ cal. ind. hr.

Collect faeces after time t_d, dry and weigh (F). Determine the calorific value of faeces (C_{gF}).

Thus, $A = \dfrac{N_{fe}\cdot C_{nf}}{t\cdot N}$ $\dfrac{F\cdot C_{gF}}{t\cdot N}$ cal. ind. hr.

N2a: Supply a known number (N_f) of food items (individual organisms or made-up pellets) of uniform weight (w_f) or calorific content (C_{nf}). After time t, count the number of food items uneaten (N_{fn}); this method cannot be used if food items are partially eaten.

Thus, $C = \dfrac{(N_f - N_{fn})\cdot w_f\cdot C_{gf}}{t\cdot N}$ or $\dfrac{(N_f - N_{fn})\cdot C_{nf}}{t\cdot N}$

or total area $\dfrac{\cdot(n_f - n_{fn})\cdot w_f\cdot C_{gf}}{t\cdot N}$ cal. ind. hr.

where n_f and n_{fu} represent the concentration of food items per unit area (e.g. number per metre square) and the total area is the area being investigated where feeding occurs.

N2b: Determine C as in *N2a*; t is longer than t_d. Collect faeces after time t_d, dry and weigh (F). Determine the calorific value of faeces (C_{gF}).

Thus, $A = C - \dfrac{F \cdot C_{gF}}{t \cdot N}$ cal. ind. hr.

R2a: Determine t_d. Determine the specific radioactivity of food (C_{rf}) Place animals in a known volume (V) or known area (TA) of medium of known concentration of food (r_f). After time t, remove animals, determine the concentration of food not eaten (r_{fn}) and check C_{rf}.

Thus, $C = V. \dfrac{(r_f - r_{fn}) \cdot C_{rf}}{t \cdot N}$ or $TA. \dfrac{(r_f - r_{fn}) \cdot C_{rf}}{t \cdot N}$ cal. ind. hr.

If C_{rf} changes during time t ,use an average value. It may be possible to employ this method in the field with predators and their marked prey.
A can be determined only from $A = P + R + U$.

R2b: For this method, it is useful but not essential to evaluate t_d. Determine the specific radioactivity of the food (C_{rf}). Place the animals in a known volume (V) or known area (TA) of medium of known concentration (r_f). After time t, remove animals, determine the concentration of food not eaten (r_{fn}), making sure that solid faeces have been separated and that dissolved faeces do not contaminate f_n.

Then, $C = V. \dfrac{(r_f - r_{fn}) \cdot C_{rf}}{t \cdot N}$ or $TA .\dfrac{(r_f - r_{fn}) \cdot C_{rf}}{t \cdot N}$ cal. ind. hr.

It may be possible to employ this method in the field with predators and their marked prey.
In order to measure A, it is necessary to determine the total radioactivity of the solid and dissolved faeces (for aquatic animals) $(R_{Fa} + R_{Fb})$ and also the sum of other losses (R_L) which will include

respiratory, urinary and other losses. The techniques involved are described for aquatic animals in Sorokin (1968).

Thus, $A = C - \dfrac{(R_{Fa} + R_{Fb}) \cdot C_{rF}}{t \cdot N}$ cal. ind. hr.

R3a: Determine t_d. Determine the specific radioactivity of food (C_{rf}). Feed the animals on radioactive food until time t. Remove animals, wash and measure the body radioactivity (where $w_t = RC_t$); w_o is represented by the background level of radioactivity (R_o). Assess whether losses of radioactivity due to respiratory losses and urinary products (R_L) are appreciable.

Thus, $C = \dfrac{(R_{ft} - R_o) \cdot C_{rf} \pm R_L \cdot C_{rL}}{t \cdot N}$ cal. ind. hr.

A can be determined only from $A = P + R + U$.

R3b: Determine t_d. Determine the specific radioactivity of food (C_{rf}). Feed animals on radioactive food until time t. Remove animals (transferring them to unlabelled food) and determine the total radioactivity of the body (R_{ft}) and background radioactivity (R_o). Determine the radioactivity of solid (R'_{Fa}) and dissolved faeces (R'_{Fb}) in aquatic animals produced during time t. Assess whether losses produced due to respiratory and urinary products (R_L) are appreciable.

$C = \dfrac{(R_{ft} - R_o) \cdot C_{rf} + (R'_{Fa} + R'_{Fb}) \cdot C_{rF} \pm R'_L \cdot C_{rL}}{t \cdot N}$ cal.ind.hr.

To measure A, continue to collect faeces produced by animals in unlabelled food for a period twice t_d (in order to obtain all labelled faeces). Determine the total radioactivity of the solid and dissolved faeces ($R''_{Fa} + R''_{Fb}$). Assess losses in radioactivity due to respiratory and urinary losses (R''_L). Then,

$A = C - \dfrac{(R'_{Fa} + R'_{Fb} + R''_{Fa} + R''_{Fb}) \cdot C_{rF} + (R'_L + R''_L) \cdot C_{rL}}{t \cdot N}$ cal.ind.hr.

Reviews of the application of radioactive isotopes to the study of nutrition can be found in Sorokin 1966, 1968, Rigler 1971, Southwood 1966.

M4a: Determine t_d. Mix food well with powdered platinum as a marker; determine the calorific equivalent of one mg platinum (C_{mf}). Feed the animal on marked food for time t and determine the platinum content of the body (M_f). The main limitation of this method is whether the method of measuring platinum is sensitive enough to determine the amount of platinum taken during time t. Then,

$$C = \frac{M_f \cdot C_{mf}}{t \cdot N} \text{ cal. ind. hr.}$$

Ivlev (1939b)

M4b: Determine t_d. As above in M4a, determine the calorific equivalent of one mg of platinum in the food (C_{mf}). Feed the animals on marked food. Collect faeces produces during time t_F (see key of symbols). Determine the total platinum content of these faeces (M_F). Then,

$$C = \frac{M_F \cdot C_{mf}}{t_F \cdot N} \text{ cal. ind. hr.}$$

To measure A, the following methods are available:

1. Dry and weigh the faeces collected during time t_F (F). Determine the calorific value of faeces (C_{gF}). Then,

$$A = \frac{M_F \cdot C_{mf}}{t_F \cdot N} - \frac{F \cdot C_{gF}}{t_F \cdot N} \text{ cal. ind. hr.}$$

2. Determine the calorific values per mg platinum of food (C_{mf}) and faeces (C_{mF}). The percentage assimilation efficiency can be calculated from the following:

$$\frac{A \cdot 100}{C} = \frac{C_{mf} - C_{mF}}{C_{mf}} \cdot 100$$

Knowing C in terms of cal. ind. hr. A may also be calculated.

3. Determine the concentrations of platinum in food (M_f) and in faeces (M_F). The percentage assimilation efficiency can be calculated from the following:

$$\frac{A \cdot 100}{C} = \frac{1 - M_f \cdot 100}{M_F}$$

Knowing C in terms of cal. ind. hr, A may also be calculated in the same terms. However, this method assumes that the calorific values of food and faeces are the same, which is unlikely to be true, and so provides only an approximation of assimilation efficiency.

4. It is possible also to apply Phillipson's 'direct' method of determining assimilation efficiency using naturally or artificially coloured food which permits the exact weight of faeces produced from a known weight of food eaten.

7B MEASUREMENT OF FOOD INGESTION IN INSECTS WITH RADIOACTIVE ISOTOPES

H. DOMINAS

7B.1. Introduction

The food ingestion rate can be determined by two different isotopic approaches:

1. The radioactive food is given to the animal until its increasing radioactivity reaches a constant value i.r. when the input of the radioactive isotope is equal to its output. At this time the animal is transferred into a medium containing unlabelled food and the rate of decrease of the radio-activity is measured.

It is assumed that this rate (expressed as the biological half-life time) corresponds to the rate of uptake of the radioactive isotope when the animal was fed on radioactive material. The specific radioactivity being known one can calculate the quantity of food consumed during a given period of time (Reichle and Crossley, 1967).

2. The radiocative food is given to the animal for a period of time too short for any radioactive faeces to have been produced. After this period the animal is killed and cleaned carefully of external radioactivity. The radioactivity of the animal is then counted and the food ratio calculated as in the method 1. (Crossley, 1963). The second method was applied in our work on the determination of the food ratio of *Tribolium* (Dominas, Klekowski, Zyromska-Rudzka, in lit). The detailed description of the method given below, as well as the experimental data, were taken from this paper.

7B.2. Equipment

A well equipped radioisotopic laboratory is indispensible to perform such kind of investigations (Schramm, 1965).

The following additional equipment should be available:
1. Thermostats
2. A Waring-blender
3. Usual laboratory glass such as pipettes, beakers, erlenmeyer flasks etc.
4. Sieve for the flour

7B.2.a. Labelling of the flour with orthophosphate ^{32}P.

Tribolium being a relatively small insect (the wet weight of an adult is about 2 mg) the flour must have a high specific radioactivity. Attention should also be paid on how to obtain labelled flour similar in all respects to the natural product (consistence, humidity etc.).

The flour containing labelled orthophosphate is prepared as follows: A 25 g sample of flour is weighed out from 500 g and mixed in a beaker with 50 ml of an aqueous solution of orthophosphate containing 2-3mC. The operation is performed from behind a 0·5 cm thick perspex shield to avoid irradiation. The dry cake is then thoroughly pulverized in a Waring-blender following which the remaining 475g of flour is added and the homogenization is continued until the radioactivity is evenly distributed in the sample. The homogenity of the flour is checked by measuring the radioactivity of two or three 50–100 mg samples.

7B.2.b. Feeding of Tribolium with flour containing orthophosphate ^{32}P

Few hundreds of individuals on a determined stage of development are placed into a jar containing 1 g of flour per individual (independently of the stage of development). The flour should have a relative humidity of 70% and a temperature of 29°C. For this reason the culture of insects is carried out in a thermostat (capacity 50 l) containing water in 2 Petri dishes of a 10 cm diameter. The thermostat is kept behind a perspex shield.

The time during which the insects are kept in the radioactive flour was established experimentally: this is the time after which 50% of the individuals have started to defecate (Dominas, Klekowski and Zyromska-Rudzka, in

press). This time established for each stage of development is in the limits of
2·5 h–3·5 h.

7B.2.c. Cleaning of the insects of the flour

After a period of time (2·5 h–3·5 h) which depends on the stage of development the insects are removed from the flour in the following way: the flour
containing the larvae is sieved and the insects are killed by ether. The sieving
should be done very carefully, with the protection of a mask and a perspex
shield. A monitor should constantly be switched on to check the contamination caused by some scattering of the flour. The insects are rinsed in water
containing some detergent, put into a beaker with the detergent solution,
stirred for few minutes and transferred to another beaker using dense gauze
cloth. This manipulation is repeated 4–5 times using usually about 2 l of
detergent solution and the larvae are finally rinsed with tap water.

The efficiency of the washing is checked in the following two ways:
1. by measuring the radioactivity of samples of the detergent solution used
to wash the insects.
The washing is repeated until the washing solutions do not contain radioactivity.
2. at the same time samples of insects (10 individuals) are counted for
radioactivity. The larvae are considered free of external radioactivity when
their radioactivity remains constant after two consecutive washings.

7B.2.d. Measurement of the radioactivity of the insects.

6–8 samples each containing 20 larvae (in the case of earlier stages of development the number of larvae in the sample should be higher: 100–200 larvae)
are placed on ebonite plates and radioactivity is determined using a Geiger-
Muller counter. There is no necessity for homogenizing the larvae before
measuring their radioactivity: our data show that homogenized or not, the
larvae labeled with ^{32}P give the same quantity of cpm. Each sample is counted
2–3 times and the mean value calculated per one individual.

7B.2.e. Calculation

An example of the calculation is given below: The radioactivity of 100 mg of
flour labeled with ^{32}P was 158,000cpm. 200 9-day larvae were placed into

this flour for 3 h. After washing as described above 6 samples each containing 30 larvae were counted twice for radioactivity. The results were:

1. 1569 cpm   1539 cpm
2. 2229 cpm   2199 cpm
3. 2018 cpm   1988 cpm
4. 2005 cpm   1975 cpm
5. 2408 cpm   2450 cpm

The total of these 12 countings was 23862 and the mean value obtained after dividing by 12 was 1988 cpm. Dividing this value by 30 we got the radioactivity per one individual: 66·2 cpm. The specific radioactivity of the flour being known we can now calculate the quantity of flour contained in the digestive tract of one insect: if the radioactivity of 100 mg of flour was 158,000 cpm one larva has taken up:

$$\frac{100 \times 66 \cdot 2}{158,000} = 0 \cdot 042 \text{ mg of flour}$$

Result: During 3 h one 9-day *Tribolium* larva has taken up 0·042 mg of flour.
Remark: One should remember that the half-life period for ^{32}P is relatively short (about two weeks), therefore the specific radioactivity of the flour should be measured preferably the same day when the experiment with the insects is performed. In other case, a correction for the decay of radioactivity should be introduced using respective tables.

7B.3. Final Remarks

The application of isotopic methods to such studies is subject to some errors:
1. A high dose of radioactivity may exert some influence on the degree of food assimilation, on the rate of food ingestion etc.
2. The mixture of the food with the radioactive compound may differ from the natural food by its chemical and physiochemical properties and its assimilation may not be identical with that of the natural product.
3. There is also a possibility that the animal takes up the food selectively i.e. the flour may be absorbed or excreted with another rate than the radioactive compound (orthophosphate ^{32}P) itself. This kind of error can be avoided by getting food bound chemically with the label (e.g. by getting flour from plants grown on radioactive substrate).

References

DOMINAS H., KLEKOWSKI R.Z. & ZYROMSKA-RUDZKA H. (in press) Food intake by *Tribolium castaneum* measured with radio-isotope P³².

PARSI O.H. & SIKORA A. (1967) Radiotracer analysis of the trophic dynamics of natural isopod populations. In K. Petrusewicz (ed.) *Secondary productivity of terrestrial ecosystems.* 741–771, Warszawa–Krakow, PWN.

REICHLE D.E. & CROSSLEY D.A. JR. (1967) Investigations on heterotrophic productivity in forest insect communities. In K. Petrusewicz (ed.) *Secondary productivity of terrestrial ecosystems.* 563–581, Warszawa–Krakow, PWN.

SCHRAMM W. (1965) *Chemistry and biology laboratories. Design, construction, equipment.* Oxford, Pergamon Press.

8

Bioenergetic Budgets and Balances

8A COMPUTATION OF ENERGY BUDGET USING ALL STAGES OF TRIBOLIUM

T. PRUS

8A.1. Introduction

This outline is to study energy flow at the special level and is based on experience with flour beetle, *Tribolium castaneum*, cI. Similar studies on bioenergetics of terrestrial invertebrates have been carried out e.g., by Phillipson (1960), Smalley (1960), Wiegert (1964, 1965), Healey (1967), Klekowski *et al.* (1967), and others. General reviews of the methods employed for estimation of energy flow in terrestrial invertebrates can be found in MacFadyen (1967), Phillipson (1967), and Varley (1967).

The species studied lives in wheat flour which constitutes both its habitat and food. Its complete development consists of the following stages and their corresponding duration at temperature of 29°C and relative humidity of 75%; egg–4·09 days, larva–17·58 days, pupa–6·45 days, adult–226·75 days (Park *et al.*, 1961). Thus the development from an egg to the adult stage lasts 28·12 days. The females start laying eggs 2–3 days after eclosion, the process is continuous, and there is usually equal proportion of males to females in the population.

The aim of this outline* is to characterize energy flow through an individual of this species, from its birth to the adult stage and also in reproducing adults. This task will be approached in two categories, in terms of instantaneous (daily) energy budget, calculated for each stage separately from the material gathered at 24 hr intervals, and cumulative energy budget describing energy flow through an individual during its whole life history, from the birth to a given stage of development. In addition, a discussion will be given of two indices that describe efficiency of a given organism in converting energy into the form available to other organisms.

The energy budget is understood here as the quantitative analysis of

---

*All numerical data, figures, and general approach (if not specified otherwise) are taken from paper by Klekowski *et al.* (1967).

energy entering an organism, its pathways within the organism, utilization, conversion, and dispersion, acc. to formulae:

$$C = P + R + FU; \; A = P + R,$$

where:

C–energy ingested (consumption), P–energy incorporated in body or reproductive products (production), R–energy used for metabolism (maintenance), FU–energy unassimilated (faeces, urine or egestion), A–energy assimilated (assimilation).

Efficiency of an organism as convertor of energy is described by indices of gross production efficiency (K_1) and net production efficiency (K_2):

$$K_1 = \frac{P}{C} \cdot 100; \quad K_2 = \frac{P}{A} \cdot 100.$$

Energy utilization by an organism is characterized by an index of Assimilation/Consumption. It is this proportion of energy which is assimilated by an organism out of a total of energy ingested. For discussion of these concepts consult Klekowski (1970).

8A.2. Elements of energy budget and their conversion to calories

Since it is author's intention to stress the computation of energy budget rather than the ways of gathering the necessary data and measurements, only a brief account of the latter can be presented here: consumption, measured by radioisotope method—final results in mg flour consumed by individual · 24 hrs (Dominas *et al.*, in press); production—by weighing method —results in mg wet weight of body growth/individual · 24 hrs; maintenance— by respirometric method—results in mg O_2 consumed per individual. 24 hrs; egestion—found by difference, in calories.

To calculate energy budget it is necessary to convert all results listed above into equal, additive units—calories. For this, additional measurements are of a necessity.

In order to convert mgs of wet weight of food consumed (C), and weight of body growth or eggs produced (P) into calories, it is necessary to known dry weight content (by drying method) and calorific value of this products (e.g. by combustion in microcalorimeter). Having these two parameters measured, further calculation of C and P is easily made by multiplying mgs of dry weight by corresponding calorific values of 1 mg dry weight.

Since in this study most of parameters were measured at 48 hr intervals, it was more appropriate to calculate first the subsequent standing crops of individual's body in calories and then find the growth rate (P) by subtraction. Conversions into calories of wet weight of substance consumed or produced and all other data involved in it, are presented in Table 8A.1.

Table 8A.1. Instantaneous production and consumption rates in chosen stages of *T. castaneum* (conversion into calories).

		Wet weight (μg)	dry wt / wet wt (%)	Dry weight (μg)	Calorific value of 1 mg dry wt, (cal)	Energy produced or consumed/ individ./ 24 hrs.
Production	Larva 9 days old Instar V	191·36	29·3	56·07	5·169	0·1458
	Larva 11 days old	364·81	27·0	98·50	5·903	
	Eggs produced per female/24 hrs	656·36	51·3	336·74	5·038	1·696
Consumption	Food consumed per larva 10 days old/24 hrs	300·00*	89·1	267·30	4·225	1·128

*Data after Dominas *et al.* (in press).

To convert ml O_2 consumed into calories, it is of necessity to measure the respiratory quotient (RQ) and use corresponding oxycalorific coefficient. By multiplying the latter by ml O_2 consumed per individual per 24 hrs the daily cost of maintenance is obtained in calories. Table 8A.2 comprises all these values necessary for conversion of ml O_2 consumed into calories.

Table 8A.2. Instantaneous oxygen consumption rate (conversion into calories).

Stage	Oxygen consumption		Respiratory quotient CO_2/O_2	Oxy-calorific coefficient (ml O_2/1 cal)*	Energy metabolized per ind. per 24 hrs /cal
	μl/ind/hr	ml/ind/24 hrs			
Larva V instar	4·1507	0·0996	0·88	4·899	0·4879
Reproducing female	9·0097	0·2162	0·82	4·825	1·0432
Adult male	6·8000	0·1632	0·80	4·801	0·7835
Average adult	7·9048	0·1897	0·81	4·813	0·9131

*After Bladergroen (1955).

Having done all of the above discussed measurements and conversions, one can proceed to computation of the energy budget.

8A.3 Types of energy balance in Tribolium

During the life history of *Tribolium* and of many holometabolic insects, 4 main types of energy balance can be distinguished (Fig. 8A.1). The first one,

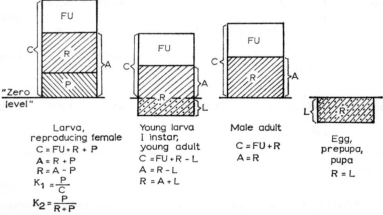

Fig.8A.1. Main types of energy balance during the development of *T. castaneum*, cI. *C*—consumption (food intake), *FU*—egestion (faeces), *R*—cost of maintenance (metabolism), *P*—production (body, exuvia, eggs), *L*—Loss of stored energy (when $R > A$), A—assimilation, K_1—index of gross production efficiency, K_2—index of net production efficiency (after Klekowski *et al.* 1967).

typical for the growing larvae and reproducing females, is characterized by a storage of energy, accumulated either as body growth or the egg production. The second type, found only in the 1st instar larvae and young adults, is characterized by the loss of energy accumulated. The assimilated food is insufficient to counterbalance the cost of maintenance and lacking energy must be taken from the storage in the body. The third type is characteristic only for adult males in which steady state of energy balance can be observed. There is no production (except for reproductive male products which were neglected here), and the whole energy consumed is used for metabolism. The fourth type, similarly as the second one, lasts shortly and is characteristic for non-feeding stages: eggs, prepupae, and pupae. Since there is no energy ingested the organism maintains its living at the expense of energy stored earlier.

Out of these possibilities the energy budgets for larvae and reproducing females (first type) as well as for adult male (third type) will be considered now. The first two are chosen on account of their great significance in the process of biomass production.

8A.4. Instantaneous (daily) energy budgets for chosen stages

Table 8A.3, being compilation of Tables 8A.1 and 8A.2 includes all parameters of daily energy budgets and production efficiency indices which will be discussed below.

Table 8A.3. Instantaneous daily energy budget for chosen stages of *T. castaneum* (cal/ind. x 24 hrs).

Stage	Consumption C	Production P	Maintenance R	Assimilation A	Efficiency indices $K_1 = \dfrac{P}{C}$ %	$K_2 = \dfrac{P}{A}$ %
Larva V instar	1·13*	0·146	0·488	0·634	13	23
Reproducing female	5·9**	1·696	1·043	2·739	29	62
Adult male	1·7**	—	0·783	0·783	—	—
Average adult	3·85**	0·848	0·913	1·761	22	48

*From Table 8A.1.
**Based on assumption that assimilation is ≃ 46% of consumption (after Evans and Goodliffe 1939).

8A.4.a. Larva (instar V, 10 days old)

Daily consumption rate over individual of this age, measured with P^{32}, amounts to 1·128 cal. Out of this quantity, 0·634 (56%) cal is assimilated and the remaining 0·495 cal (44%) is egested as faeces. Of the energy assimilated, 0·146 cal (23%) is incoporated in the body tissues and the remaining 0·488 cal (77%) is lost in the respiration process. Thus of the total energy ingested only 13% (K_1) is retained in the larva, or 23% (K_2) of the energy assimilated. The budget equation is as follows:

1·128 cal (C) = 0·146 cal (P) + 0·488 cal (R) + 0·494 cal (FU)

8A.4.b Adults

Since daily consumption rates measured by radioisotope methods (Dominas *et al.*, in press) were often considerably underestimated it was necessary to calculate the energy budgets for this stage depending on literature data, pertaining to a closely related species, *Tenebrio molitor*. Evans and Goodliffe (1939) have reported that assimilation in this species was 46% of the consumption. Assuming the same percentage of assimilation in *T. castaneum* adults, the following, tentative energy budgets can be calculated.

8A.4.c Reproducing female

Daily consumption rate would be 5.9 cal from which 3.2 cal were egested and 2.7 cal assimilated. Of this latter amount, 1.7 cal (62%) is produced as eggs (there is no increase in body weight and in the calorific value of the adults), and 1 cal (38%) is the cost of maintenance. Thus 29% (K_1) of the energy ingested, or 69% (K_2) of the energy assimilated would be accumulated as egg production.

The equation would read:

5·9 cal (C) = 1·7 cal (P) + 1.0 cal (R) + 3·2 cal (FU)

8A.4.d. Male

In the male of *T. castaneum*, daily food intake would be much lower than that of the female, amounting to about 1·7 cal. Cost of maintenance is 0.8

cal and the remaining energy is egested since there is no perceptible production.

The equation would be as follows:

$$1{\cdot}7 \text{ cal } (C) = 0{\cdot}8 \text{ cal } (R) + 0{\cdot}9 \text{ cal } (FU)$$

8A.4.e. Average reproducing adult

Since males and females in *Tribolium* populations usually occur in equal proportions, one can assume that the average reproducing adult will consume 3·85 cal, assimilate 1·76 cal, produce 0·85 cal as eggs, and use 0·9 cal for maintenance of its life. On account of very high production of eggs by the female, the efficiency indices for an average individual are still high: K_1—22%, K_2—48%.

The equation would be:

$$3{\cdot}85 \text{ cal } (C) = 0{\cdot}85 \text{ cal } (P) + 0{\cdot}9 \text{ cal } (R) + 2{\cdot}1 \text{ cal } (FU).$$

It can be inferred from the above that *Tribolium castaneum* cI strain is an organism which is rather effective in energy transformation.

8A.5. Cumulative energy budget

From the ecological point of view it is also interesting to ascertain cumulative form of the budget and its elements, i.e., the amount of food consumed by an organism from its birth to a given moment of life, as well as the expenditure and fate of this energy. Similarly, cumulative indices of gross and net production efficiency can be calculated. In order to estimate the amount of energy that has flown through the organism to a given moment it is necessary to sum up instantaneous values for each of the elements of energy budget and then calculate the indices.

The cumulative forms of all budget elements (except consumption) plotted against time (development, or age) are presented in Fig. 8.A2. They were calculated for an average individual, sex ignored. The amount of energy which has been assimilated by one individual during its life-time reveals first rapid increase during the last two larval stages due to an extremely high respiration. The second increase in assimilation occurs in adult stage, when females are producing eggs. These increased rates coincide with increased production of body tissue (first one) and of eggs (second one).

Since in *Tribolium* adults all the parameters of the cumulative budget were found to increase in a constant rate (Fig. 8A.2) and the body biomass remained

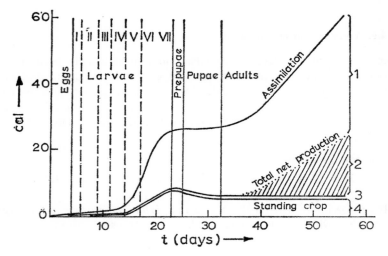

Fig.8A.2. Cumulative energy budget of *T. castaneum*, cI (without consumption). 1—cumulative cost of maintenance, 2—cumulative egg production, 3—cumulative exuvia production, 4—standing crop of body (after Klekowski *et al.* 1967).

constant, it was possible to develop some equations which would allow to estimate the cumulative elements of energy budget for any particular moment of adult life. These equations are presented in Table 8A.3 (Klekowski *et al.*, 1967).

In adults, after 36th day of life, the cumulative production of body remains constant 6·169 cal; egg production from that time on is also constant and its rate is 0·854 cal per average individual per 24 hrs. The cost of maintenance accumulated up to 36th day is 22·944 cal per average individual. Constant is also the rate of respiration: 0·869 cal per average adult per 24 hrs. The introduction of these constants will make it possible to evaluate for a given day the cumulative values of total production, cost of maintenance, assimilation, and indices of efficiency.

The net production efficiency index, K_2, in its cumulative form as presented in Fig. 8A.3, is increasing rapidly during the larval development, decreasing

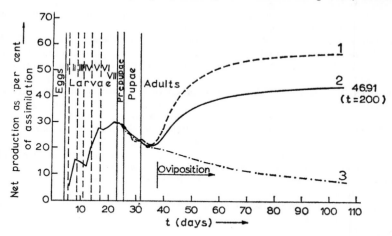

Fig.8A.3. Index of cumulative total net production efficiency of *T. castaneum*, cI. 1—reproducing female, 2—average individual, 3—adult male (after Klekowski *et al.*, 1967).

in the pupal stage and young adults, and is increasing again asymptotically in reproducing adults. This last increase is entirely due to the egg production, since its value for male tends to zero. The average value of this index for reproducing adult, as calculated from equation, will reach the level of 46 per cent, on 200th day of the adult life. The cumulative form of net production efficiency index points to *T. castaneum* as a very efficient transformer of energy, both in the last larval instars, or still more in the reproductive period of life.

References

BLADERGROEN W. (1955) *Einfuhrung in die Energetik und Kinetik biologischer Vorgange.* Wepf and Co., Basel, 368 pp.

DOMINAS H., KLEKOWSKI R.Z. & ZYROMSKA-RUDZKA H. (in press) Food intake by *Tribolium castaneum* measured with radioisotope P^{32}.

EVANS A.C. & GOODLIFFE E.R. (1939) The utilization of food by the larva of the meal-worm *Tenebrio molitor* L. *Proc. R. ent. Soc. Lond.*, Ser. *A.* **14**, 57–62.

HEALEY I.N. (1967) The energy flow through populations of soil Collembola. In Petrusewicz K. (ed.) *Secondary Productivity of Terrestrial Ecosystems.* PWN, Warszawa-Krakow, **2**, 695–708.

KLEKOWSKI R.Z. (1970) Bioenergetic budgets and their application for estimation of production efficiency *.Pol. Arch. Hydrobiol.*, **17**, 55–80.

KLEKOWSKI R.Z., PRUS T. & ZYROMSKA-RUDZKA H. (1967) Elements of energy budget of *Tribolium castaneum* (Hbst) in its development cycle. In (Petrusewicz K. ed.) *Secondary Productivity of Terrestrial Ecosystems.* PWN, Warszawa–Krakow, **2**, 859–879.

MACFADYEN A. (1967) Methods of investigation of productivity of invertebrates in invertebrates in terrestrial ecosystems. In (Petrusewicz K. ed.) *Secondary Productivity of Terrestrial Ecosystems.* PWN, Warszawa–Krakow, **2**, 383–412.

PARK T., MERTZ D.B. & PETRUSEWICZ K. (1961) Generic strains of *Tribolium:* Their primary characteristics. *Physiol. Zool.* **34**, 62–80.

PHILLIPSON J. (1960) The food consumption of different instars of *Mitopus morio* (T.), (Phalangidea) under natural conditions. *J. Anim. Ecol.* **29**, 299–307.

PHILLIPSON J. (1967) Secondary productivity in invertebrates reproducing more once in a lifetime. In: (Petrusewicz K. ed.) *Secondary Productivity of Terrestrial Ecosystems.* PWN, Warszawa–Krakow, **2**, 459–475.

SMALLEY A.E. (1960) Energy flow of a salt marsh grasshopper population. *Ecology*, **41**, 672–677.

VARLEY G.C. (1967) Estimation of secondary production in species with annual life-cycle. In (Petrusewicz K. ed.) *Secondary Productivity of Terrestrial Ecosystems*, PWN, Warszawa–Krakowa, **2**, 447–457.

WIEGART R.G. (1964) Population energetics of meadows spittlebugs (*Philaenus spumarius* L.) as affected by migration and habitat. *Ecol. Monogr.* **34**, 225–241.

WIEGERT R.G. (1965) Energy dynamics of the grasshoppers 'populations in old field and alfalfa ecosystems. *Oikos* **16**, 161–176.

Vertebrate Bioenergetics

9

Calorimetry

9A CALORIMETRY IN ECOLOGICAL STUDIES

A. GÓRECKI

9A.1 The calorie as an ecological unit

Ecological energetics, which is the study of energy transfer and energy transformation from one form to another within ecosystems, uses the calorie or kilogramcalorie, as its basic unit of measurement (Phillipson, 1966). If the calorific value of natural plant and animal material is determined then biomass can be expressed in energy units. Similarly energy budgets and balances for other ecological parameters can be computed in terms of calories. Thus the calorie is becoming a generally accepted unit, allowing a better comparison of energetic processes than numbers or biomass.

From the point of view of energy transfer the usefulness of the calorie, a unit of heat energy, is due to the fact that all forms of energy can be transformed into heat; they cannot be wholly transferred into any other form of energy. However, in the international unit system 'SI' (Système International) which has been adopted by many countries, the calorie has been cancelled and the joule, a unit of mechanical energy, has been introduced instead. The direct link between mechanical energy and heat energy is expressed in Joule's Law, namely, that the equivalent of one calorie in units of mechanical energy is $4 \cdot 187 \times 10^7$ ergs or $4 \cdot 187$ joules; (or alternatively a joule is the equivalent of $2 \cdot 388 \times 10^{-4}$ kcal). In all probability the calorie will continue to be used by dietitians, cooks and ecologists for a very long time before the joule is in common use.

9A.2 Determination of calorific values

There are a large number of tables in which the nutritive values of various plant and animal material are recorded and tabulated. However, the usefulness of these tables to ecological studies is limited. They deal chiefly with human and animal food and only occasionally is data concerning natural ecological materials included. Moreover dietitians, and breeders are mainly interested in the composition of particular foods and not in their total calorific value. They base their studies therefore on recalculations from gross food composition instead of on direct calorimetric determinations. The amount of carbohydrate, fat and protein contained in a given product is multiplied by the appropriate coefficient. Atwater's coefficients used in human dietetics have been rounded off to 4 kcal/g for proteins and carbohydrates, and 9 kcal/g for fats of both plant and animal origin. The coefficients given are different for domestic animals, for example, values of 3·5; 3·4; 7·6 for chicken protein, carbohydrate and fat respectively (Spector, 1956; p. 230). Thus nutrition tables provide only approximate levels for the assimilated part of food which is lower than its total calorific value.

In studies on productivity and energy flow, we are interested in this total calorific value. Sometimes this is later reduced by coefficients of assimilation or digestibility, which may be different for various trophic levels. Therefore, ecologists have to do the necessary calorimetric measurements themselves, as a result the bomb calorimeter is becoming a common tool in a large number of ecological laboratories (Long, 1934; Slobodkin and Richman, 1960; Golley, 1961; Phillipson, 1964; Górecki, 1965a).

Three measures of calorific value are commonly used. The first is the value obtained directly from the combustion of dried material in the calorimeter which gives the total calorific value per unit dry weight. The second is the calorific value of dry weight after subtracting the ash content; this is the ash-free dry weight calorific value and exceeds the former by the percentage of ash in the combusted sample. The third value which is used directly for calculations of productivity is the calorific value of fresh biomass. This value can also be obtained from the above by adding the water content to the value of dry weight. These different values are used in calculations depending on the object in view.

9A.3 Energy values of terrestrial animals

Nowadays more and more calorific values for different plant and animal material can be found in ecological literature. Golley (1959, 1961) made an attempt to collect and tabulate the scattered data. His tables contain the calorific values for 64 plant species and 14 animal species; among the latter only two refer to vertebrates.

It can be assumed that the ash-free calorific values for vertebrates will range between 4·5 and 8·0 kcal/g (Golley, 1967). The highest values will be represented by fat hibernating mammals or birds before a long migration.

Odum, Marshall and Marples (1965) studied the calorific value of body tissues and fat-free tissues of more than 20 bird species. They found that the energy content of birds' bodies before migration was much higher than after migration, the calorific value of non-fat tissues being very much the same in both cases. For non-migrating birds the calorific values were lower than those for birds returning from migration (Table 9A.1).

Table 9A.1. Energy values of some selected terrestrial vertebrates and invertebrates

Species	cal/g dry weight	kcal/g fresh biomass	References
Mammals			
Sorex minutus	5·025†	1·63†	Myrcha, 1969
Sorex araneus	4·555†	1·50†	Górecki, 1965b
Sorex avaneus	4·991†	1·53†	Myrcha, 1969
Apodemus flavicollis	5.002†	1·40†	Górecki ,1965b
Apodemus agrarius	5·179†	1·64†	Górecki, 1965b
Clethrionomys glareolus	4·934†	1·45†	Górecki, 1965b
Microtus arvalis	4,856†	1·47†	Górecki, 1965b
Mus musculus	5·675	1.70	Golley, 1959
Mus musculus	6·205	2·25	Myrcha and Walkowa, 1969
Microtus pennsylvanicus	4.650	1·40	Golley, 1959
Lepus europaeus	7·500	1·87	Myrcha, 1967
Oryzomys palustris	5·840	1·90	Davis and Golley, 1963
Capreolus capreolus	6·065	2·16	Weiner, 1973
Sigmodon hispidus	5·355†	—	Golley, 1970
Birds			
Parula americana before migration	7·680	—	Odum *et al.*, 1965
Parula americana after migration	6·075	—	Odum *et al.*, 1965
Piranga olivacea	7·575	—	Odum *et al.*, 1965
Piranga olivacea after migration	6·645	—	Odum *et al.*, 1965
Hylocichla mustellina before migration	7·725	—	Odum *et a.l*, 1965
Hylocichla mustellina after migration	5·990	—	Odum *et al.*, 1965
Zenoidura macroura	—	2·61	Brisbin, 1968
Passer domesticus	5·535	—	Odum *et al.*, 1965

Table 9A.1. continued

Species	cal/g dry weight	kcal/g fresh biomass	References
Passer domesticus	5.512	1·91	Górecki, after Pinowski, 1967
Perdix perdix	5.253†	1·90†	Szwykowska, 1969
Cyanocitta cristata	5·090	—	Odum *et al.*, 1965
Richmodena cardinalis	5·530	—	Odum *et al.*, 1965
Telmatodytes palustris	6·170*	—	Kale, 1965
Amphibians			
Bufo bufo	5·003†	1·25†	Mazur, 1967
Rana arvalis	4·628†	1·16†	Mazur, 1967
Arthropods			
Crematogaster clara	6.130*	—	Kale, 1965
Chaeptopsis sp.‡	5.520*	—	Kale, 1965
Trigonotylus sp.	5·300*	—	Kale, 1965
Prokelisia marginata	5·380*	—	Kale, 1965
Orchelinum fidicinum	5·432	—	Golley, 1959
Philaenus spumarius	5·807	—	Wiegert, 1965
Schistocerca sp.	5·363	—	Golley, 1959
Tenebrio molitor	6·314*	—	Slobodkin and Richman, 1961
Tribolium castaneum	6·000	—	Klekowski *et al.*, 1967
Myrmica sp.	6·309	1·92	Pętal, 1967
Arachnidea‡	5·820	—	Gibb, 1957
Arachnidea‡	5·620*	—	Kale, 1965
Tyroglyphus lintneri	5·808*	—	Slobodkin and Richman, 1961

*ash-free caloric value
†annual mean
‡average values of some (2—7) species

The body calorific value of non-hibernating mammals also changes throughout the year. The problem was studied using small rodents and shrews of the European temperate zone as examples (Górecki, 1965). Seasonal changes in voles and mice as a group appear to be much greater than differences between particular species. The average increase in the energetic value of mice and voles was from 4·5 at the end of winter to 5·2 in late summer. The annual mean values for five different species together ranged from 4·6 to 5·2 kcal/g (Table 9A.1). The changing calorific values for some amphibians (frogs and toads) from spring to autumn were determined by Mazur (1967).

In Table 9A.1 there are also listed the calorific values for some representative terrestrial invertebrates taken from numerous papers (e.g. Golley, 1959; Klekowski *et al.*, 1967; Connell after Golley, 1961; Kale, 1965). The list does not include all the available calorimetric determinations which are now available in the literature. Cummins *et al.* (1971) has recently produced a very comprehensive list of calorific values in a wide variety of organisms.

9A.4 Body composition and calorific value

The calorific value of an animal is connected with its gross body composition. It depends, to a considerable degree, on the fat content and, to a lesser extent on the mineral content. The animal fat has a high calorific value and usually exceeds 9.0 kcal/g (Odum *et al.*, l.c.; Mazur, l.c., Myrcha, 1967; Sawicka-Kapusta, 1968). The calorific value of non-fat dry weight, that is protein and carbohydrate, is normally much lower ranging from 4–5 kcal/g (Odum *et al.*, l.c; Sawicka-Kapusta, l.c.). Therefore, even a small change in fatness may be responsible for some significant changes in the calorific value of the total body (see Chapter 3). For that reason some ecologists determine the fat content with the help of the Soxhlet apparatus (see Chapter 9C), instead of determining the total calorific value. Later the energetic value of the whole body is calculated from the proportion of fat to fat-free body (Kale, 1965; Sawicka-Kapusta, 1968).

The majority of ecologists, however, prefer to use direct calorimetric determinations to eliminate the time-consuming extraction and calculation. The fatness of an animal is a good indicator of its physiological condition and accordingly both methods can be recommended in ecological studies.

The mineral content of animals is closely connected with their skeleton, exoskeletons being comparatively larger than internal ones. It may also change with age and with the different developmental stages of the animal (Górecki, 1965b). The total calorific values for animals with heavy skeletons may be highly diverse, yet, when expressed in terms of the ash-free dry weight, become very similar to each other (Slobodkin and Richman, 1961). If the calorific values of the ash-free dry weight are to be used, however, in determinations of the energy flow then the ash content has to be known.

References

BRISBIN I.L. (1968) A determination of the caloric density and major body components of large birds. *Ecology*, **49**, 792–794.

CUMMINS, K.W. & WAYCHECK, J.E. (1971) Caloric equivalents for investigations in ecological energetics. *Mitt. int. Ver. Limnol.* **18**, 1–158.

DAVIS D.E. & GOLLEY F.B. (1963) *Principles in mammalogy*. Reinh. publ. Corp. 335 pp. London, New York.

GIBB J. (1957) Food requirements and other observations on captive tits. *Bird study*, **4**, 207–215

GOLLEY F.B. (1959) Table of caloric equivalents. *Mimeogr. Univ. Georgia.* 7 pp.

GOLLEY F.B. (1961) Energy values of ecological materials. *Ecology*, **42**, 581–584.

GOLLEY F.B. (1967) Methods of measuring secondary productivity in terrestrial vertebrate populations. In Petrusewicz K. (ed.) *Secondary Productivity of Terrestrial Ecosystems.* 99–124 pp. Warszawa, Kraków.

GÓRECKI A. (1965a) The bomb calorimeter in ecological research. *Ekol. pol. "B"*, **11**, 145–158. (In Polish with English summ.).

GÓRECKI A. (1965b) Energy values of body in small mammals. *Acta theriol.* **10**, 333–352.

KALE H.W. (1965) Ecology and bioenergetics of the long- billed marsh wren in Georgia Salt Marshes. *Publ. Nutt. Ornithol. Club.* 142 pp. Cambridge, Massachusettes.

KLEKOWSKI R.Z., PRUS T. & ŻYROMSKA-RUDZKA H. (1967) Elements of energy budget of *Tribolium castaneum* (Hbst.) in its developmental cycle. In Petrusewicz K. (ed.) *Secondary Productivity of Terrestrial Ecosystems.* 859–894 pp. Warszawa, Kraków.

LONG F.L. (1934) Application of calorimetric methods to ecological research. *Plant Physiol.* **9**, 323–337.

MAZUR T. (1967) Seasonal variations in the energy reserves of *Bufo bufo* (L.) and *Rana arvalis* Nilss. (Anura) in Poland. *Ekol. pol. "S.A."*, **15**, 607–613.

MYRCHA A. (1967) Caloric value of the body and internal organs of *Lepus europaeus*. *Small Mammals Newslett.* **3**, 10.

MYRCHA A. (1969) Seasonal changes in caloric value, body water and fat in some shrews. *Acta theriol.* **14**, 211–227.

MYRCHA A. & WALKOWA W. (1968) Changes in caloric value of the body during the postnatal development of white mice. *Acta theriol.* **13**, 391–400.

ODUM E.P., MARSHALL G.S. & MARPLES T.G. (1965) The caloric content of migrating birds. *Ecology.* **46**, 901–904.

PĘTAL J. (1967) Productivity and the consumption of food in the *Myrmica laevinodis* Nyl. population. In Petrusewicz K. (ed.) *Secondary Productivity of Terrestrial Ecosystems.* 841–855 pp. Warszawa, Kraków.

PHILLIPSON J. (1964) A miniature bomb calorimeter for small biological samples. *Oikos.* **15**, 130–139.

PHILLIPSON J. (1966) *Ecological energetics*. E. Arnold Ltd. 57 pp. London.

PINOWSKI J. (1967) Estimation of the biomass produced by a tree sparrow (*Passer m. montanus* L.) population during the breeding season. In Petrusewicz K. (ed.) *Secondary Productivity of Terrestrial Ecosystems.* 357–368 pp. Warszawa, Kraków.

SAWICKA-KAPUSTA K. (1968) Annual fat cycle of field mice, *Apodemus flavicollis* (Melchior, 1834). *Acta theriol.* **13**, 329–339.

SLOBODKIN L.B. & RICHMAN S. (1960) The availability of miniature bomb calorimeter for ecology. *Ecology.* **41**, 784–785.

SLOBODKIN L.B. & RICHMAN S. (1961) Calories/gm. in species of animals. *Nature.* **191**, 299.

SPECTOR W.S. ed. (1956) *Handbook of biological data.* Saunders Comp. 584 pp. Philadelphia, London.

Szwykowska M.M. (1969) Seasonal changes of the caloric value and chemical composition of the body of the partridge (*Perdix perdix* L.). *Ekol. pol.* "S.A.". **17**, 795–809.

WEINER J. (1973) Dressing percentage, gross body composition and caloric value of the roe deer.

WIEGERT R.G. (1965) Intraspecific variation in calories/g of meadow spittlebugs (*Philaneus spumarius* L.). *BioScience.* **15**, 543–545.

9B   THE ADIABATIC BOMB CALORIMETER

A. GÓRECKI

The calorimeter has been a traditional tool in physical, chemical and nutrition laboratories for a long time. Recently, more and more ecologists have begun to employ it in their studies of productivity and the calorie (cal or kcal) has become a common and very useful unit of measurement. Now biomass and production can also be expressed in terms of energy.

9B.1 Apparatus and procedure

A calorimeter measures the heat of combustion (net caloric value) produced during the burning of a definite amount of fuel in a steel calorimetric bomb. The products of combustion then are carbon dioxide, liquified water, sulphuric acid and nitrogen. Such a determination of heat of combustion was introduced by Berthelot—and today a variety of calorimeters are in use.

The Polish calorimeter KL-3 is an adiabatic calorimeter of the Berthelot type (Fig. 9B.1). Its main parts are: (a) water calorimeter, (b) calorimeter bucket with top, (c) water jacket with stirrer and thermometer, (d-e) electric stirrer and ignition unit, (f) certified thermometer with the scale covering a range of 15-25°C graduated in 0·002°C divisions. The burning of samples is

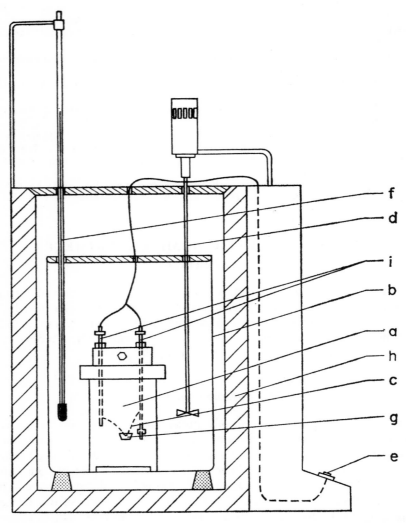

Fig. 9B.1. Cross section of the calorimeter KL-3 type: a—bomb calorimeter; b—calorimeter bucket; c—fuse wire; d—electric stirrer; e—ignition contact; f—calorimeter special thermometer; g—combustion capsule; h—water jacket; i—electrodes.

carried out in an oxygen bomb which is a thick-walled, pressure steel container. The bomb is sealed with a screw-on bomb head in which there are two electrodes (i) as well as oxygen filling connections (inlet and outlet valves, Fig. 9B.1).

Five to ten millilitres of distilled water are poured into the bomb and the sampled material is inserted into a combustion capsule (crucible) made of acid-proof steel. The sample is connected with the electrodes by ignition fuse wire (h). Next the bomb is filled with oxygen under a pressure of 20-25 atm. and the prepared bomb is submerged into the calorimeter which should contain a definite amount of water (approx. 2·5 litres) at a slightly lower (0·5–1·5°) temperature than that in the thermostat jacket. The bomb has to be immersed up to the electrodes. Measurements are started after the temperatures of the water in the bucket and in the jacket have been balanced, that is 10–15 minutes later. When, after a few readings, the temperature remains constant, the ignition contact is switched on and the sample material is ignited. Initially the rise in temperature of the water surrounding the bomb is very rapid, but gradually becomes slower and, in 15–20 minutes, the temperature stabilizes and it will begin to fall slowly after a further 10–15 minutes period. The whole measurement takes about 30 minutes during which time temperatures are read from the thermometer every 30 seconds. The installation of an alarm bell signalling the time of each subsequent thermometer reading considerably facilitates the operation of the calorimeter by one man. The changes of temperature which occur are illustrated in Fig. 9B.2 which gives the combustion of a vole (*Clethrionomys glareolus*) sample (Fig. 9B.2). In the process of combustion of any sample three periods can be distinguished: (1) the introductory period of stabilization of temperature, (2) the central period when the temperature increases and (3) the terminal period when temperature is stabilized at a high level or starts slightly falling (because of radiation of heat).

9B.2 Calculation

Results of the determinations are calculated according to various formulae derived from the heat balance equation:

$$Q = G \times Wg \times (t_n - t_o).$$

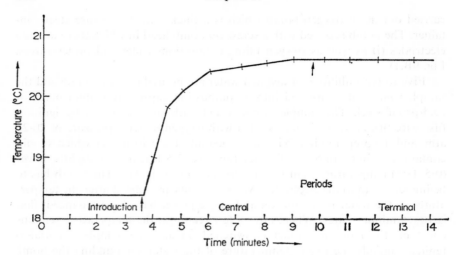

Fig. 9B.2. Temperature readings during the combustion in a calorimeter of vole sample (*Clethrionomys glareolus*).

The following formula is one of them:

$$Wg = \frac{Ww\,(t_n + C - t_o) - \Sigma b}{G}$$

where: Wg is the caloric value of the sample, Ww—is the net-caloric value of the calorimeter system, that is the water-equivalent of the calorimeter; t_n—the final temperature of the central period; t_o—the initial temperature of the central period; C—the correction for radiation; Σb—the sum of the corrections; and G—weight of samples in grams.

Since the caloric value of the calorimeter (Ww) has to be standardized for each calorimeter, a thermochemical standard, that is a substance of a known caloric value, usually benzoic acid (6·324 cal/g) or β naphthol (8·244·5 cal/g) is burned. The water equivalent has to be established every few months for the value Ww is subject to some small changes.

Correction (C) can be calculated from various formulae; the most accurate is the Regnault-Pfaundler formula (Kleiber, 1961).

$$C = n\,\Delta p + \frac{\Delta k - \Delta p}{t_k - t_p}\left(\sum_{2}^{n-1} t + \frac{t_n + t_o}{2} - nt_p\right)$$

CALORIMETRIC DETERMINATION DATA SHEET No...4......

Date Feb. 27'68 Sample .Tissue of Cl. glareolus..... Burning Nr ...2.....

Water eqivalent (Ww) ..3,036.'7.... Temp. of water jacket....19.0...°C

SAMPLE:

Total wt of sample,
combustion capsule
and fuse wire .... 5.900 g

Total wt. of comb caps.
and wire .... 4.363 g

Wt. of fuse wire .... 0.006 g

Wt. of comb. capsule ....4.357...g

Weight of sample ....1.537...g

Wt. of combust.
caps. with ash.................. 4.585 g

Wt. of comb. caps......4.357...g

Wt. of ash ....0.229...g

Wt. of ash-free sample .1.308...g

% of ash ....14.87.....

BURNING: (thermometer readings)

1.	17.67	16.	19.78	31.	20.03	46.	
2.	17.67	17.	19.82	32.	20.03	47.	
3.	17.67	18.	19.87	33.	20.03	48.	
4.	17.67	19.	19.90	34.	20.04	49.	
5.	17.67	20.	19.93	35.	20.04	50.	
6.	17.67	21.	19.95	36.	20.04	51.	
7.	17.67	22.	19.96	37.	20.04	52.	
8.	17.78	23.	19.99	38.	20.04	53.	
9.	18.19	24.	19.99	39.	20.04	54.	
10.	18.70	25.	20.00	40.	20.04	55.	
11.	19.07	26.	20.01	41.	20.04	56.	
12.	19.32	27.	20.01	42.	20.04	57.	
13.	19.49	28.	20.02	43.	20.04	58.	
14.	19.61	29.	20.02	44.		59.	
15.	19.70	30.	20.02	45.		60.	

CALCULATIONS:

Final temperature (Tn) ...20.04.....°C

Initial temperature (To)....17.67.....°C

Difference of temp. (ΔT) .2.37.....°C

Correct. for radiation (C)...0.0......°C

Ww x (ΔT+C) ....7,196.98....cal

Sum of corrections (Σb) ...24.95.... cal

Ww x (ΔT + C) – Σb ....7,172.03....cal

Ww x (ΔT+C) – Σb ....4,666.23.. cal

Wt. of sample (G)

Ww x (ΔT + C) – Σb ......5,483.18...cal

Wt. of ash-free sample

Amount of NaOH used for titra–

tion: ....15.3.....ml x .1.50...

= ....22.95........cal

REMARKS ....

....

....

....

....

The Langbein formula is much simpler, though the results obtained are a little overestimated.

$$C = n \cdot \Delta k + \frac{\Delta p + \Delta k}{2}$$

where: Δk — is the mean decrease in temperature of each reading during the final period; n the number of reading during the central period; t_o—the initial temperature of the central period; t_n—the final temperature of the central period; Δp—the mean increase in temperature of each reading in the initial period; t_p—the mean temperature of the initial period; $\overset{n-1}{\underset{2}{\Sigma}} t$— the sum of temperatures of the central period without the first and the last temperature; t_k—the mean temperature of the final period.

The sum of corrections comprises corrections for the heat of fuse wire combustion and the heat of acid formation.

$$b = b_D + b_N$$

where b_D—is the correction for the heat of fuse wire combustion and b_N—correction for the heat of formation of HNO_3.

The heat of combustion for steel fuse wire with 0·1 diameter is 1·600 cal/g, or for a fuse wire composed of 54% copper, 26% nickel and 20% tin with 0·1–0·08 diameter is 775 cal/g. The heat of formation of acids (HNO_3 and H_2SO_4) is usually recalculated in terms of HNO_3. The acids appear in the bomb as a result of nitrogen oxides and sulphur left after the combustion dissolving in the water. They are titrated using 0·1 N NaOH with phenolphtalein or methyl orange as an indicator. Calculations can be derived from the equation:

$$b_N = a \times 1·43$$

where a—is the amount of NaOH used.

9B.3 Preparation of ecological materials for burning

The preparation and combustion of ecological materials (plant and animal tissue) varies somewhat from the usual treatment of solid combustible materials. Consequently, it presents a number of additional difficulties.

Some methods applicable to animal tissues will be presented using small vertebrates as examples (Golley, 1961; Górecki, 1965a, b; Lieth, 1968). The preparation of a frog, sparrow or mouse for combustion in a calorimeter means drying and grinding the animal tissues. After weighing and labelling, but before drying, the animal is dessicated. Then it can be dried in an ordinary drying oven at high temperature or in a freezer-drier at low temperatures. In both cases the process of drying is considerably shortened by lowering the pressure (vacuum oven or freeze-drying apparatus). The drying time of a mouse in a vacuum oven is three times shorter than in an ordinary oven (Górecki, 1965a), since a 'vapour point' is exceeded at the temperature of 75–80°C. Drying as rapidly as possible is essential because of the progressive decomposition of fats to fatty acids with a different caloric value. Lyophilisation is the best methd for it eliminates the possibility of any changes in the chemical constitution of fats; it requires, however, a freeze-drying apparatus.

After drying the animal to a constant dry weight and careful weighing, the sampled material is ground. A Waring blender was found to be the simplest and sufficiently precise method since these blenders do not crumble the dry tissue but grind it to dust. The ground material is additionally kneaded and divided into a number of samples. The weight of a sample for burning should be 0·5 to 1·5 g; however, to minimize the error of calculations heavier samples of 1·2 to 1·5 g can be used.

With whole animals, tissues are sufficiently compact (fat content) to avoid powdering during combustion therefore they may not need to be formed into pellets.

Animal tissues can be combusted thoroughly and yet still leave a great deal of ash after the process. The ash of small vertebrates consists mainly of $Ca_3(PO_4)$ and after combustion the remaining ashes should be carefully weighed. This is necessary for the calculation of the caloric value of ash-free tissue (Slobodkin and Richman, 1961). In some cases, the ash content can be determined additionally by burning a sample in a muffle oven.

The preparation of plant tissues is somewhat different from the preparation of animal matter. Plants can first be dried in an ordinary laboratory oven at a temperature of approximately 60°C. Then they are ground in a Wiley mill or Waring blender and, in some cases of small samples, in a small Waring blender; the powder is made into pellets in a pellet press. The pellets are then placed in combustion capsules in weighing bottles and re-dried for a few hours at temperatures of 102°C–105°C, to obtain a constant weight. The ashes left after combustion are weighed to 0·001 g. The ash content, used in the determination of the ash-free caloric value, is normally somewhat lower in plant tissues than in animal tissues.

The number of determinations is a vital point since two kinds of errors occur in the process. One is an error of method itself (the number of samples), the other results from the individual variability (the number of animals from which the samples were taken). Usually samples are burned in duplicate or triplicate (Golley, 1961), yet whenever the statistical error of method is insignificant, one sample is sufficient. However, on account of the considerable variability of ecological materials the determination of the caloric value of at least 10 animals or samples is strongly recommended.

References

GOLLEY F.B. (1961) Energy values of ecological materials. *Ecology*, **42**, 581–584.
GÓRECKI A. (1965a) The bomb calorimeter in ecological research. *Ekol. pol.* "S.B.", **11**, 145–158. (In Polish with English summ.).
GÓRECKI A. (1965b) Energy values of body in small mammals. *Acta theriol.* **10**, 333–352.
KLEIBER M. (1961) *The fire of life: an introduction to animal energetics*. J. Wiley Inc., New York, London, 454 pp.
LIETH H. (1968) The measurement of caloric values of biological material and determination of ecological efficiency. In Eckardt F.E. (ed.) *Functioning of terrestrial ecosystems at the primary production level. Proc. Copenhagen Symp.* 233–242 pp. UNESCO, Paris.

9C   FAT EXTRACTION IN THE SOXHLET APPARATUS

K. SAWICKA–KAPUSTA

In the analysis of gross body composition in regard to bioenergetics, a determination of the proportion of fats in relation to other organic compounds is important (Hayward, 1965). The caloric value of fats is twice as high as that of proteins and carbohydrates; in the case of animal fat it is often more than 9·0 kcal/g (Kleiber, 1961). For this reason the average caloric value of the body depends mainly on its fat content (Odum *et al.*, 1965; Mazur, 1967; Sawicka-Kapusta, 1968; see also Chapter 9A). This is important when the net production is expressed in term of energy units. Determination of fat reserves in migrating birds is a reliable method for studying the bioenergetics of their flights. Energy requirements of passerine birds for overnight survival in winter (roosting period) may be also estimated from their fat reserves.

Hence fat extraction was employed in many bird studies (Odum, 1960; Dol'nik & Bljumental, 1964; Evans, 1969).

A widely known, simple and efficient technique for fat extraction utilizes the Soxhlet apparatus. Its description can be found in many handbooks of chemical analyses, and therefore only details concerning the treatment of ecological materials will be discussed here.

9C.1 Procedure

The Soxhlet apparatus consists of an extractor with an extraction thimble inside, a condenser and a flask with the solvent. The apparatus is manufactured in various sizes, the capacity of the flasks usually ranging from 100 to 1000 millilitres. For the extraction of fat from an animal the size of a frog, sparrow or mouse the capacity should be between 250 to 500 ml.

The following organic solvents have been used to extract fat from small mammals and birds: ether, toluene, and benzene as well as mixtures, such as benzene-alcohol 4:1; ethanol-ether 3:1; chloroform-methanol 2:1. Ether has proved to be the best and the most convenient solvent due to its low boiling point 34·6°C. Fat free paper filters are sufficient for ecological purposes.

In preparing a small vertebrate for extraction, the body is cut into pieces and placed in a paper thimble. The thimble with the sample is dried to constant weight, in a vacuum oven at a temperature of $+60°C$; this takes about 48 hours. A freeze-dry lyophilisation apparatus is much quicker. Before extraction the dry sample is placed for 30 min. in ethyl alcohol to denature the proteins to prevent them from being washed out with the fat during the extraction period. Though the size of the thimble matches the size of the extractor, the former should be filled with ether to no more than half its height. The time of extraction depends mainly on the amount of fat in the body. For example, a medium fat mouse has to be extracted for at least 8–10 hours during which the solvent will complete about 25 cycles through the apparatus. The extraction is usually carried out in three series (e.g. $10+10+5$ cycles), the solvent (ether) being replaced after each series. All of the fat will not be removed if the period of extraction is too short. After the extraction the sample is dried at a temperature of 60–70°C and weighed. Extracted fat and the fat-free body left after the extraction can be burned in a bomb calorimeter (see Chapter 9B).

U

9C.2 Calculation

As a result of extraction an ether extract sometimes called 'crude-fat' is obtained. The amount of fat is usually expressed as a percentage in proportion to the dry weight of the animal. The difference between the amount of fat and the dry weight of the body before extraction gives fat-free body weight which can be used for metabolic studies. Fat content in the body can also be expressed in the form of other indices (Jameson & Mead, 1964; Odum *et al.*, 1965).

Energy values of the whole body can be accurately calculated if caloric values of both fat and fat-free body are determined in a bomb calorimeter. In this case, fat content plus fat-free body weight have to be multiplied simply by their caloric values. If these caloric values are not known, an equivalent of about 9·2 kcal/g of fat and 4·7 kcal/g of fat-free tissues can be adopted in order to obtain rather crude estimation of caloric content of the whole animal.

All these simple calculations are shown on fat content data sheet (Fig. 9C.1) for a field mouse (*Apodemus flavicollis* Melch.). This example, as well as caloric determinations derive from the study by Sawicka-Kapusta (1968). Attention should be drawn to the fact that with the used procedure the stomach content was removed from the mouse before fat was extracted. Therefore the caloric content of mouse does not include ingested food.

Fig. 9C.1. (right) A fat content data sheet with an example of fat extraction and caloric content calculations for a field mouse.

FAT CONTENT DATA SHEET

Species __FIELD MOUSE (APODEMUS FLAVICOLUS)__ Code: __MW-75 x__

Age, sex and reproduction state : __AD ♀, LACTATING__

Date and place of capture : __JULY 12, 1966 – OJCOW NATIONAL PARK__

DRYING:

Fresh body weight/ biomass/	31.35 g
Stomach content	1.70 g
Biomass without stomach content	29.65 g
Dry weight	8.05 g
Per cent of water	73.0 %

EXTRACTION :

Dry weight after first 10 cycles	7.20 g
Dry wt. after next 10 cycles	7.14 g
Dry wt. after final 10 cycles	7.14 g
Fat–free body weight	→ 7.14 g
Fat content/ 8.05 – 7.14 g /	0.91 g
Per cent of fat	11.0 %

CALORIC VALUES OF WHOLE ANIMAL :

Caloric value of fat __9.124__ kcal/g and fat-free body __4.544__
kcal/g as determined in the bomb calorimeter

Fat	0.91×9.124	=	8.303 kcal
Fat–free body	7.14×4.544	=	32.444 kcal
Whole animal		=	40.747 kcal
Dry weight	$\dfrac{40.747 \text{ kcal}}{8.05 \text{ g}}$	=	5.062 kcal/g
Fresh weight/biomass/	$\dfrac{40.747 \text{ kcal}}{29.65 \text{ g}}$	=	1.374 kcal/g

REMARKS:

References

DOL'NIK V.R. & BLJUMENTAL T.I. (1964) Bioénergetika migracij ptic [Bioenergetics of bird migrations]. *Uspehi Sovrem. Biol.*, **58**, 280–301 [in Russian].

EVANS P.R. (1969). Winter fat deposition and overnight survival of yellow buntings (*Emberiza citrinella* L.). *J. Anim. Ecol.* **38**, 415–423.

HAYWARD I.S. (1965) The gross body composition of six geographic races of *Peromyscus*. *Canada. J Zool.* **43**, 297–308.

JAMESON E.W. & MEAD R.A. (1964). Seasonal changes in body fat, water and basic weight in *Citellus lateralis, Eutamias speciosus* and *E. amoenos. J. Mammal.* **45**, 359–365.

KLEIBER M. (1961). *The fire of life—an introduction to animal energetics.* J. Wiley & Sons, Inc. New York. p. 454.

MAZUR T. (1967). Seasonal variations in the energy reserves of *Bufo bufo* (L.) and *Rana arvalis* Nilss. (Anura) in Poland. *Ekol. pol. A.* **15**, 607–613.

ODUM E.P. (1960). Lipid deposition in nocturnal migrant birds. *Proc. XII Intern. Ornithol. Congr.*, Helsinki, pp. 563–578.

ODUM E.P., MARSHALL S.G. & MARPLES T.G. (1965). The calorific content of migrating birds. *Ecology* **46**, 901–904.

SAWICKA-KAPUSTA K. (1968). Annual fat cycle of field mice, *Apodemus flavicollis* (Melchior, 1834). *Acta theriol.* **13**, 329–339.

10

Respirometry

10A METHODS OF MEASURING RESPIRATORY EXCHANGE IN TERRESTRIAL VERTEBRATES*

P. MORRISON AND G. C. WEST

Respiratory exchange provides a convenient measure of the energy utilization of an animal according to basic relations of the form

$$(CHOH)_n + nO_2 \rightarrow nCO_2 + nH_2O + ENERGY$$

Since the proportions between the several terms of the equation are fixed, each will be related to the other by a constant, thus (Kleiber, 1961):

energy : carbohydrate = 4·0 kcal/g

energy : O_2 = 5·0 cal/cc

energy : CO_2 = 5·0 cal/cc

CO_2 : O_2 = 1·00 (R.Q.)

Accordingly, the energy utilization of an animal consuming carbohydrates may be computed as 5·0 calories for each cc of oxygen consumed.

Similar relations but with different constants describe the utilization of the other principal foodstuffs, fat and protein. In animals living on a mixed diet, the proportions of each constituent must be known in order to calculate the energetic equivalent with full accuracy. These proportions may be computed if the output of urinary nitrogen, defining protein catabolism, and the respiratory quotient are known. However, the caloric equivalent for oxygen, unlike that for carbon dioxide, does not differ markedly between the three classes of foodstuffs so that it is a common and not unreasonable practice to use a mean value of 4·8 cal/cc of oxygen for animals on a mixed diet. This procedure seems quite satisfactory in the ecological bioenergetics

*Publication No. 89 from the Institute of Arctic Biology, University of Alaska, College, Alaska. Supported in part by NIH Grant GM-10402.

under consideration here, although in species known to have very specific diets such as hibernators metabolizing stored fat, some modification of the constant may be called for.

10A.1 Direct Calorimetry

A direct measurement is to be preferred, in principle, over an indirect one, and indeed the first calorimetric determinations of Lavoisier (Lavoisier and LaPlace, 1780) measured animal heat directly as water produced by melting ice. Although early measurements were often crude and inexact, Lavoisier's procedure in which the layer of ice surrounding and receiving heat from the animal was protected thermally by a second layer of ice, is elegant and precise. However, as a practical arrangement it is clumsy and restricts the measurement to a single ambient temperature near zero.

Subsequent direct calorimeters, usually for man or large animals, isolated the subject thermally and carried away the heat produced in a fluid stream (Lusk, 1928). The volume flow and the change in temperature of the fluid defined the heat production. Such systems were again cumbersome and due to their inherent heat capacity were slow to reach equilibrium and thus to respond to changes in rate. New materials such as stryofoam which allow the construction of chambers with high insulation and negligible heat capacity may make direct calorimetry on small animals feasible.

Another principle employed recently for measuring heat production in man and smaller animals is the gradient calorimeter (Benzinger and Kitzinger, 1949) which measures the temperature difference (and therefore the heat flow) across the walls of the animal chamber by means of a "net" of thermocouple junctions (thermopile) arranged in series with alternate junctions inside or outside thus integrating the heat flow in all directions. In the aggregate however, direct calorimetry has been employed rarely as compared to the great number of indirect calorimetric studies.

10A.2 Closed Circuit Systems

The essential operational principle of closed-circuit systems is the basic gas law which states that at constant temperature the amount of gas in an enclosed space is directly proportional both to volume and to pressure. In closed-circuit systems, carbon dioxide is continuously absorbed as it is produced by the animal so that the net reaction is to remove oxygen from the system. Thus, if either volume or pressure is held constant, changes in the other term will define the consumption of oxygen. Such apparatus can therefore be divided into volumetric (constant pressure) or manometric (constant volume) systems. In both types it is essential that the temperature be carefully controlled.

The adsorbant may be placed directly within the chamber and the carbon dioxide adsorbed by passive convection or it may be contained in a separate tube or canister and supplied actively by a pump. Liquid adsorbants such as sodium or potassium hydroxide or porous solid adsorbants such as soda lime, baralime or Ascarite may be employed. Provision may be made, periodically or at the end of the experiment, to estimate the amount of carbon dioxide produced either gravimetrically or by titration.

Perhaps the best known volumetric system is that of Benedict used for the clinical determination of basal metabolic rates (Carpenter, 1915) in which the pressure is held constant by a balanced spirometer which, during its descent, records the consumption of oxygen on a kymograph. Here the subject is connected to the system by a mouthpiece. A more relevant system for animals (Krogh, 1916) encloses the subject within a chamber which is connected to a spirometer. With animals, such as small mice, the spirometer must be very delicate in order to maintain constant pressure while still recording directly and this presents mechanical limitations. A recent modification overcomes this problem by means of a 'servo' system comprising a sensitive pressure transducer and a mechanically driven bellows or a 'waterless spirometer'. Another volumetric system employs a small spirometer but avoids continuous recording by using only two points in the excursion to define a fixed and reproducible volume (Morrison, 1947). This comprises an automatic system in which the time required to use the volume aliquot defines the rate of oxygen consumption. The Kalabukov-Skvortzov system described in Chapter 10C maintains a closed system at constant pressure by the admission of water from a measuring burette whose meniscus is read at regular intervals.

Although manometric systems such as the Warburg apparatus are convenient for smaller organisms, the principle has not been widely used for larger animals. Werthessen (1937) described a system which supplied oxygen electrolytically essentially in relation to the position of an aqueous manometer. An automatic apparatus which times the consumption of equal manometric aliquots of oxygen (Morrison, 1951) is described in chapter 10B.

10A.3 Open-Circuit Systems

In open-circuit systems a stream of air is passed through the animal chamber to maintain the oxygen level and carry away the carbon dioxide produced. Such systems have the advantage that the ambient temperature need not be so closely controlled, and so they may be employed at low temperatures which are impractical for thermostated baths.

In the earliest studies (Regnault and Reiset, 1849) the carbon dioxide alone was measured either gravimetrically or by titration after absorption on alkali. Later with the development of accurate analytical procedures for gases, it was possible to collect the exhaust air, measure its volume, and analyze for oxygen and carbon dioxide, (Haldane system, Scholander analyzer). Alternatively, samples could be collected periodically or a small fraction of the measured flow could be collected continuously. Such systems can even be made portable to be carried by man or larger animals.

With the development of direct sensors for oxygen and carbon dioxide it has become possible to avoid manual analysis and, at constant flow, to obtain a continuous record of either or both entities. As sensors for oxygen, the Pauling paramagnetic principle has been commonly used and the oxygen electrode now offers another approach. For carbon dioxide, infrared sensors and thermal conductivity sensors have been used.

While the oxygen and carbon dioxide analyzers utilized in open flow systems offer a very simple approach in principle, it should be remembered that they are all expensive and often troublesome in operation, requiring constant attention to insure proper function and accuracy. Gas flow volume also present problems in accurate measurement and in stability and must be continuously recorded.

10A.4 A Specific Open-flow System

Open-flow systems for metabolism employing oxygen recorders are in use in many laboratories, and since this technique was not demonstrated or des-

Fig. 10A.1. (right) Schematic diagram of an eight-channel open-flow respirometer system. Solenoids are numbered 1-10A, hand operated needle valves are designated by $\otimes$. Tubes between chambers and solenoids contain water absorbant. For flow pattern, see text.

cribed elsewhere in this volume a description of a multiple channel system is provided (Fig. 10A.1).

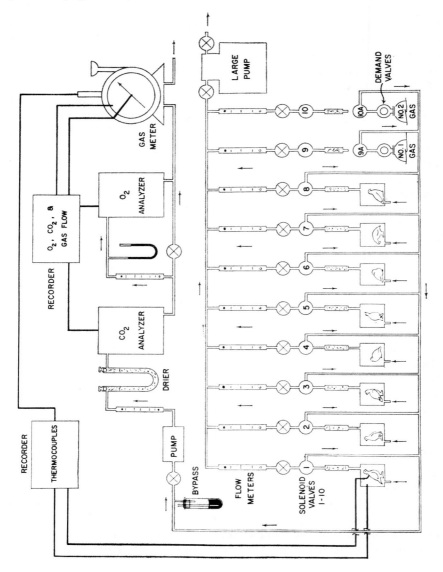

The animal chambers are all connected through a manifold so that they may be ventilated by a single large pump. Individual flow meters and needle valves allow balancing of the flow in each channel to a standard value. Drying towers containing calcium chloride or Anhydrone (magnesium perchlorate) may be provided at the inlet of each metabolic chamber in order to reduce the humidity in the chamber, especially at high ambient temperatures and are always provided at the outlet in order to eliminate the condensation of water in the sampling line.

Sampling from any channel is accomplished through individual solenoid valves (No. 1–8) which shunt the flow into the sample line and may be opened in sequence as often as every 5 minutes. The sampling line is ventilated by a separate smaller pump and also supplied with a flow meter and needle valve so that its flow can also be adjusted to the standard rate. Thus, when a chamber is switched to the sample line there is no change in the flow rate and equilibrium of the respiratory gases is maintained. Additional drying tubes ensure the complete removal of water from the sample before passing through the carbon dioxide analyzer (Beckman No. IR215, ranges 0–5% and 0–1%) which is sensitive to water vapor. Thermal tempering coils may also be provided if the animals are maintained at very low temperatures. The sample then goes through the oxygen analyzer (Beckman No. F–3, paramagnetic, ranges 16–21%, 19–21% and 20–21%) but a shunt is provided so that the flow through the sensing cell is limited to about 150 cc/min. The recombined flow is finally passed through a wet test gas meter which accurately measures the total sample line flow.

Signal voltages from the two analyzers go to a strip chart recorder (Honeywell No. 17, electronic, dual pen) to provide continuous records of both carbon dioxide and oxygen. On the same chart a sidemarker pen provides a record of the sample gas flow as revolutions of the gas meter (litres). Channel selection is accomplished by an automatic stream selector (Beckman) which programs the flow through the 8 channels. It may be noted that the simultaneous use of 8 channels will often not be possible, 4–6 being a reasonable limit under ordinary conditions. Individual channels may be omitted or left on for longer than the minimum cycle. The system is designed to identify channels on the chart by a brief deflection of the oxygen pen at the beginning of each recording period (i.e. Ch. 1 = 10, Ch. 2 = 20, etc.). Two additional channels (9 and 10) supply standard gases which are admitted manually or programmed into the sample line from their pressure tanks through demand valves as required. The two gases usually employed are 17%

oxygen in nitrogen and 2% carbon dioxide in air which allow 'zero' and 'scale' adjustments of the two analyzers. These concentrations need not be exact but the actual value must be determined by analysis. Body temperatures, chamber temperatures, and the temperature inside the gas meter are recorded on a thermocouple recorder (Leeds and Northrop Speedomax, G, Model S).

The flow rate is ordinarily set so that the carbon dioxide concentration does not exceed 1%, or roughly 100 times the maximum oxygen consumption. The minimum flow rate is limited by the minimal requirement of the oxygen analyzer (50 cc/min). However, the normal flow through the chambers for most animals is 1000 cc/min. For larger animals whose requirements exceed 10–20 cc O_2/min. two or more channels may be attached to a single chamber thus multiplying the flow rate by the number of channels used.

The calculation of results is straightforward. First, the flow as recorded from the gas meter is corrected to standard temperature and pressure. Since the chart has been calibrated and confirmed by the use of the standard gas mixture, the values for oxygen (removed) and for carbon dioxide (produced) may be read from the chart, and used directly if they are equal (R.Q. = 1·0). If they are not equal, a correction must be applied to the oxygen value because of 'shrinkage' of the total gas volume during transit (incomplete replacement of oxygen by cardon dioxide). The maximum correction (R.Q. = 0·70) is by a factor of 1·064 (Depocas & Hart, 1957). This system has been used successfully on a variety of small birds and mammals including humming-birds (Trochilidae), finches (Fringillidae, Ploceidae), voles (*Clethrionomys, Microtus*), ptarmigan (*Lagopus*), crows (*Corvus*), rabbits (*Lepus*), and beaver (*Castor*). It has also been modified for use with larger animals such as man and reindeer (*Rangifer*).

For use in ecological bioenergetics this system could certainly be simplified. The measurement of oxygen alone would be adequate and, with care in flow design and in monitoring, the wet gas meter could be eliminated. A mean factor can be computed to correct for the volume change during passage.

10A.5 Discussion

A variety of systems have been described for measuring respiratory exchange in animals. Indeed, there are almost a plethora of systems but most have embodied some improvement in technology or design for a particular application. In choosing a system for ecological bioenergetics, as opposed to physiological measurements, the special circumstances that should be considered are: (1) measurements should be continued over a full day (24 hours) so that an automatic or recording system is desirable; (2) most measurements will be made at ordinary temperatures (10–30°C) so that thermostated baths are practical; (3) larger chambers will be required to give the animal natural freedom of movement so that response time (lag) will be longer and the sensitivity reduced; (4) individual measured periods can be fairly long (0·5–1·0 hr.) so that less sensitivity is required and the lag in response will be less important; and (5) a lower precision will be required (perhaps 5% *vs* 2%) particularly since the field observations to which these measurements are matched will always be considerably more uncertain.

References

BENZINGER T.H. & KITZINGER C. (1949) Direct calorimetry by means of the gradient principle. *Rev. Sci. Instr.* **20**, 849–860.

CARPENTER T.M. (1915) *A comparison of methods for detremining the respiratory exchange of man.* Carnegie Institution of Washington, Washington, D.C., Publication No. 216.

DEPOCAS F. & HART J.S. (1957) Use of the Pauling oxygen analyzer for measurement of oxygen consumption of animals in open-circuited systems and in a short-lag, closed-circuit apparatus. *J. Appl. Physiol.* **10**, 388–392.

KLEIBER M. (1961) *The Fire of Life.* J. Wiley & Sons, New York.

KROGH A. (1916) *The Respiratory Exchange of Animals and Man.* London.

LAVOISIER A.L. & DE LAPLACE P.S. (1780) *Memoir sur la chaleur, Memoires de l'Academie Royale.* p. 355.

LUSK G. (1928) *The Elements of the Science of Nutrition.* Saunders, Philadelphia.

MORRISON P.R. (1947) An automatic apparatus for the determination of oxygen consumption. *J. Biol. Chem.* **169**, 667–679.

MORRISON P.R. (1951) An automatic manometric respirometer. *Rev. Sci. Inst.* **22**, 264–267.

REGNAULT V. & REISET J. (1849) Recherches clinique sur la respiration des animaux. *Ann. de chim. et de phys. Ser* 3., **26**, 299–519.

10B MORRISON RESPIROMETER AND DETERMINATION OF ADMR

P. MORRISON AND W. GRODZINSKY

One or another of the various systems for measuring metabolic activity may provide special advantage depending upon the animal material or the circumstances. This closed system, manometric respirometer provides a continuous record of oxygen consumption and locomotory activity of homeothermic animals with body weight of 5–2000g (Morrison, 1951). A fixed manometric aliquot of oxygen ranging from 1cc to 300cc or larger is admitted through a solenoid valve controlled from an attached mercury manometer, and the time required to consume this aliquot is recorded on a continuous record. Because of its automatic design which obviates attendance it has proved very useful in applications involving extended measurements such as those describing the daily metabolic cycle or the (average) daily metabolic rate. For ecological bioenergetics the ADMR, the mean value of a run over 24-hours during which time the animal is kept in a large chamber at a temperature similar to its natural habitat, is of great value, for construction of DEB of small rodents (Grodziński & Górecki, 1967; see also Chapter 3).

The special conditions involving longer measured periods and therefore larger aliquots of oxygen allow some simplification on the original design which emphasized the use of small aliquots for the determination of minimum metabolism.

10B.1 Principle

A closed system (chamber) is connected to a manometer containing mercury whose position reflects the pressure within the system. As the animal consumes oxygen the pressure falls in the system and the mercury rises in the closed arm. The excursion of the mercury column is limited on each side by a platinum contact. When the mercury reaches the contact on the closed side a relay system acts to open a solenoid valve which supplies oxygen to the chamber through another lead. As oxygen enters, the mercury rises in the open arm. When it reaches the contact on that side the relay system shuts off the solenoid valve. During this process a definite amount of oxygen has been added to the chamber.

It is essential that the temperature of the system be carefully maintained in the water bath, particularly in this application where very large chambers are employed for small animals. However, it is possible to attach a thermo-barometer chamber to the 'open' side of the manometer which will compensate for any such changes in temperature and as well isolate the system from barometric pressure changes.

10B.2 Apparatus

This system draws oxygen directly from a high pressure reserve tank through a standard reducing valve to a second more sensitive pressure controller which further reduces the supply pressure below 20 cm of water. Oxygen then passes, on demand, through the solenoid valve into the chamber. The chamber is connected to the closed arm of the manometer which is constructed of 10 to 15 mm tubing with a connecting bottom section of 1–mm capillary tubing. The constricted section prevents oscillation of the mercury in response to vibration or deep respiration by the animal but does not hamper the steady flow of mercury from one arm to the other. The manometer is supplied with platinum contacts to provide upper and lower pressure limits. An adjustable mercury reservoir connected below the capillary section permits change to the mercury level to increase or decrease the pressure differential.

In the basic control circuit (Fig. 10B.1) pulses from the manometer actuated an electronic relay (ER) drawing about 1 microampere which in turn operated the ratchet relay (RR) controlling the solenoid valve (V). In the original system for most precision the signal (S) was actuated when either circuit was closed through the mercury column, As simplified in Poland by W. Maczek, J. Pindel, A. Górecki & W. Grodzinski for measuring ADMR, only a single pulse is recorded while the solenoid valve is open. This circuit using simple relays is diagramed in Fig. 10B.2. The use of less sensitive relays results in some sparking and oxidation at the mercury surface which can reduce sensitivity. This corrosion may be suppressed by the addition of a buffer zone of an inert gas such as nitrogen which transmits the pressure through a thin membrane.

10B.3 Chambers

The choice of a chamber for ADMR involves two opposing factors: the need for sufficient size to allow an adequate freedom of movement to the animal and the limitation of accuracy resulting from the increased volume. Fortunately, the longer measuring periods which are acceptable for 24-hour measurements (20–60 min.), and the fact that errors in individual periods will average out in summing the ADMR make a reasonable compromise possible. For small mammals ranging in size from the shrew to the squirrel

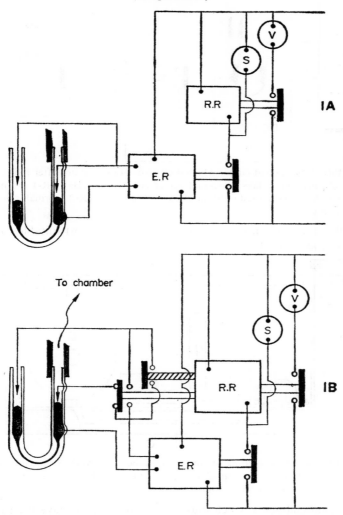

Fig. 10B.1. Schematic circuit diagram for the systems. 1A, basic circuit; **1B**, with interlocking to prevent phase reversal. R.R. = Ratchet Relay; E.R. = Electronic Relay; S = Signal Pen; V = Solenoid Valve.

chambers of 5–40 litres capacity have been suitable. Demonstrated were a 9–litre chamber (23 × 19 × 16 cm) appropriate for most mice and voles

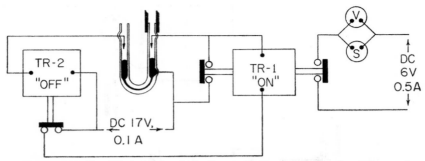

Fig. 10B. 2. Simplified circuit which disregards Hg overshoot after filling and using 2 telephone relays: a second set of contacts on relay **TR-1** (normally open) 'holds' **TR-1** closed until the circuit is broken by **TR-2** (normally closed) after filling.

(Fig. 10B.3 and a 17–litre chamber (43 $\times$ 27 $\times$ 21 cm) suitable for larger voles, rats and squirrels.

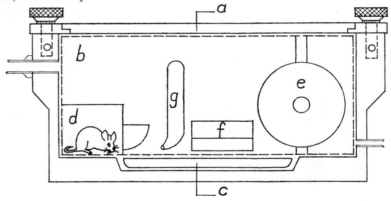

Fig. 10B. 3. Cross section of animal chamber for **ADMR** measurement: a—plastic lid (top); b—animal cage; c—CO_2 absorber; d—nest box; e—activity wheeel; f—feeder; g—water bottle.

The metal chambers which must be absolutely tight have been made of either aluminium alloy or stainless steel. A plastic top screwed on (a in Fig. 10B.3), provides visibility for inspection. An animal cage (b) made of metal mesh and a carbon dioxide absorber (c) are tightly fitted inside the chamber. The cage is also equipped with a nest box (d) as well as with an activity wheel, a feeder (f) and a drinking bottle (g). The chamber is connected with the respirometer by means of two tubes, while electrical connections for the activity indicator (Górecki and Hanusz, 1968) run directly to the recorder.

The four-channel apparatus used for ADMR recorded on a 12–channel Jaquet operation recorder using a wax-coated chart (g & g—). Each respirometer channel (chamber) was represented on the chart by three traces describing (1) oxygen consumption, (2) presence in or out of the nest box, and (3) exercise on the running wheel; Fig. 10B.4 illustrates a representative metabolic data sheet.

10B.4 Procedure

In determination of ADMR, wild animals from the field should be kept in the laboratory for no more than the several days necessary for their psychological adjustment to captivity. Directly before the run and immediately after it the rodents are weighed and their rectal body temperature taken (note: 'Metabolism Data Sheet ADMR'). The animals are given natural food, water and some nesting material (cotton). The lids (tops) are tightly screwed on the chambers with cages and dishes of potassium hydroxide as an absorber of CO_2 inside. During the course of measurement of ADMR which usually lasts 26–28 hours, the chambers are entirely submerged in a large thermostated water bath and connected with the respirometer and recorder. During the initial hour, the mercury manometers are adjusted so that the periods between the fillings were not shorter than 30–40 min.

10B. 5 Calibration

Since the effective volume of the system depends on the size of the animal and the volume of the absorbants, calibration must be carried out during the run. This is accomplished by withdrawing gas from the system into a calibrated burette or syringe of appropriate size, thus simulating the utilization of oxygen by the animal. During this time the measuring syringe must be submerged in the water bath (or jacketed) to define the temperature. Since the animal is consuming oxygen during calibration the true volume of the aliquot of oxygen (V) is somewhat greater than the measured volume (V_m) according to the relation:

$$V = V_m(t/t–\mathbf{t});$$

where t is the average consumption period at calibration (before and after) and **t** is the time used in withdrawal during calibration. The value (**t**) may be

DATE: Start __10·00(AM)__ End __13·00(PM)__   Run duration __27^h [12-12]__

July 11/12, 67

METABOLISM DATA SHEET (ADMR)

Species, sex and No. __MICROTUS ARVALIS ♂ #82__

Code __MSS-125__   Run No. __33__

Chamber No. __1__   Manometer No. __1__   Record No. __125__

	Gross Wt.	Tare	Net Wt.	TB-°C
Initial			19.7g	38.8°C
Final			19.3g	38.4°C
Average			19.5g	

Water bath temp. __20°C__   Period of light __L 4-20^h__

Food and nest __OATS, CARROTS E POTATOES__

Daily food consumption __———__ / faeces / __———__

Calibration:

Time	Vol.ccm	Duration/sec/	B.Pressure	T°C
7/11/67 – 10^h	50.0	14	753.8mm	22°C
18^h	49.5	12	753.4mm	
8/11/67 – 13^h	49.5	12	751.2mm	21°C
	49.5	13		
Average	49.6	13'	752.8mm ⟶ ×0.9018	

$$V_{stp} = \underline{44.81 + 0.37}(corr.) = \underline{45.18} \, ccm \, O_2$$

Activity wheel counter : start __5541__ end __6762__ sum. __1221__

Respirometer time counter: start __1870.2__ end __1897.2__ diff. __27^h__

Pressure in oxygen tank: start __118__ end __117__ diff. __1 Atu.__

Fig. 10B. 4. Representative chart showing metabolism and activity response in six channels.

OXYGEN CONSUMPTION

No. of fillings **56·27**   ADMR **5.4317** ccmO$_2$/g/hr

Bioenergetics: **0·6256** Kcal/g/day   **12.20** Kcal/animal/day
(2542·3)

12	14	16	18	20	22	24	2	4	6	8	10	12
$\frac{2}{20}+$ $5\frac{38}{108}$	$\frac{70}{108}+$ $4\frac{0}{0}$	$\frac{0}{0}+$ $4\frac{33}{70}$	$\frac{27}{70}+$ $4\frac{35}{50}$	$\frac{15}{50}+$ $5\frac{35}{50}$	$\frac{15}{50}+$ $4\frac{27}{70}$	$\frac{43}{70}+$ $3\frac{45}{100}$	$\frac{55}{100}+$ $4\frac{14}{100}$	$\frac{86}{100}+$ $3\frac{35}{55}$	$\frac{20}{55}+$ $3\frac{85}{100}$	$\frac{15}{100}+$ $4\frac{20}{100}$	$\frac{80}{100}+$ $3\frac{15}{85}$	
5.45	4.65	4.47	5.08	6.00	4.68	4.06	4.69	4.50	4.21	4.35	3.98	
246.23	210.08	201.95	229.51	271.08	211.44	183.43	211.89	203.31	190.20	196.53	179.81	
6.31	5.38	5.17	5.88	6.95	5.42	4.70	5.43	5.21	4.87	5.03	4.61	

ACTIVITY (NEST–RUN)

Total daily $\frac{1452}{3}$ = 484.0 min.

	12	14	16	18	20	22	24	2	4	6	8	10	12
mm	120 86	— —		18 13 10 20	85 42 8 18	180 54 60	58 74	130 20	18 60 15 30	25 42 18	50 25	20 23 89	18 13
mm	206	—		61	153	294	132	150	123	95	75	132	31
min	68.6	—		20.3	51.0	98.0	44.0	50.0	41.0	31.6	25.0	44.0	10.3

RUNNING WHEEL

Total daily $\frac{824}{3}$ = 274.6 min

	12	14	16	18	20	22	24	2	4	6	8	10	12
mm	0	180 146	33	50 13	100 13	25 62	10 8	65 10	13 25	20 7	16 20	5 3	
mm	0	326	33	63	113	87	18	75	38	27	36	8	
min	0	108.6	11.0	21.0	37.6	29.0	6.0	25.0	12.6	9.0	12.0	2.6	

timed directly with a stop watch, but a more convenient method of sufficient accuracy is to measure it on the chart, expressing both **t** and *t* in mm.

10B. 6 Calculations

The basic value for calculation of ADMR is the volume of oxygen consumed by the animal in each cycle. The corrected volume is additionally adjusted to Standard Temperature and Pressure (STP—0°C and 760 mm Hg) using the temperature of water bath (chambers) and the atmospheric pressure using standard gas tables or the formula given in Chapter 10C. In calculations of daily metabolic rate short periods of actual filling are ignored because their total time amounts to less than 10 minutes per day (e.g. about 50 fillings, less than 10 sec.).

The number of periods (N) as counted on the chart over a full 24-hour period is multiplied by the corrected volume of one filling (V_{stp}) to give the daily oxygen consumption for the animal as cc/day (A). This rate may be converted into other units as desired:

For physiological comparisons, to the BMR for example, the specific oxygen consumption as cc/g hr is common and represents value (A) divided by the body weight in g and the constant, 24 hr/day (**B**).

To relate to food consumption, value (A) is multiplied by the constant 4·8 cal/cc O_2 to give cal/day (C).

For ecological calculations involving biomass the specific caloric consumption is useful and represents value (C) divided by the mean body weight in g to give cal/g day (D).

The chart with the records can provide a number of other kinds of information. We can read out and calculate the metabolic rate (oxygen consumption) in 2-hour intervals and plot the daily rhythm of such metabolism. We can also find out a maximum and minimum value of daily metabolism (in full run). The record of activity gives a pattern of daily (circadian) activity, as well as duration of activity outside the nest in minutes and motility on treadmill (see Data Sheet); it is also possible to estimate a metabolic cost of the activities by relating them to the oxygen consumption record.

References

GÓRECKI A. & HANUSZ T. (1968) A simple indicator for registering the activity of small mammals. *Ekol. pol.* "S.B.", **14**, 33–37, (In Polish with English summ.).

GRODZIŃSKI W. & GÓRECKI A. (1967) Daily energy budget of small rodents. In Petruse-wicz K. (ed.) *Secondary Productivity of Terrestrial Ecosystems.* 295–314 pp. Warszawa, Kraków.

MORRISON P.R. (1947) An automatic apparatus for the determination of oxygen consumption. *J. Biol. Chem.* **169,** 667–679.

MORRISON P.R. (1951) An automatic manometric respirometer. *Rev. Sci. Instr.* **22,** 264–267.

10C KALABUKHOV-SKVORTSOV RESPIROMETER AND RESTING METABOLIC RATE MEASUREMENT

A. GORECKI

A closed-system manometric respirometer was constructed by Kalabukhov (Kalabukhov, 1940) and then modified by Skvortsov (Skvortsov, 1957; Kalabukhov, 1962). This respirometer facilitates measurement of the resting metabolism rate (RMR) of birds and mammals of different body sizes. RMR includes the minimal cost of maintenance plus a value for the specific-dynamic action of food (SDA) together with the cost of thermoregulation at temperatures below the thermoneutral zone. Metabolic rates obtained at various ambient temperatures can be used to construct a temperature-metabolism curve and to estimate the cost of thermoregulation to homeothermic animals (Górecki, 1966). The energy requirements for thermoregulation represent a considerable part of the warm-blooded animal's daily energy budget (see Chapter 3).

The Kalabukhov-Skvortsov respirometer has been adapted for measurements of the average daily metabolic rate (ADMR), but the process is time-consuming because the respirometer lacks full automation (Gębczyński, 1963). However, it has become widely used in many ecological and physiological laboratories becaues it is simple, compact and convenient to use even under field conditions.

10C. 1 Apparatus and Procedure

The main parts of the Kalabukhov-Skvortsov respirometer are: an animal chamber (a glass desiccator—a) containing a CO_2 absorbant, an oxygen container (b), a calibrated burette (c) and a water-type pressure valve (d) with manometer (Fig. 10C. 1).

The apparatus operates on the following principle: oxygen is consumed by the animal and the expired carbon dioxide is immediately absorbed. The decrease in gas pressure causes oxygen to be drawn from the container into the animal chamber (Fig. 10C. 1). Then an equal amount of water flows from the burette into the container through the water-type pressure valve to replace the oxygen. Oxygen consumption can be read from the water level in the burette. During the experimental period the pressure inside the respirometer is slightly lower than the atmospheric pressure.

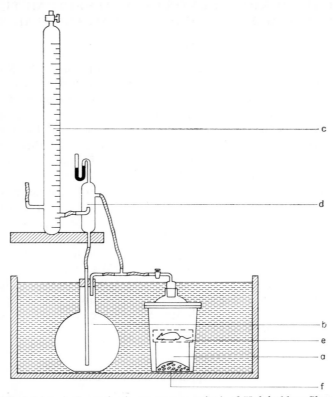

Fig. 10C. 1 (above) General scheme (cross-section) of Kalabukhov-Skvortsov respirometer: a—glass desiccator—animal chamber, b—oxygen container, c—calibrated burette, d—water type pressure valve with manometer, e—wire-mesh cage, f—carbon dioxide absorber (modified from Skvortsov, 1957). **Fig. 10C. 2** (right) A representative metabolic data sheet.

METABOLIC DATA SHEET (RMR)

Species ..Apodemus agrarius.. Bath temperature.....20°C...........
Date and time .....Feb.29 '72.... B Pres .740.0.. mm Hg. Corr.xO .886..

Animal No and sex	Gross wt.	Tore/tage/	Average bw	Body temp. in °C
1 0 ♀	79.8	57.5	22.3	37.3
2 0 ♂	76.5	54.8	21.7	38.1
3 0 ♀	76.4	54.7	21.7	37.8
4 0 ♂	77.8	59.7	18.1	37.5

OXYGEN CONSUMPTION

Cage No Time in Min.	1	2	3	4	Remarks
0	26	19	5	10	
5	34 / 8	30 / 11	18 / 13	26 / 16	
10	46 / 12	38 / 8	28 / 10	37 / 11	
15	55 / 9	47 / 9	32 / 4	43 / 6	
20	64 / 9	62 / 15	46 / 14	48 / 5	
25	74 / 10	77 / 15	57 / 9	58 / 10	
30	85 / 11	86 / 9	65 / 8	73 / 15	
35	94 / 9	96 / 10	75 / 10	84 / 11	
40	102 / 8	108 / 12	87 / 12	94 / 10	
45	112 / 10	117 / 9	95 / 8	102 / 8	
50	121 / 9	125 / 8	102 / 7	110 / 8	
55	129 / 8	136 / 11	111 / 9	120 / 10	
60	138 / 9	148 / 12	119 / 8	129 / 9	
S_{30}	—	—	—	—	
S_{60}	112	129	114	119	
S_{60} x Corr.	99.23	114.29	101.00	105.43	
ccm O_2/g/h	4.45	5.27	4.65	5.82	
kcal/g/h	0.0214	0.0253	0.0223	0.0280	

The animal is put into a small wire-mesh cage to restrict its movements. For small rodents with body weight of 20–30 g, cages with overall dimensions of about 40 × 40 × 85 mm are used; for bigger animals (up to 150–200 g) cages of 55 × 55 × 100 mm are needed. The cage is placed into the chamber (volume equal to 1·5–3 litres) on a tripod. Approximately 15–20 g of granulated potassium hydroxide are placed at the bottom of the chamber to absorb CO_2. The desiccator is covered with a ground glass top and sealed with vaseline. The chamber is placed into a constant temperature water bath held within ± 0·2°C of the desired level. Usually 4–6 respirometers are placed into the same water bath. After constant temperature is reached, the chambers are connected to the rest of the apparatus by means of rubber tubing and ground glass joints (Fig. 10C. 1). The calibrated burettes of 400 to 500 ccm have been previously filled with distilled water, and the containers with medicinal oxygen with no more than 0·5% impurities. The manometers are also filled with distilled water.

A period of at least 20 to 30 minutes should be allowed for the animal to adapt to its new environment before readings are taken. The oxygen consumption is determined by the difference in the water level, using the bottom of the meniscus, in the burette at 5-minute intervals. Variation in oxygen consumption per 5-minute interval can be correlated with changes in the animal's behaviour. All the readings are recorded on the Metabolic Data Sheet —RMR (Fig. 10C.2). The animal is weighed at the end of the experiment and its rectal body temperature taken.

10C. 2 Calculations

The total volume of oxygen consumed by the animal during a run can be calculated from the difference between the initial and final readings of the water level in the burette taken over a 30 or 60-minute period of time. If the animal showed increased activity during any of the 5-minute intervals, then the volume of oxygen consumed during that 5 minutes is subtracted from the half-hourly and hourly total in order to obtain RMR. The volume of oxygen per hour has to be adjusted to standard temperature and pressure (0°C and 760 mm Hg) either by using a correction factor from a table, or by calculating it from the formula:

$$V_{stp} = \frac{(B - W)\, T\, V}{760 \text{ mm of Hg} \,.\, 273°C}$$

where:

V_{stp} = the volume of oxygen (ccm) consumed at standard temperature and pressure

B = barometric pressure in mm of Hg

W = vapour pressure

T = ambient temperature in °C

V = difference in the water level (ccm) in the burette at the beginning and end of experiment

The barometric pressure is read before and after each RMR measurement and the average reading is used in the above calculations.

The corrected volume (V_{stp}) divided by the weight of the animal gives the amount of oxygen consumed per unit body weight per hour (ccm O_2/g/h). In order to express the metabolism in terms of calories per gram per day (kcal/g/day), it is sufficient to multiply the former value by 115·2 (4·8 × 24 hrs.), assuming that the respiratory quotient (RQ) = 0·82, and the caloric value of 1 litre of oxygen as equal to 4·8 kcal (Kleiber, 1961).

References

GĘBCZYŃSKI M. (1963) Apparatus for daily measurement of oxygen consumption in small mammals. *Bull. Pol. Acad. Sci. Cl.II.*, **11**, 433–435

GÓRECKI A. (1966) Metabolic acclimatization of bank voles to laboratory conditions. *Acta theriol.* **11**, 399–407.

KALABUKHOV N.I. (1940) Wliyanie temperatury na potreblenie kisloroda lesnymi i zheltogorlymi myshami. (The effect of temperature on the oxygen consumption in field mice.) *Dokl. A.N. SSSR.* **26**, 89–90.

KALABUKHOV N.I. (1962) Seasonal changes in the organism of mammals as the indicators of environmental effects. *Symp. theriol.* 156–174. *Publ. House Czech. Acad. Sci. Praha.*

KLEIBER M. (1961) *The fire of life—an introduction to animal energetics.* I. Willey Inc., New York, London, 1–454.

SKVORTSOV G.N. (1957) Ulutshennyi metodika opredeleniya intensivnosti potrebleniya kislorodau gryzuna drugich melkich zhyvotnych. (Improved method of determination of the oxygen consumption rate in rodents and other small animals.) *Gryzuny i borba z snimi.* **5**, 424–432.

10D METHODS OF ESTIMATING THE METABOLIC RATE IN AMPHIBIANS WITH SOME REFERENCES TO THE APPLICATION OF EXPERIMENTAL DATA IN ECOLOGICAL STUDIES

P. POCZOPKO

10D. 1 Metabolic rate measurements

Many devices can be used to determine respiratory exchange in amphibians. Here we will discuss a very simple apparatus with a closed air system, which can be easily constructed in almost every ecological laboratory. Despite its simplicity the apparatus is quite sensitive and accurate. A slightly more complicated model has been extensively used by Whitford and Hutchison (1963, 1965). The determination of respiratory exchange with either apparatus is time consuming because the oxygen supply is operated manually. When considering the possibilities of introducing automatic oxygen supply and recording one must remember that the metabolic rate of amphibians is low. A frog (*Rana temporaria*) weighing 25 g consumes only 4 ml O_2 per hour (Dolk and Postma, 1927) whereas a rodent of similar size (*Clethrionomys rutilus*, with a body weight of 24 g) consumes about 100 ml O_2 per hour (Grodziński, 1966). A mistake in the determination of oxygen consumption amounting to 2 ml/hr would cause an error of 2% for the rodent and of 50% for the frog.

A result with an error of 50% is useless, but such an error can easily be made when an apparatus with a closed air system is subjected to poor thermoregulation. Temperature fluctuations of commonly used thermostats are $\pm$ 0·5°C, thus giving a range of 1°C. If the capacity of the respiratory chamber is 1 litre, a 1°C change in temperature alters the air volume in the chamber (at constant pressure) by 3·66 ml. This is almost equal to the hourly oxygen consumption of a frog weighing 25 g. The type of experimental error is greater for larger respiratory chambers. Better thermostats would improve the accuracy but they are relatively expensive. It is more convenient and cheaper to use a compensating chamber which makes the system independent of small temperature changes.

Due to the low respiratory exchange in amphibians some types of automatic oxygen supply are excluded. For instance Morrison's respirometer

(1951) in which the oxygen supply is operated by mercury manometer is not sensitive enough. With some improvements, Kalabukhov's respirometer (1951) can also be used (Chapters 10B, 10C). However, the most promising way of using an automatic oxygen supply seems to be the electrolytic method as described by Macfadyen (1961).

Open circuit systems for respiratory exchange measurements are not suitable for investigations on amphibians. Only the carbon dioxide production can be easily measured by passing air through the respiration chamber which is preceded and followed by CO_2—absorbing vessels containing $Ba(OH)_2$. The carbon dioxide produced is determined by titration (see Chapter 10E). Measurements made with modern gas analyzers, based on physical principles such as thermoconductivity, magnetism, or infra-red light absorption, require the oxygen deficit and carbon dioxide surplus in the air leaving the respiration chamber to be 0·4% or more. In the commercial diaferometer (Kipp and Zonen) the minimum flow rate is 0·8 1/min. The minimum detectable oxygen consumption must, therefore, be 3 to 4 ml per minute. Connecting the respiration chamber directly to the measuring system of the diaferometer a smaller air flow rate can be used (150 ml/min.). This enables us to measure the oxygen consumption at a rate of 0·6 ml/min. which is almost 10 times higher than in the quoted example of *Rana temporaria* (4 ml/hr = 0·067 ml/min).

10D. 2. The significance of what is being measured

In homoiothermic animals one usually determines: the basal metabolic rate, the resting metabolic rate or the average daily metabolic rate. All of these can be determined under well standardized conditions and have a more or less clear biological meaning. The basal metabolic rate is the minimum heat production associated with processes necessary to sustain the life of the animal. In poikilothermic amphibians it is difficult to establish the basal metabolic rate and normal metabolism can vary over a wide range. Some of the factors affecting the metabolic rate in homoiothermic animals also influence the metabolic rate of amphibians. Other factors, however, have different qualitative effects on the metabolic rates in these two types of animals.

The difference between the metabolic rate of fed and fasting amphibians is similar to that in homoiotherms (Dolk and Postma, 1927), but the calorigenic effect of food seems to be a small item in the overall energy expenditure in amphibians due to relatively low food intake in these animals.

The effect of muscular activity on metabolism is probably much greater than the effect of food. This effect is hard to measure due to the difficulty of establishing the environmental conditions for normal levels of activity for animals in respiratory apparatus. Field observations on activity may be misleading. Watching the slow movements of a toad (*Bufo bufo*) and the quick jumps of a frog (*Rana esculenta*), one gets the impression that the frog spends more energy for activity than the toad. There are, however, good reasons to believe that just the opposite is true. *Bufo bufo* spends more time actively seeking food than *Rana esculenta* which waits for its prey to pass by and uses its ability to jump primarily to avoid danger. This does not hold true for every species of *Rana*, for example *Rana temporaria* is known to be an active hunter. Also the sluggish jumps of toads may be less efficient energetically than those of frogs.

Ambient temperature has a most pronounced effect on the metabolic rate in amphibians. The range of environmental temperature harmless to these animals lies between the critical thermal maximum and minimum as used by Cowles and Bogert (1944). The maxima and minima are not constant in different environments. Hutchison and Kosh (1965) found that the critical thermal maximum (CTM) for painted turtles acclimated to a 16:8 hr light: dark photoperiod is higher than that of those acclimated to an 8:16 hr light: dark photoperiod. Since acclimation to photoperiod affected in a similar manner the gas exchange in *Ambystoma maculatum* (Whitford & Hutchison, 1965a) one may conclude that acclimation to photoperiod can shift the CTM of this and other species of amphibia. If short term acclimation (14 days) is capable of shifting the CTM, seasonal changes in day length and average daily temperature may be even more effective.

The metabolic rate of amphibians within the range between the critical thermal minimum and maximum follows Van't Hoff's law, but Van't Hoff's quotient (Q_{10}) may be not constant even for the same species. The fact that the metabolic rate of the spotted salamanders (*Ambystoma maculatum*) acclimated to 16:8 light: dark photoperiod was significantly higher than in those acclimated to 8:16 light: dark photoperiod (in both groups determined at 15°C) suggests that either Q_{10} in these groups of animals was different or metabolic rate in one group was uniformly higher even though the Q_{10} was equal. The effect of season on metabolic rate, achieved by altering Q_{10}, or some other manner, must be greater than that of the short term acclimation in laboratory. Dolk and Postma (1927) found that the rate of gas exchange in

Rana temporaria during the breeding season was at least three times that found during the winter even though the ambient temperature during the experiments was the same, namely 24·6°C. Similar seasonal changes in metabolic rate were found by Fromm and Johnson (1955) in *Rana pipiens* tested at 22 to 25°C. Thus it is not enough to establish the Q_{10} of a given amphibian species during one season and then use this quotient for estimating the metabolic rate of that species during other seasons. For ecological studies on amphibian energetics more detailed information is needed. Since the ecologist is interested in knowing the metabolic rate of an animal in its natural habitat he must imitate in his experiments the conditions of that habitat. This concerns particularly the ambient temperature. Results on the seasonal variability of respiratory exchange in frogs (Dolk & Postma, 1927; Fromm & Johnson, 1955), although very interesting, have no proper meaning for ecologists since they were obtained at temperatures considerably higher (at least throughout 3 seasons) than the mean temperature of the natural habitats of these frogs.

The effect of metabolic acclimation of amphibians to environmental conditions suggests that only freshly caught animals should be used in laboratory studies of metabolism. This effect of acclimation suggests, however, that by appropriate treatment it might be possible to evoke hibernation in frogs during the summer time. This possibility would be very convenient but it probably does not exist. Holzapfel (1937) investigated hibernation in frogs (*Rana pipiens*) by placing them in a refrigerator at 0°C in each month of the year and found that only those frogs which were placed in the refrigerator during the period from October to April developed the typical features of hibernation. This shows that the estimation of respiratory exchange in hibernating amphibians should be done only during the winter time. Since some amphibian species hibernate on the bottom of ponds or streams they should be submerged in water when determining the winter metabolic rate. Many amphibians spend much time in water even during the summer time but only a few of them stay for a long time under water without using their lungs. Prolonged submersion decreases metabolic rate in *Bufo bufo* (Leivestad, 1960), *Rana esculenta* (Poczopko, 1959/60) and presumably in other species.

10D. 3 Presenting the experimental data

This part of my chapter consists partly of Kleiber's recommendations (1947) and partly of my own views.

Metabolic rate is usually expressed in terms of oxygen consumption or heat production per animal, per unit of weight or per unit of surface area. As Kleiber (1957) writes 'none of these various units of body size in which metabolic rate may be expressed is absolutely superior to all others'. For the ecologist the most instructive is the expression of the metabolic rate of a given species per animal and unit of time. Usually a day is chosen as the time unit but in the case of amphibians, due to the considerable seasonal variations, the annual metabolic rate may be more meaningful. When considering thermoregulation, it is necessary to express the metabolic rate per unit of surface area. Whereas, when dealing with rates of enzymatic processes in cells or tissues, the expression of the metabolic rate per unit of body weight is most proper. It is clear therefore, that the best way to express the metabolic rate depends on the problem being investigated. The investigator should remember, however, that his experimental data, quite often obtained with great effort, is likely to be used in connection with various problems and not only his own particular one. The experimental data should therefore be published in a way enabling easy recalculation and comparison.

The question of whether a metabolic rate of 51 kcal/kg–day is high or low is impossible to answer without knowing the size of the animal. The rate of 51 kcal/kg–day is normal for a 3 kg cat, but it would be very low for a rat weighing 210 g (normal 104 kcal/kg–day) and very high for a cow of 435 kg (normal 19 kcal/kg–day). Metabolic rate in amphibians is usually expressed in ml of O_2 consumed per g of body weight per hour. This way of expressing the metabolic rate of 14 amphibian species was used by Whitford and Hutchison (1963, 1965, 1966) and Vinegar and Hutchison (1965). The very interesting papers of these authors, on cutaneous and pulmonary respiration, containing the data obtained with a good technique, should be regarded as very valuable contributions to the knowledge on physiology of respiration in amphibians. On the other hand these papers are of limited value for bioenergetical considerations as the body weight of their animals is not given. The same objection applies to my own papers (Poczopko, 1959/1960, 1963).*

Kleiber (1947, 1961) proposes to use metabolic body size as body weight in $kg^{3/4}$. The interspecific average metabolic rate of adult homoiotherms, as calculated by Kleiber, is 70 $kcal/kg^{3/4}/day$. This average value enables us to

---

*When this chapter was completed a new paper by Hutchison *et al.* (1968) was published. The relationship between body size and metabolic rate is discussed in this paper in a way that provides more material (data) for a bioenergetic approach.

predict roughly the metabolic rate of adult homoiotherms from their body weight. The metabolic rate of poikilotherm animals cannot be predicted on the basis of Kleiber's unit of metabolic body size. Nevertheless this unit can be very useful in comparative physiology and ecology. It may be regarded as a base line which enables easy comparison of the metabolic rates of different animals under different conditions. For instance, the metabolic rate of a 25 g frog may be 4 ml O_2/hr and equal 7·77 kcal/kg$^{3/4}$–day; this metabolic rate is approximately 10 times lower than the interspecific mean for homoiotherms.

Kleiber (1947) suggests that the minimum requirements for the publication of data on metabolic rates should include the body weight and the total metabolic rate expressed per unit of body weight. Due to the important influences of external factors (such as temperature, season, photoperiod) on metabolic rate in amphibians, the conditions of the experiments should be clearly described. For easy interspecific comparison of metabolic rates it is desirable to express the data obtained in proportion to Kleiber's unit of metabolic body size.

References

COWLES R.B. & BOGERT C.M. (1944) A preliminary study of the thermal requirements of desert reptiles. *Bull. Amer. Mus. Nat. Hist.* **83**, 265–296.

DOLK P.O. & POSTMA N. (1927) Ueber die Haut und Lungenatmung von *Rana temporaria*. *Z. Verg. Physiol.* **5**, 417–444.

FROMM P.O.& JOHNSON R.E. (1955) The respiratory metabolism of frogs as related to season. *J. cell. comp. Physiol.* **45**, 234–240.

GRODZIŃSKI W. (1966) Bioenergetics of small mammals from the Alaskan taiga forest. *Lynx.* **6**, 51–55.

HOLZAPFEL R.A. (1937) The cyclic character of hibernation in frogs. *Q. Rev. Biol.* **12**, 65–84.

HUTCHISON V.H. & KOSH R.J. (1965) The effect of photoperiod on the critical thermal maxima in painted turtles (*Chrysemys picta*). *Herpatologica*, **20**, 233–238.

HUTCHISON V.H., WHITFORD G. & KOHL M. (1968). Relation of body size and surface area to gas exchange in anurans. *Physiol. Zool.* **41**, 65–85.

KALABUKHOV N.I. (1951). Metodika eksperimentalnyh issledovanii po ekologii nazemnyh pozvonočnyh. *Gos. Izd. Sov. Nauka*, Mokva. p. 177.

KLEIBER M. (1947) Body size and matabolic rate. *Physiol. Rev.* **27**, 511–541.

KLEIBER M. (1961) *The fire of life: an introduction to animal energetics* J. Wiley Inc 454 pp. New York, London.

LEIVESTAD H. (1960) The effect of prolonged submersion on the metabolism and heart rate in the toad (*Bufo bufo*). *Acta Univ. Bergensis. S. Math.* Rerumque Natrual. **5**, 1–15.

MACFADYEN A. (1961) A new system for continuous respirometry of small air-breathing invertebrates under near-natural conditions. *J. exp. Biol.* **38**, 323–343.

MORRISON P.R. (1951) An automatic respirometer. *Rev. Sci. Instr.* **22**, 264–267.

POCZOPKO P. (1959/1960) Respiratory exchange in *Rana esculenta* L. in different respiratory media. *Zool. Pol.* **10**, 45–55.

POCZOPKO P. (1963) Oddychanie plazów. (Respiration in amphibia.) *Przegl. Zool.* **7**, 5–18. (In Polish with English summ.).

VINEGAR A. & HUTCHISON V.H. (1965) Pulmonary and cutaneous gas exchange in the green frog, *Rana clamitans*. *Zoologica* **50**, 47–53.

WHITFORD W.G. & HUTCHISON V.H. (1963) Cutaneous and pulmonary gas exchange in the spotted salamander, *Ambystoma maculatum*. *Biol. Bull.* **124**, 344–354.

WHITFORD W.G. & HUTCHISON V.H. (1965) Gas exchange in salamanders. *Physiol. Zool.* **38**, 228–242.

WHITFORD W.G. & HUTCHISON V.H. (1965a) Effect of photoperiod on pulmonary and cutaneous respiration in the spotted salamander, *Ambystoma maculatum*. *Copeia* **1**, 53–58.

WHITFORD W.G. & HUTCHISON V.H. (1966) Cutaneous and pulmonary gas exchange in *Ambystomoid salamanders*. *Copeia* **3**, 573–577.

10E A RESPIRATORY CHAMBER FOR MEASURING METABOLIC RATE IN AMPHIBIANS

P. POCZOPKO

Due to the low metabolic rate of amphibians, instruments for investigating respiratory exchange in these animals should be very sensitive. The apparatus described here, in spite of its simplicity, is sensitive enough to be satisfactorily used for bioenergetical studies on amphibians. It is a simplified model of the apparatus described by Whitford and Hutchison (1963). Oxygen consumption is determined by direct reading and carbon dioxide production (if necessary) by chemical analysis.

The apparatus consists of two chambers of equal capacity made of vacuum jars tightly closed by metal or plastic covers (Fig. 10E.1). One of the jars (a) serves as a respiratory chamber, the other one, (i) is a compensation chamber. The chambers are connected by a manometer (h) containing a fluid of low density and viscosity such as Brodie's or Krebs' manometer fluid (Umbreit *et al.*, 1957, p. 66). Both chambers are placed in a constant temperature water bath. The syringe used for supplying oxygen to the respiratory chamber should be calibrated with mercury. All connexions and stopcocks should be checked for leaks prior to each run.

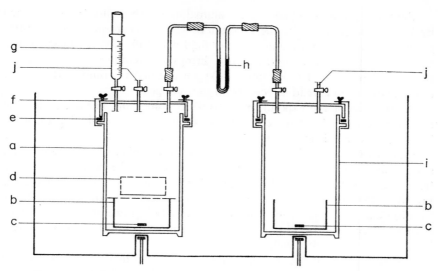

Fig. 10E.1. Diagram for the apparatus for investigating gas exchange in amphibians; a—respiratory chamber; b—absorption vessel; c—magnetic stirring bar; d—animal cage; e—rubber ring; f—sealing device; g—oxygen supplying syringe; h—manometer; i—compensation chamber; j—equilibrating stopcocks.

To measure respiratory exchange, vessels with known amounts of barium hydroxide solution (0·1 N) are put into the respiration chamber and compensation chamber. In each of the vessels a plastic or glass-coated magnetic bar is placed and rotated at regular intervals by magnetic stirrers to ensure effective carbon dioxide absorption. If sodium or potassium hydroxide pellets are used (when only oxygen consumption is determined) no stirring is necessary. The absorption vessel in the respiratory chamber is covered with a perforated plastic plate on which the cage with the experimental animal is placed. The exact time of sealing the animal in the respiration chamber should be recorded. Care should be taken at this time to keep the equilibrating stopcocks (j) open and the stopcocks leading to the manometer closed. After the system reaches temperature equilibrium, the equilibrating stopcocks are closed and those connecting the chambers with the manometer opened. If there is no shift in the manometer fluid the apparatus is sufficiently equilibrated to start measuring oxygen consumption.

Since oxygen is consumed by the animal and carbon dioxide is absorbed by the barium hydroxide, the amount of gas in the respiratory chamber gradually decreases and the manometer fluid will shift. Oxygen injected into the respiratory chamber by the syringe brings the manometer fluid back to the normal position. The amount of oxygen injected to compensate for the oxygen consumed is read directly from the syringe.

At the end of the experiment the vessels containing barium hydroxide are removed from the chambers, and the amount of carbon dioxide absorbed is determined by titration with 0.1 N hydrochloric or sulphuric acid. The barium hydroxide solution in the compensation chamber serves as a control, since it absorbs carbon dioxide at the same rate as the barium hydroxide from the respiratory chamber both prior to the experiment and during titration. Details for the procedure of carbon dioxide determination by titration can be found in most manuals of analytical chemistry (e.g. Handbook of Chemistry and Physics, Cleveland, Ohio, Chemical Rubber Publishing Co., 41st ed., 1955–1960).

The respiratory quotient is calculated from the volumes of carbon dioxide produced and oxygen consumed (corrected to STP). Finally the heat production of the experimental animals is calculated from the calorific equivalent of oxygen. Since carbon dioxide produced by the animal is absorbed by the barium hydroxide during the time of equilibrating the apparatus but oxygen consumption is not measured at this time an adjustment must be made in the calculations of the respiratory quotient.

In many cases the determination of carbon dioxide production can be omitted and heat production calculated directly from oxygen consumption by assuming that the respiratory quotient is 0.8 and 1 litre of oxygen has a heat equivalent of 4.8 kcal. The procedure is then considerably easier.

RESPIRATORY EXCHANGE DETERMINATION

Problem... *Spring metabolic rate* ...   Date... *8th April 1968*

No. of experiment... *T-2*

Animal sp... *Rana temporaria*   Barometric pressure ... *752*

Body weight ... *25*   Temperature ... *20°C*

Sex ... *♂*

- -

CO₂ determination	O₂ determination

CO_2 determination

Start... *9¹⁵*

End... *12¹⁵*

hr., min. ... *3 hr*

Amount of Ba/OH_2, ml. *50, 0.0989N*

Molarity of acid ... *0.1031*

Experimental-... *40.60 ml*
control   ... *2.15 ml*

Difference ... *38.45 ml*

CO_2 produced, g. *0.0216* ml. *10.986*
per unit of time ... *3.662 ml/hr*

O_2 determination

Start... *9⁴⁵*

End ... *12¹⁵*

hr., min. ... *2 hr 30 min*

O_2 consumption from direct
reading ... *11* ... × *0920*

O consumption reduced to
STP... *10.12*

per unit of time... *4.048 ml/hr*
= *9715 ml/24hr = 0.0097 l/24hr*

- -

Calculations

$$RQ = \frac{CO_2}{CO_2} = \frac{3.662}{4.048} \approx 0.90$$

Calorific equivalent of oxygen/kcal/l/from table/ ... *4.912*

Daily heat production = Daily O_2 consumption/l/x calorific
equivalent of oxygen = *0.097 × 4.912 = 0.476 kcal/24hr*

Additional calculation which was made for the metabolic rate of the frog (Rana temporaria)
see Respiratory Exchange Determination above.

CO_2 produced by the frog combines with water:

$$CO_2 + H_2O = H_2CO_3 \qquad\qquad 1$$
$$44 + 18 = 62$$

When 0.1 N $Ba(OH)_2$ is used to bound H_2CO_3 and O.1 N HCl for titrating $Ba(OH)_2$
not bound with H_2CO_3 (equation 2) we get the following reactions:

$$Ba(OH)_2 + H_2CO_3 = BaCO_3 + 2 H_2O \qquad\qquad 2$$
$$\text{and } Ba(OH)_2 + 2HCl = BaCl_2 + 2 H_2O \qquad\qquad 3$$

The above equations show that when x ml $Ba(OH)_2$ combined with H_2CO_3 then the
number of ml 0.1 N H_2CO_3 is the same.

When 50 ml $Ba(OH)_2$ is used the amount of $Ba(OH)_2$ combined can be calculated from
the equation: 50 − a = x, where a = number of ml 0.1 N HCl used for titrating free
hydroxide.

Because in 1000 ml   1 N H_2CO_3 is 62g H_2CO_3
then in 1000 ml 0.1 N H_2CO_3 is $\dfrac{62}{2 \cdot 10}$ g

and in   1 ml 0.1 N H_2CO_3 is $\dfrac{62}{2 \cdot 10 \cdot 1000}$ g

therefore in   x ml 0.1 N H_2CO_3 is $\dfrac{62 \cdot x}{2 \cdot 10 \cdot 1000}$ g

Equation 1 shows that 62 g H_2CO_3 corresponds to 44 g CO_2
the amount of CO_2 produced can be then calculated:

$$\frac{62 \; x}{2 \cdot 10 \cdot 1000} \cdot \frac{44}{62} = \frac{44 \; x}{2 \cdot 10 \cdot 1000} \text{g } CO_2 \qquad\qquad 4$$

Our case:
   Amount of $Ba(OH)_2$ used = 50 ml, 0.0989 N.

   correction factor $= \dfrac{0.0989}{0.1} = 0.989$

   Molarity of HCl = 0.1031
   correction factor $= \dfrac{0.1031}{0.1} = 1.031$

   x = (50 . 0.989) − (38.45 . 1.031) = 49.45 − 39.65 = 9.82
   according to formula 4.
   $\dfrac{44 \cdot 9.82}{2 \cdot 10 \cdot 1000} = \dfrac{432.08}{20000} = 0.0216$ g CO_2

Since 44 g CO_2 = 22400 ml
then $\dfrac{0.0216 \cdot 22400}{44} = \dfrac{483.84}{44} = 10.986$ ml $CO_2 \simeq 11$ ml CO_2

References

UMBREIT W.W., BURRIS R.H. & STAUFFER J.F. (1957) *Manometric techniques, a manual describing methods applicable to the study of tissue metabolism.* Burgess Publ. Co. 338 pp. Minneapolis, Minn.

WHITFORD W.G. & HUCTHISON V.H. (1963) Cutaneous and pulmonary gas exchange in the spotted salamander, *Ambystoma maculatum. Biol. Bull.* **124,** 344–354.

11
Feeding and Nutrition

11A FOOD HABITS AND FOOD ASSIMILATION IN MAMMALS

A. DRODZ

In the study of energy flow through mammal populations some questions concerning feeding ecology have to be answered. The most important are: (1) what part of the primary net production of a given ecosystem is available as food to mammals; (2) what part of the energy consumed by mammals is assimilated and (3) what fraction of nutrients is returned as faeces and urine to the cycle. The following chapter will deal chiefly with these problems.

11A. 1 Available food

As a rule the food available to rodents is only part of what the plants have produced and primary production cannot be regarded as being available in its total bulk. The energy available to rodents was defined by Grodziński (see Chapter 3) as 'that food which is easy to find and is being chosen and eaten by these animals'. Consequently, the estimation of the food available is usually based on some knowledge of mammalian food habits.

In a study of the natural food of mammals, several methods can be employed: (1) an analysis of stomach contents (Holišova, *et al.*, 1962; Williams, 1962); (2) an analysis of faeces (Godfrey, 1953; Steward, 1967, Watts, 1970); and (3) a determination of the natural feeding grounds and food stores (Carleton, 1966; Turček, 1967; Gentry & Smith, 1968). Furthermore, numerous experiments on food preferences, both under laboratory and field conditions, have been carried out recently. In order to determine the food habits of small mammals it is sufficient to use two methods only: the analysis of stomach contents and choice tests. Both methods are qualitative, supplement each other and give fairly similar results. (Drożdż, 1966; see also Chapter 11C and 11B.2).

The food habits of two rodents that are predominant in a beech forest, namely, the bank vole (*Clethrionomys glareolus*) and the yellow-necked field mouse (*Apodemus flavicollis*) have been estimated in this way. The bank voles are polyphagous and can live on both concentrated and bulky food, whereas the field mouse prefers concentrated food, such as seeds and animal matter.

Estimation of the available food to these rodents was compared to the primary production present, that is, to the annual plant increment (growth). An example of such an estimate for a beech forest from the point of view of a vole or a mouse is given in Table 11A. 1. The standing crop of wood and

Table 11A.1 Net primary productivity of beech forest and food available to rodents. All values in kcal/ha/year x 10^3 (After Droźdź, 1967).

Kind of food	Net productivity of beech forest	Food of voles	Food of mice
1. Harb layer vegetation	1,083	918	512
2. Tree leaves	13,427	671	324
3. Tree twigs (trunks and branches)	3,584 (25,557)	107	?
4. Tree seeds	225 (28·5–456)	202 (22–360)	202 (22–360)
5. Fungi and invertebrates	74	51	47
6. Total food supply (1–5)	44,000 (= 10.3 ton)	1,949 (1,769–2,007)	1,085 (905–1,234)

new growth is omitted because it cannot be eaten by rodents. They eat tree seeds and some leaves readily, but bank voles also consume the buds and bark of available twigs. The production of edible plants among the ground flora is lumped together. Those plants which were preferred in the choice tests and were also found in the stomachs may be regarded as edible plants. When estimating the total food available, fungi and food of animal origin have also to be taken into consideration.

The total primary production of the beech forest in the Ojców National Park has been estimated as being equal to over 10 tons/hectare/year which corresponds to 44,000,000 kcal/ha/year (Droźdż, 1966). The food available to either rodent (bank voles and field mice) does not exceed 2,000,000 kcal/ha/year. Consequently, the food available to voles or to mice corresponds to only 4·4% of the annual primary net production in the forest studied (Table 11A. 1).

In contrast, it seems that for herbivorous rodents (voles) living in grassland ecosystems almost the whole vegetation biomass may serve as potential food (Golley, 1960; Pearson, 1964; Grodziński *et al.*, 1966). Similarly granivorous rodents (mice) usually find more food in grasslands than in forests (Odum *et al.*, 1962; Pearson, 1964). Thus, although forests have a much higher total production than grasslands they appear to offer less food.

West (1967) presented an excellent estimation of the food available to birds by analysing the contents of bird stomachs and evaluating seed production.

An entirely different approach was suggested by Newsome (1967), who merely studied the body weight of rodents kept in enclosures set in a given habitat. He considered a given body weight to be a sufficient indicator of a natural food supply.

It seems that the food available to the mammals in a given ecosystem can be estimated; but later it can be calculated fairly precisely as to how much of that food is to be consumed by animals. The question which remains unanswered is what is the proportion of food utilized to the food additionally destroyed. Some rodents, for instance, only partially eat the procured food. On the other hand, grazing frequently stimulates the growth of pasture and thus of meadow production.

11A. 2 Food digestibility and assimilation

In order to estimate which part of the energy consumed is utilized by mammals it is necessary to determinate the digestibility and assimilation of their natural foods.

For small mammals and birds such determinations are easily made in metabolic cages using the balance method (see Chapters 11D and 11E). However, the available data on the nutrition of wild animals (rodents) are minimal. Table 11A. 2 shows the tabulated results of several authors concerning the digestibility and assimilation of various natural diets of rodents. Energy losses in faeces and urine depend chiefly on the kind of food consumed. For example, mice and voles (*Apodemus agrarius*, *Apodemus flavicollis*, *Clethrionomys glareolus*, *Microtus arvalis*) who were offered concentrated and mixed food consisting mainly of tree seeds and grain, lost 7 to 12% of the food energy in the faeces and 1 to 4% in the urine. The voles (*Microtus arvalis*, *Microtus pennsylvanicus*, *Clethrionomys glareolus*) which were given

Table 11A.2. Digestibility and assimilation of food in small rodents. The range of coefficients of different diets is given; the mean values of natural diets are in brackets. (All values in per cent of energy intake).

Species	Digested energy	Metabolizable energy	References
Rice rat *Oryzomys palustris*	88–95		Sharp, 1967
Cotton rat *Sigmodon hispidus*	91·2	86·5	Golley, 1962
Old-field mouse *Peromyscus polionotus*	64–93		Golley, 1962
Meadow vole *Microtus pennsylvanicus*	82–90		Golley, 1960
Common vole *Microtus arvalis*	70·4–92·3 (81·3)	65·2–89·7 (77·5)	Droždž, 1968
Bank vole *Clethrionomys glareolus*	77·4–92·0 (86·8)	72·0–88·7 (82·9)	Droždž, 1968
Yellow-necked field mouse *Apodemus flavicollis*	81·3–92·2 (88·2) 90·0	78·7–90·9 (86·1)	Droždž, 1968 Turček, 1956
Striped field mouse *Apodemus agrarius*	89·6–90·4 (90·0)	88·9–89·0 (88·9)	Droždž, 1968

bulky food lost 22 to 30% of the food energy in the faeces and approximately 5% in the urine. The energetic value of faeces was 4·5 to 5·5 kcal/g (ash-free dry weight), while that of urine ranged from 48 to 331 cal/g of liquid (Droždž, 1967, 1968).

Similarly, carnivorous mammals assimilated 80 to 90% of their energy intake, the amount depended upon the kind of food eaten (Golley *et al.*, 1965). Digestibility of the dry mass and the caloric value of faeces are determined by the amount of indigestible components of the food, such as bones, hair, feathers etc. (Davis & Golley, 1963; Pearson, 1964; Golley *et al.*, 1965). Four species of insectivorous shrews were found to have a very high level of food assimilation (Buckner, 1964), reaching peak values of 78 to 95% of the food energy intake.

Digestibility appears to fluctuate considerably in ruminants. The faeces of cattle may contain 10 to 70% of the energy intake (Blaxter, 1962). Similarly, in deer, the coefficient of digestibility fluctuates from 49 to 86% in accordance with the diet (Davis & Golley, 1963: Ullrey *et al.*, 1968; Mautz, 1971; Drożdż & Osiecki 1973).

The mammals discussed above show considerable plasticity in the utilization of various foods. Although their assimilation level is restricted by the structure of their alimentary tract, it depends mainly on the quality of the food. The digestibility of herbivorous animals is limited to a great extent by the fibre content of the food; while with carnivores it is usually bones, feathers and hair that are the indigestible portions. Finally, the energy flow (assimilation) through a population is lower than the actual consumption by an amount equal to the energy value of the faeces plus urine.

11A. 3 Cost of maintenance based on metabolizable energy

Studies of digestibility and assimilation of food allow ecologists to estimate not only the amount of energy returning to the cycle in the form of faeces and urine but also to determine daily energy requirements (Drożdż, 1968). It is obvious that food is the only source of energy. Measurement of the amount of metabolizable energy by the method of food rations enables us to estimate animal energy requirements. Energetic utilization of food intake can be presented schematically as follows (Kleiber, 1961; slightly modified by ecological symbols from Petrusewicz, 1967):

Food energy (C)
(gross energy)

Energy in faeces (F)   Digested energy (D)

Energy in urine and methane (U)   Metabolizable energy (A) ($R + P$)

SDA   Net energy
(calorigenic effect of food)

                          Energy for maintenance (R)

                          Energy for production (P)

Energy contained in food is called gross energy, energy intake or consumption (C) (Petrusewicz, 1967). After subtracting the caloric value of the faeces (F) from the food energy the remainder is digested energy (D), hence the formula:

$$D = C - F$$

Energetic losses through urine and methane (U) further reduce the amount of energy that can be utilized by the animal. This quantity is called metabolizable energy, assimilation or energy flow (A), and is calculated from the following:

$$A = C - FU$$

All of the metabolizable energy cannot be utilized for existence and production requirements. A part of it is lost as the heat of calorigenic effect of food (SDA); such losses range from 3 to 20% of the gross energy and are dependent upon the diet of the animal.

After subtracting SDA from the metabolizable energy, net energy is obtained. This energy comprises the energy for body maintenance and includes the cost of thermoregulation, musclework etc. It also includes the energy of production, namely, that used by the animal for the synthesis of its own body. Ecologists divide the metabolizable energy into respiration, or cost of maintenance (R) and into production (P), which is equal to the energy for production. Respiration, in the ecological sense of the word, includes total energy transformed into heat and utilized in the life processes (existence energy + SDA). This may be expressed by:

$$A = R + P$$

When the weight of an animal living on an existence diet does not change then the total metabolizable energy is equal to the costs of maintenance (respiration), or:

$$A = R \text{ when } P = 0$$

Therefore the cost of maintenance calculated from metabolizable energy can be compared to the respiration measured by oxygen consumption (see Chapter 10).

Such an evaluation was made for rodents (Drożdż, 1968), by comparing the average daily metabolic rate—ADMR (Gębczyński, 1966; Grodziński,

1966; Górecki 1968; see also Chapter 3) with the energetic requirements calculated from the metabolizable energy. This showed that the results
obtained by the method of food balance were very close to respirometric
determinations.

Changes in body weight during the experiment can cause some trouble.
In order to calculate the real cost of maintenance some corrections must be
introduced. The changes in weight may be the result of gains or losses of fats,
proteins, carbohydrates and minerals, as well as from variations in water
content, or even the contents of the alimentary tract. Differences in fat
content are most pronounced during periods of malnutrition or overfeeding
(Cumming & Morrison, 1960). Therefore, when a loss in body weight occurs,
the energetic equivalent of fat tissue, 7 kcal/g, should be added to the calculated cost of maintenance (King, 1961). When a gain in weight is observed,
the costs of deposition should be reduced by an amount equal to 14 kcal of
metabolizable energy per 1 g of fat deposited (Kielanowski, 1965). Most
frequently the fat constitutes about 2/3 of the weight gain, the rest being due
to water, carbohydrates and proteins (Blaxter, 1962). Therefore, for each
gram of gain of biomass 9 kcal should be subtracted. When the correction
is calculated in this way it cannot be too large. Of course such estimations are
only approximate if the composition of the lost or gained body tissue is not
known precisely.

In ecological bioenergetics the cost of maintenance can be studied both
by gaseous exchange determinations and by food utilization (Golley, 1967).
Respirometry (see Chapter 6 and 10) is more precise but the method of
food balance is technically simpler and allows the estimation not only of the
cost of maintenance but also of two important parameters in the energy flow
balance, namely, digestibility and assimilation of food. Last but not least
the method allows respirometer determinations to be checked.

References

BLAXTER K.L. (1962) The energy metabolism of ruminants. *Hutchinson Scient. and
Techn.* 380 pp. London.
BUCKNER C.H. (1964) Metabolism, food capacity, and feeding behaviour in four
species of shrews. *Can. J. Zool.* **42**, 259–279.
CARLETON M.W. (1966) Food habits of two sympatric Colorado sciurids. *J. Mammal.*
47, 91–103.

CUMMING M.C. & MORRISON S.D. (1960) The total metabolism of rats during fasting and refeeding. *J. Physiol.* **154**, 219–243.

DAVIS D.E. and GOLLEY F.B. (1963) *Principles in mammalogy.* Reinhold Publ. Corp. 335 pp. New York.

DROŻDŻ A. (1966) Food habits and food supply of rodents in the beech forest. *Acta theriol.* **11**, 363–384.

DROŻDŻ A. (1967) Food preference, food digestibility and the natural food supply of small rodents. In Petrusewicz K. ed. *Secondary Productivity of Terrestrial Ecosystems.* 323–330 pp. Warszawa, Kraków.

DROŻDŻ A. (1968) Digestibility and assimilation of natural foods in small rodents. *Acta theriol.* **13**, 367–389.

DROŻDŻ A., OSIECKI A. (1973) Intake and digestibility of natural feeds by roe-deer. *Acta theriol.* **18**: 81–92.

GENTRY J.B., SMITH M.H. (1968) Food habits and burrow associates of *Peromyscus polionotus. J. Mammal.* **49**, 562–565.

GĘBCZYŃSKI M. (1966) The daily energy requirement of the yellow-necked field mouse in different seasons. *Acta theriol.* **11**, 391–398.

GODFREY G.K. (1953) The food of *Microtus agrestis hirtus* (Bellamy, 1839) in Wytham, Berkshire. *Säugetier. Mitt.* **1**, 148–151.

GOLLEY F.B. (1960) Energy dynamics of a food chain of an old-field community. *Ecol. Monogr.* **30**, 187–206.

GOLLEY F.B. (1962) *Mammals of Georgia.* Univ. Georgia Press. 218 pp. Athens.

GOLLEY F.B. (1967) Methods of measuring secondary productivity in terrestrial vertebrate populations. In Petrusewicz K. ed. *Secondary Productivity of Terrestrial Ecosystems.* 99–124 pp. Warszawa, Kraków.

GOLLEY F.B., PETRIDES G.A., RAUBER E.L. and JENKINS J.H. (1965) Food intake and assimilation by bobcats under laboratory conditions. *J. Wild. Manag.* **29**, 442–447.

GÓRECKI A. (1968) Metabolic rate and energy budget of bank vole. *Acta theriol.* **13**, 341–65.

GRODZIŃSKI W. (1966) Bioenergetics of small mammals from the Alaskan taiga forest. *Lynx.* **6**, 51–55.

GRODZIŃSKI W., GÓRECKI A., JANAS K. and MIGULA P. (1966) Effect of rodents on the primary productivity of alpine meadows in Bieszczady Mountains. *Acta theriol.* **11**, 419–431.

HOLIŠÓVA V., PELIKAN J. and ZEJDA J. (1962) Ecology and population dynamics in *Apodemus microps* Krat. and Ros. (Mamm.: Muridae). *Acta Acad. Sci. Czechosl.* **34**, 493–540.

KIELANOWSKI J. (1965) Estimates of the energy cost of protein deposition in growing animals. In Blaxter K.L. ed. *Energy Metabolism.* Acad. Press. 13–20 pp. London, New York.

KING J.R. (1961) The bioenergetics of vernal premigratory fat deposition in the white-crowned sparrow. *Condor.* **63**, 128–142.

KLEIBER M. (1961) *The fire of life: an introduction to animal energetics.* J. Wiley Inc. 454 pp. London, New York.

MATUSZEWSKI G. (1966) Studies on the european hare. XIII. Food preference in relation to trees branches experimentally placed on the ground. *Acta theriol.* **11**, 485–496.

MAUTZ W.W. (1971) Confinement effects of dry-matter digestibility coefficients displayed by deer. *J. Wildl. Mgmt.*, **35**: 366–368.

NAUMOV N.P. (1948) *Ocherki sravnitelnoĭ ekologii myshevidnykh gryzunov. (Outline of comparative ecology of small rodents).* Izd. A.N. SSSR. 204 pp. Moskva.

NEWSOME A.E. (1967) A simple biological method of measuring the food supply of house mice. *J. Anim. Ecol.* **36**, 645–650.

ODUM E.P., CONNELL C.E. and DAVENPORT L.B. (1962) Population energy flow of the primary consumer components of old-field ecosystems. *Ecology.* **43**, 88–96.

PEARSON O.P. (1964) Carnivore-mouse predation: an example of its intensity and bio-energetics. *J. Mammal.* **45**, 177–188.

PETROV O.V. (1963) Pitanie mysnevidnykh gryzunov lesostepnykh dubrav v laborator-nykh usloviyakh. Laboratory studies on the small rodents from oak-woods. *Vopr. Ekol. Biocenol.* **8**, 119–173.

PETRUSEWICZ K. (1967) Suggested list of more important concepts in productivity studies (definitions and symbols). In PETRUSEWICZ K. ed. *Secondary Productivity of Terrestrial Ecosystems.* 51–58 pp. Warszawa, Kraków.

SHARP H.F. JR. (1967) Food ecology of the rice rat, *Oryzomys palustris. J. Mammal.* **48**, 557–563.

STEWARD D.R.M. (1967) Analysis of plant epidermis in faeces: a technique of studying the food preference of grazing herbivores. *J. Appl. Ecol.* **4**, 83–111.

TURČEK F.J. (1956) Quantitative experiments on the consumption of tree seeds by mice of the species *Apodemus flavicollis. Arch. Soc. 'Vanamo'.* **10**, 50–59.

TURČEK F.J. (1967) Ökologische Beziehungen der Saugetiere und Gehölze. *Slow. Akad. Wiss.* 210 pp. Bratislava.

ULLREY D.E., YOUATT W.G., JOHNSON H.E., FAY L.D., BRENT B.E., KEMP K.E. (1968) Digestibility of cedar and balsam fir browse for the white tailed deer. *J. Wildl, Mgmt.*, **32**, 162–171.

WATTS C.H.S. (1970) The food eaten by some Australian desert rodents. *S. Austr. Nat.*, **44**, 4.

WEST G.C. (1967) Nutrition of tree sparrows during winter in central Illinois. *Ecology.* **48**, 58–67.

WILLIAMS O. (1962) A technique for studying Microtine food habits. *J. Mammal.* **43**. 365–368.

11B ESTIMATION OF FOOD PREFERENCES

J. PINOWSKI AND A. DRODZ

The food preferences and food habits of animals are normally studied in order to estimate the potential food supply available to them in natural ecosystems (see Chapter 3). This is usually accomplished, under laboratory conditions, by a simple test of choice also called the 'cafeteria test' in ecological jargon. Only two tests of choice will be described here: one, a test for gran-ivorous birds (Section 11B. 1) and the other, a test for small rodents (Section 11B. 2). It should be noted however, that similar feeding experiments can be adapted for other terrestrial vertebrates.

11B. 1 Choice tests for granivorous birds

The best choice test is a method well known in behavioural studies of birds (Hespenheide, 1966; Kear, 1962). The test described here can have practical application in the study of energy flow from primary production to a bird population. It provides a quantitative determination of the food preferred by granivorous birds and a crude estimate of their daily food consumption (Turček, 1967).

This feeding experiment will be described for the European tree sparrow (*Passer montanus* L.) which is distributed over most Eurasia and feeds mainly on seeds of weeds. The birds are offered a free choice of six weeds that are commonly found in the fields where tree sparrows feed, for example *Echinochloa crusgalli*, *Setaria glauca*, *Amaranthus retroflexus*, *Polygonum convovulus*, *Polygonum persicaria* and *Chenopodium album*.

The birds, caught in the wild, are placed in individual cages (105 × 60 × 52 cm). They are then left for two or three days to become accustomed to the cage, after which they are weighed (to the nearest 0·1 g).

The seeds are dried to a constant weight and 4 gram of each seed species are weighed out. Litre jars, one for each of the seed species to be tested, are placed in the cage and the order of their arrangement is noted on a data sheet using special symbols. The seeds are placed on the bottom of the jars and the covered jars are carefully placed into the cage after dark in the evening.

During the following days the jars are uncovered at a fixed hour (e.g. 8 a.m.) and the behaviour of the birds is observed for 60 min. The choice of jar and the duration of food consumption are noted. During the whole procedure extreme caution must be exercised so as not to frighten the birds. Each evening the food remaining in each jar is collected, dried to constant weight and weighed separately, after the excrements and other incidental dirt have been removed. An experiment with one bird should be continued for a period of 10 consecutive days, the arrangement of jars being changed daily in order to prevent the bird from becoming conditioned to a particular arrangement of seeds. During the whole experiment the conditions of temperature and humidity should be as nearly constant as possible.

The edible content of the seeds can be determinated with the assistance of the granivorous birds themselves. For this purpose an additional experiment is carried out in which the tree sparrows are given about 2 g of seeds to consume completely. The proportion of edible parts (endosperm) present can be determined by subtracting the weight of the seed-shells (pericarp) remaining from the original total weight.

The results from the main experiment can be expressed as weight of seed eaten minus the weight of the seed-shells (pericarp). It can also be presented as a percentage seed eaten of total available thus expressing a measure of food preference. Visual observations during the first hour of food intake can also be given as the proportion of time spent by the tree sparrow on the jar containing a particular seed species. Direct observations will indicate the most preferred food. The experiment should normally be carried out with 10 animals to encompass individual variability of food habits. If two or more birds are kept together in one cage visual observations will not give clear results because usually only one bird can feed in one jar during a given time interval. On the other hand the results based on seed weights can be more distinct and closer to a natural situation in a population.

Both the natural food supply of granivorous birds and their food habits change drastically throughout the year (West 1967) therefore the experiment has to be related to a definite season and should be repeated in various seasons.

Acknowledgement

The technical details of the feeding experiments were tested and developed by W. Tomek M.Sc.

References

HESPENHEIDE H.A. (1966) The selection of seed size by finches. *Wilson Bull.*, **78,** 191–197.
KEAR J. (1962) Food selection in finches with special reference to interspecific differences. *Proc. Zool. Soc. Lond.*, **138,** 163–204.
TURČEK F.J. (1967) Some methods of the food habits of *Passer montanus* and *Passer domesticus. Intern. Stud. on Sparrow.* **1,** 23–25.
WEST G.C. (1967) Nutrition of tree sparrows during winter in Central Illinois. *Ecology*, **48,** 58–67.

11B.2 Choice Tests for Small Rodents

Tests involving choice for small rodents introduced by Chitty (1954) has been subsequently employed and modified quite independently by a number of ecologists (e.g. Miller 1954; Górecki and Gębczyńska 1962; Petrov 1963; Drożdż, 1966). For the test, rodents in the laboratory are offered a choice of several kinds of food and the degree of consumption is estimated. The experiment should be carried out with animals captured in a studied area and natural foods actually available in different seasons be used. In forest areas these are herb layer vegetation, tree- and shrub-seeds, fruit, leaves, buds and tree twigs. Replication of the experiment in successive seasons is necessary because food available to rodents in the growing season differs considerably from that obtainable in winter. Some plants disappear, others take their place, seeds fall down in the autumn and so on. The natural diet of rodents changes according to what food is available during the year.

The procedure described here has been successfully employed for mice and voles (Drożdż, 1966). During the experiment rodents are placed in metal cages or in large jars. In three consecutive days each animal is offered 3 to 5 kinds of food available in nature and, in addition, is supplied with pellets of some standard food together with water *ad libitum*. The stalks of fresh herbs can be wrapped in wet cotton to prevent withering. A small amount of granulated food is placed in the cage as a reserve in case the test food is inedible. Wild rodents usually eat pellets only in extreme conditions when the food offered is unacceptable. During the subsequent 3-day periods, other groups of foods are tested. Finally, tests with the foods previously rejected are repeated.

The degree of consumption of particular components of the diet is estimated daily using a scale 0, 1, 2, 3 (Petrov 1963; Drożdż, 1966). The numbers in this scale correspond approximately to the following percentages: $0 = 0\%$ (food was not touched), $1 = 0 - 30\%$, $2 = 30 - 60\%$ and $3 = 60 - 90\%$ of food was consumed.

The experiment should be done with about 10 animals to encompass individual variability in food habits. The results can be presented in the form of arithmetic means and percentages. The method of calculation of the mean values is presented in Table 11B.2.1. using bank voles (*Clethrionomys glareolus*), which were offered three plants from the herb layer vegetation in a beech forest as an example (Drożdż, 1966).

Table 11B.2.1

Kind of food (plants)	*Clethrionomys glareolus* individual animals											
	1	2	3	4	5	6	7	8	9	10	$\bar{x}$	%
Mycelis muralis	3	3	3	3	2	3	2	3	3	3	2.8	80
Galeobdolon luteum	0	3	3	2	1	1	0	2	1	1	1.4	40
Carex silvatica	0	1	1	1	0	2	0	0	0	0	0.5	10

For a closer examination of food habits it is strongly recommended that stomach content analysis of animals trapped in the field should be used (see Chapter 11C).

References

CHITTY D. (1954) The study of the brown rat and its control by poison. In *Control of Rats and Mice*. Clarendon Press. 160–299 pp. Oxford.

DROŻDŻ A. (1966) Food habits and food supply in the beech forest. *Acta theriol.*, **11**, 363–384.

GÓRECKI A. & GĘBCZYŃSKA Z. (1962) Food conditions for small rodents in a deciduous forest. *Acta theriol.*, **6**, 275–295.

MILLER R.S. (1954) Food habits of the Wood-mouse, *Apodemus sylvaticus* (Linné, 1758), and bank vole, *Clethrionomys glareolus* (Schreber, 1780) in Wytham Woods, Berkshire. *Saugetierk. Mitt.*, **2**, 109–114.

PETROV O.V. (1963) Pitanie myshevidnykh gryzunov lesostepnkyh dubrav v laboratornykh usloviyakh. (Laboratory studies on the food of small rodents from oak-woods). *Vopr. Ekol. Biocenol.*, **8**, 119–173.

11C ANALYSIS OF STOMACH CONTENTS OF SMALL MAMMALS

A. DRODZ

Analysis of stomach contents or of the whole digestive tract is the best and most direct way of investigating an animal's food habits. Identification of various foods from gut contents is comparatively easy with amphibia and reptiles (Darevskij & Terentev, 1967) and with most birds. With mammals however, and particularly with small rodents and insectivores, it can be achieved only by microscopic techniques because the foodstuffs are thoroughly ground by the teeth. Such a microscopic method of rodent stomach food analysis was developed by Williams (1955), Holišova *et al.*, (1962) and then was modified and improved by several authors (Fleharty & Olson, 1969; Hansen, 1970; Hansson, 1970; Zemanek, 1972). It permits the identification of the majority of plant and animal foods that are consumed by polyphagous animals. It also enables one to calculate the frequency of the various components of ingested food. This is sufficient to estimate the natural food supply of rodent population in the ecosystem (see Chapter 11A).

Quantitative determinations obtained through the application of this method are, however, very crude and in spite of some efforts in this direction (Gębczyńska & Myrcha, 1966; Hansson, 1970), they have not yet been evaluated precisely.

The following procedure is suggested. Stomachs are dissected from fresh voles and mice that have been caught in snap-traps at various seasons of the year. After weighing the stomachs are dried or fixed in 80% alcohol. The contents of the dried stomachs are then examined and analysed microscopically after being soaked in water for 24 hours.

A good method of analysis of material from stomach of herbivorous animals was proposed by Hansen (1970). The stomach samples are washed over a 200 mesh screen (Fig. 11C.1), spread on slide (Fig. 11C.2) using Hertwig's and Hoyer's solutions (Figs. 11C.3, 11C.4) and dried in an oven at 60°C for 3 days. The material prepared in such way (Fig. 11C.5) is compared with earlier-made reference slides.

The useful diagnostic features of green parts of plants are the shape of epidermal cells (Figs. 11C.6, 11C.7, 11C.8, 11C.9), the tracheal tissue and its accessory cells (Figs. 11C.10, 11C.11) and the trichomes (Figs. 11C.12, 11C.13). With seeds, it is the general structure and shape of the starch granules that is identifiable. Animal material can be recognized from the presence of striated muscles and chitin fragments, but these are difficult to classify further. Permanent histological slides and drawings of epidermal cells should be prepared for the majority of the plant species occurring in the study area. These drawings can then be used as a key for the identification of the various constituents of the sample.

The analysis can be carried out in the following order: (1) the contents of each stomach are divided under the microscope into three main components: green plants, seeds and animal matter (2) the approximate volume of the components are estimated, (3) five slides are made of each group mentioned above, (4) 10 microscopic fields are checked on each slide, (5) the

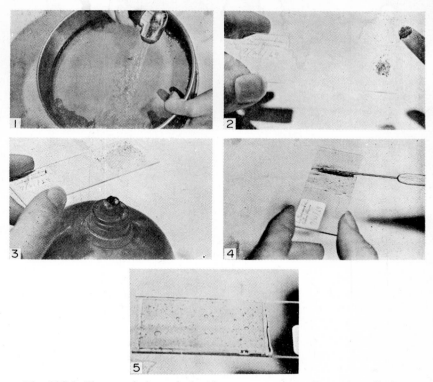

Fig. 11C.1. The sample is washed with water to remove dirt and small plant fragments over a 0.1 mm (200 mesh) screen.

Fig. 11C.2 A small amount of the sample is placed on a slide.

Fig. 11C.3 Hertwig's solution is boiled off over an alcohol burner.

Fig. 11C.4 Hoyer's solution is mixed with sample and spread evenly over the slide.

Fig. 11C.5 A finished slide showing the desirable approximate density of plant fragments and ring of Hoyer's solution around the cover slip. The air bubbles may be pressed out with a teasing needle while the slide is cooling after being heated.

Figures 1 to 5 were reproduced from Technical Report No. 18 of the US IBP Grassland Biome Program thanks to personal permission of the author, Dr. Richard M. Hansen.

[*facing page 338*]

frequency of occurrence of particular plant species, seeds and invertebrates are estimated from the percentage of stomachs containing a given food constituent.

It is recommended that seasonal analyses be carried out throughout the yearly cycle or at least during the main seasons when abundant material for investigation is available.

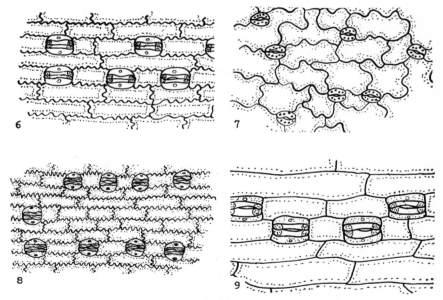

Fig. 11C.6 *Carex silvatica*—an example of Monocotyledones plants
Fig. 11C.7 *Stellaria holostea*—an example of Dicotyledones plants
Fig. 11C.8 *Luzula pilosa*—walls of epiderma cells smooth
Fig. 11C.9 *Carex pilosa*—walls of epiderma cells sinuous

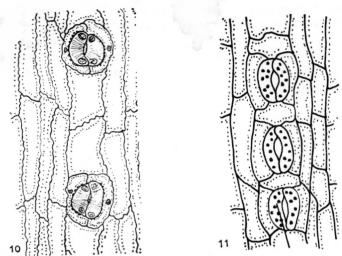

Fig. 11C.10 *Equisetum pratense*—two accessory cells present
Fig. 11C.11 *Abies alba*—four accessory cells present

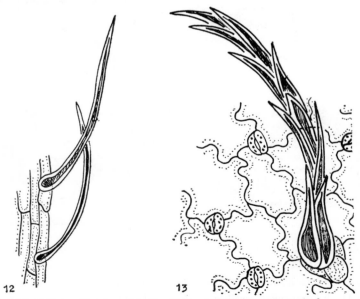

Fig. 11C.12 *Rubus idaeus*—protective hairs, unicellular, unbranched
Fig. 11C.13 *Pulmonaria obscura*—multicellulate, branched hairs. Figures 6 to 13 were drawn by Dr. K. Urbańska-Worytkiewicz.

References

DAREVSKIJ I.S. & TERENTEV P.V. (1967) Estimation of energy flow through amphibian and reptile populations. In K. Petrusewicz (ed.) *Secondary Productivity of Terrestrial Ecosystems*. 181–197 pp. Warszawa-Kraków.

FLEHARTY E.D. & OLSON L.E. (1969) Summer food habits of *Microtus ochrogaster* and *Sigmodon hispidus*. *J. Mammal.*, **50**, 475–486.

GĘBCZYŃSKA Z. & MYRCHA A. (1966) The method of quantitative determining of the food composition of rodents. *Acta theriol.*, **11**, 385–390.

HANSEN R.M. (1970) The microscope method used for herbivore diet. *Technical Report No. 18*, U.S. International Biological Program. Grassland Biome.

HANSSON L. (1970) Methods of morphological diet micro-anyalsis in rodents. *Oikos*, **21**, 255–266.

HOLIŠOVA V., PELIKAN J. & ZEJDA J. (1962) Ecology and population dynamics in *Apodemus microps* Krat & Ros. (*Mamm.: Muridae*). *Práce Br. Zákl. CSAV.*, **34**, 493–539.

WILLIAMS O. (1955) The food of mice and shrews in a Colorado montane forest. *Univ. Color. Stud. Biol.*, **3**, 109–114.

ZEMANEK M. (1972) Food and feeding habits of rodents in a deciduous forest. *Acta theriol.*, **17**, 315–325.

11D MEASUREMENT OF EXISTENCE ENERGY IN GRANIVOROUS BIRDS

S. C. KENDEIGH

Basal or standard metabolism of birds is commonly measured by the rate of oxygen consumption when the bird is quiet and in a post-absorptive state. Such data are of physiological significance but scarcely applicable to birds in the wild that are seldom in a standard metabolic state except after several hours on a roost at night. The activity of a bird in a respiratory chamber is necessarily limited and to permit it also to feed presents several difficulties, as does continuing the measurements over long periods of time.

Of much greater ecological practicality is the measurement of energy utilization of caged birds, free to move about and feed, over long periods of time, as such measurements approximate their energy requirements under free-living natural conditions. This is possible by measuring their rate of food intake and utilization.

The chapter is reproduced from 'International studies on sparrows' **1**: 26–33 (1967) ed. by Working Group on Granivorous Birds—PT Section. Some changes were introduced by the author.

Each bird, the size of a house sparrow, should be placed in an individual cage, approximately 18 centimetres wide, 30 cm high, and 30 cm deep, equipped with a perch near the centre, and a door (d) at one end large enough for insertion of a man's arm (see Fig. 11D.1). The entire cage could be

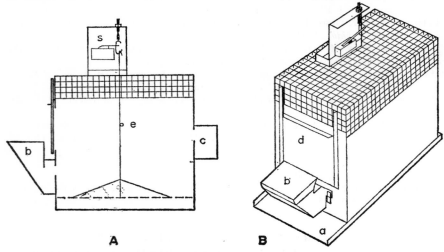

Fig. 11D.1 Cage for measuring existence energy

constructed of wire mesh and placed in a pan (a) to collect scattered food and excreta or the sides of the cage to near the top could be made of thin sheet metal. Separate food (b) and water (c) cups should be fastened outside and at opposite ends of the cage against openings made in the wall of the cage through which the bird can feed. A small perch (e) should be placed just below each opening.

The food that we are using is a chick starter mash (No. 521) prepared by the Poultry Department of the University of Illinois. Its contents are as follows:

Yellow corn (maize)		41·7%
Wheat		15·0%
Oats		10·0%
Soybean oil meal	(50% crude protein)	15·0%
Fishmeal	(60% crude protein)	5·0%
Meatscraps	(50% crude protein)	5·0%
Alfalfa meal	(17% crude protein)	5·0%

Dried whey	2·0%
Limestone	1·0%
Iodized salt	0·25%
A and D vitamins, auromycin	0·25%
Magnese sulphate	0·05%

It has a caloric value of 4·3–4·4 kcal/g and a protein content of 21 per cent. We have it ground sufficiently fine to pass through a wire screen of 1·6 mm mesh. The chick mash is stored in a cool place until used. House sparrows adjust to this food readily and experiments may start as soon as they maintain a constant weight from day to day.

An experimental period lasts for three days during which time the birds are allowed to feed *ad libitum*. At the beginning of each period approximately 75 grams of food are supplied to each cage and a similar sample is dried at 65°C for determination of water content. At the end of a period the scattered food that has fallen through the bottom of the cage into the pan is separated from the excreta using a 1·6 mm sieve and along with the unused food from the cup is dried at approximately 65°C. The excreta are also collected and dried at this time. We often cover the bottom of the pan with a sheet of aluminium foil so that the excreta can be lifted off, the waste food brushed off, and the excreta and foil put directly into an oven. Three days are ample for food and excreta to attain a constant dry weight. Particles of waste food clinging to the pellets of excreta can usually be separated more easily after drying, and food and excreta are weighted separately. Representative samples of excreta and fresh food are put in vials at low temperature for later caloric determinations.

Photoperiod has only a small effect on the amount of food consumed but in order to have experiments in various parts of the world and at various seasons strictly comparable, we suggest that they be done on a standard of 12 hours light per day.

It is always desirable to run outdoor controls by placing the caged birds where they are exposed to natural fluctuations of photoperiod light intensity and temperature. It is best to protect the cages by overhead shelter from precipitation and excessive wind. We have run birds out-of-doors for months at a time, dividing the group into three sub-groups so that one sub-group could be serviced each day.

One can expect some mortality in getting wild birds acclimated to the cages and the ground food. In some species this may amount to 50%. House sparrows appear to do best when placed directly into the individual cages, out-of-doors, where they can see and hear other caged sparrows. Usually several days are required before they are able to maintain a constant weight. They continue to be nervously active and to become alarmed at the approach of man.

Birds are easily weighed in a metal box, approximately $20 \times 6 \times 6$ cm with a tight-fitting lid. In experimental work at constant temperatures and photoperiods, the food consumption and excreta voided are measured over consecutive three-day periods only while the birds maintain a constant weight within plus or minus 2·5 per cent.

Gross energy intake is the difference in calories between the amount of food given a bird and the amount unused; excretory energy is the caloric value of the total excreta for the same period; metabolized energy is the gross energy intake minus the excretory energy; and existence energy is the metabolized energy when the birds maintain a constant weight. With constant weight the birds are eating and utilizing only sufficient energy to maintain an energy balance, no more or no less. In experimental work under constant conditions we strive to measure existence energy; in out-door studies under fluctuating conditions we measure only metabolized energy. Metabolic efficiency is metabolized energy divided by gross energy intake, often an interesting factor to measure. You cannot easily measure digested energy in birds, as in mammals, since the excreta contain not just undigested energy from the ingested food but also the unmetabolized energy from the kidneys.

Existence energy, of course, varies with temperature and it is desirable to measure it at a sufficient number of temperatures (minimum 3) to estabilsh a significant regression line and equation. Existence energy is determined at intervals of 10–15°C in the medium temperature range but at progressively smaller intervals of temperature towards the extremes as the birds show discomfort or greater difficulty in maintaining constant weight. It is also desirable to determine the lower and upper limits of temperature tolerance where only 50 per cent of the individuals can survive or maintain constant weight. We commonly start with two groups of 10–12 birds each and subject one group to progressively lower temperatures and the other to progressively higher temperatures.

Since birds, especially permanent resident species, may exhibit seasonal acclimatization or their response to ambient temperatures may vary with moult, regression lines should be determined in the middle of extreme seasons, i.e. winter (Dec.-Feb.) and summer (May-July) in the north temperate zone; or wet and dry seasons in the tropics.

Existence energy requirements increase to a maximum at the lower limit of temperature tolerance. We have good evidence that this represents the maximum potential intake that they can and will accumulate for activities at other temperatures. The difference between the maximum potential intake and existence energy intake at any ambient temperature represents productive energy that is available for free existence, flight, nestling, moulting, etc., or if not immediately used may be deposited as fat reserves on the body.

Accessory equipment required in this study are constant temperature chambers with adjustable thermo-regulators and a thermostat that will maintain temperatures within $\pm$ 1·0°C. A range of $\pm$ 2·0°C should be the maximum tolerated. In our experiments we have made no attempt to control relative humidity and it varies inversely with temperature roughly similar to what it does under natural conditions. Walk-in cabinets are preferred to prevent a deficiency of oxygen; small cabinets will require forced ventilation. Lighting should give an illumination of at least 50 foot-candles. We use fluorescent lighting which is turned on and off automatically by a time clock.

Bird, food, and excreta should be weighed to at least 0·1 gram, on a balance with a minimum precision of $\pm$ 0·05 gram. You will need large drying ovens.

Caloric determinations are made on one gram samples in duplicate in an adiabatic bomb calorimeter. When duplicate values differ by more than 2·5%, additional determinations are made. These samples are weighed to 0·1 milligram on an analytical balance with minimum precision of 0·05 mg. For necessary rapidity in weighing oven-dried material, a single pan substitution rather than a double pan equal arm balance is recommended. Recent investigations in our laboratory indicate that dried excreta stored at low temperatures do not deteriorate in caloric value. We obtain values of excreta for each bird and at different temperatures until we are assured that there are no significant differences between them. We also obtain caloric values of each new supply of chick starter mash used as bird food.

References

DAVIS E.A., JR. (1955) Seasonal changes in the energy balance of English sparrow. *Auk.*, **72**, 385–411.

KENDEIGH S.C. (1949) Effect of the temperature and season on energy resources of the English sparrows. *Auk.*, **66**, 113–127.

MARTIN E.W. (1967) An improved cage design for experimentation with passeriform birds. *Wilson Bull.*, **79**, 335–338.

11E   METABOLIC CAGES FOR SMALL RODENTS

A. DRODZ

Metabolic cages are used for the determination of the digestibility and assimilation of natural food and of the daily energy requirements of the organisms under study. The measurements are based on a method whereby ingestion is carefully measured and compared with detailed determinations of egestion in terms of faeces and urine production. The rodent cages demonstrated here (Drożdż, 1968) constitute a modification of metabolic cages commonly used in the laboratory for mice and rats.

11E.1 Description of cages

The cage is shaped like a cylinder or a cube (Fig. 11E.1). Its bottom (a) and removable top (b) are made of 7–10 mm mesh wire screen. A water bottle (c) and a feeder are fitted on the outside of the cage. The feeder has an adjustable canopy to regulate the diameter of the feeding tunnel (h) so that a feeding rodent is prevented from turning around and carrying food away from the food cup (d). The floor of the feeding tunnel (h) is also constructed of wire-mesh and covers a collection tray (i) from which excess food, spilled by the animal, may be easily retrieved. The faeces and urine are separated in the funnel below the wire mesh floor by two cones placed as shown in Fig. 11E.1. Because of the construction design the faeces bounce off the walls of the funnel or slide down the incline of the top cone (f) and roll into the faeces-cup (g). The urine flows down between the wall of the funnel and the faeces-cup and into a calibrated glass tube (j). We recommend that the cages be made from stainless steel and the funnels from vinyl or glass. A whole series of cages may be mounted on mobile racks to provide mobility within the laboratory.

For small rodents metabolic cages of two sizes are recommended. One, a small 12×13 cm., in which the area of the floor approximates 110 cm^2 and has a capacity of approx. $1 \cdot 5$ dcm^3, and a larger one, $20 \times 17 \times 14$ cm. having a floor area of approx. 340 cm^2 and capacity exceeding 45 cc. The first type is very suitable for mice having a body weight not exceeding 40 g while the other will accommodate bigger voles and rats.

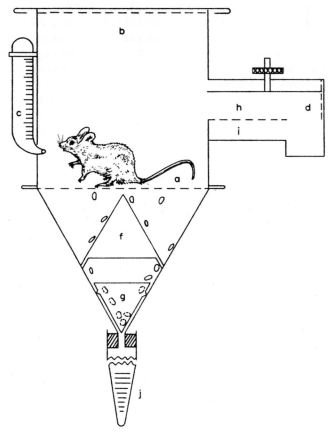

Fig. 11E.1 Cross section of the rodent metabolic cage: a—bottom; b—top; c—drinking tube; d—feeder; e—funnel; f—baffle; g—faeces cup; h—adjustable tunnel; i—food collector; j—calibrated tube for urine.

11E.2 Procedure

One experiment may be completed in about 14 days and is divided into two periods. During the first or introductory period, the animals are given test food until the alimentary tract is completely filled with new food. With mice and voles this takes two days or less. However, it is recommended that the introductory period be lengthened to 7–10 days when wild rodents are used so that the animals may become adapted to the new diet. During this stage of the experiment the rodents can be kept in normal cages and they are placed in the metabolic cages only during the second period. This usually lasts about a week (5–7 days). During this interval of time the animals are provided daily with food *ad libitum*, after which the total food consumed is determined and all the faeces and urine carefully collected. The food intake and the excretory production (excrements) are then summed and the mean values are calculated (see Energy Data Sheet).

ENERGY DATA SHEET

Date: Start... 15 I 68 ... End ... 20 I 68 ... Measurement duration... 5 ... days

Species, sex ... *Clethrionomys glareolus* ♂ ... No. of cage... 1

Initial b.wt... 24·5

Final b.wt... 24·5

Difference: gain... O ... or loss... O

Kind of food ... *beechmast*

Calorific value kcal/g dry weight

food ... 6·914

faeces ... 4·154

urine ... 0·170 ... kcal/g of liquid

Item	Wet wt. g	Dry matter g	Kcal	% of energy intake
Consumed food (C)	11·11	10.58	73·58	100.00
Faeces (F)		1.23	5·11	6·99
Digested (D)		9·35	68·05	93·01
Urine (D)	13·70		2·33	3·18
Metabolizable energy (A)			65·72	89·83
Kcal/animal–day			13·14	
Kcal/g–day			0·53	
Kcal/kg$^{0·75}$			215	

In research of this kind there is usually one animal in each metabolic cage. However, groups of 2–3 animals can also be used provided they are accustomed to each other. In this way conditions more closely resembling their natural ecological situations are created.

Diets, whose composition corresponds closely to the animal's natural food should be prepared. The content of the food is subject to considerable changes in a yearly cycle and therefore determination of digestibility and assimilation for a number of seasonal diets is advisable. The food supplied should be a homogenous mixture to prevent an animal from exercising a preference for some specific ingredients or carrying them out of the feeders. For this reason mixed foods and some seeds have to be finely ground or compressed into pellets.

Usually an abundant food supply is prepared in advance of the experiment. The water content is determined prior to storing in a tightly closed container. Faeces and urine that are collected during the experiment should be kept in a refrigerator in closed containers. The urine should be slightly acidified to pH 6 with a few millilitres of 1% H_2SO_4. In samples that are prepared in this manner neither changes in caloric value nor in nitrogen content have been noted (Fuller & Cadenhead, 1968).

The caloric value of food, faeces and urine are determined with a bomb calorimeter. Combustion of food and faeces does not require any deviation from the usual procedure recommended for other ecological materials (see Chapter 5A). Urine is combusted together with some other material of definite caloric value such as cellulose pellets or preferably on polyethylene-foil (Nijkamp, 1965). In the latter method a definite amount of urine is poured into a foil disc 12 cm in diameter and the water is completely evaporated in a vacuum oven or in a freeze-drying apparatus. Owing to the high caloric value of the foil (11.068 kcal/g) the dry urine residue is then perfectly combusted.

11E.3 Calculation

The measurements made in the metabolic cages can be used to calculate the digestibility coefficient which is expressed as a ratio namely

$\dfrac{\text{digested components of food}}{\text{consumed components of food}} \times 100$. This can be done using weights of dry matter, organic substances or other particular components of food or with units of energy. In studies of the energy flow, the major interest is in the digestibility and assimilation of food intake energy.

Digested energy (D) also called gross energy, is obtained by subtracting the caloric value of faeces (F) from the food intake (C—consumption). Metabolizable energy is the difference between digested energy (D) and the energy of the urine (U).

Daily energetic requirements can be calculated from metabolizable energy. If body weight is not changed during the experiment this means that the whole metabolizable energy is used for maintenance. If a gain or loss of body weight is observed an appropriate correction has to be made (see Chapter 11A).

An example of the determination of digestibility and food assimilation, as well as daily energetic requirements, is presented in the data sheet for the bank vole (*Clethrionomys glareolus*). The body weight of this animal, fed on beechmast, was 24·5 g (Drożdż, 1968). The amounts of food, faeces and urine are given as totals for the whole period of the experiment.

References

Drożdż A. (1968) Studies on the disgestibility and assimilation of natural foods in rodents *Ekol. pol.* S.B., **14.**

Fuller F.M. & Cadenhead A. (1968) The preservation of faeces and urine to prevent losses of energy and nitrogen during metabolism experiments. In *Energy Metabolism.* Proc. of IVth Symp. of EAAP. Jablonna, Sept. 1967.

Nijkamp H.J. (1965) Some remarks about determination of the heat of combustion and the carbon content of urine. In Blaxter K.L. (ed.) *Energy Metabolism.* 147–157 pp. Acad. Press. London, New York.

FURTHER READINGS
IN VERTEBRATE BIOENERGETICS

DAVIS D.E. & GOLLEY F.B. (1963) *Principles in Mammalogy.* Reinhold, New York. 335 pp.

FOLK G.E. (1966) *Introduction to Environmental Physiology.* Lea & Febiger, Philadelphia. 308 pp.

GESSAMAN J.A. (Ed.) (1973) *Ecological Energetics of Homeotherms—a view compatible with ecological modeling.* Monograph Series USU, Vol. 2. Utah State University Press, Logan, Utah. 155 pp.

KLEIBER M. (1961) *The Fire of Life: an Introduction to Animal Energetics.* Wiley, New York. 454 pp.

MARSHALL A.J. (1960, 1961). *Biology and Comparative Physiology of Birds.* Vol. 1, 2. Academic Press, New York and London. 518, 468 pp.

MAYNARD L.A. & LOOSLI J.K. (1962) *Animal Nutrition.* McGraw-Hill Book Company, Inc., New York. 533 pp.

MOORE J.A. (Ed.) (1964) *Physiology of the Amphibia*, Academic Press, New York and London. 654 pp.

PETRUSEWICZ K. (Ed.) (1967). *Secondary Productivity of Terrestrial Ecosystems. (Principles and Methods)*, Vol. 1, 2. PWN, Warszawa-Kraków. 379, 879 pp.

PHILLIPSON J. (1966) *Ecological Energetics.* Edward Arnold, London. 57 pp.

ŠILOV I.A. (1968) *Reguljacija teploobmena u ptic.* Izd. Mosk. Univ., Moskva. 251 pp.

SLONIM A.D. (1971) *Ēkologičeskaja fizilogija životnyh.* Izd. 'Vyvšaja škola', Moskva. 448 pp.

WHITTOW G.C. (Ed.) (1970, 1971). *Comparative Physiology of Thermo-regulation.* Vol. 1, 2. Academic Press, New York and London. 333, 410 pp.

Author Index

Subject Index

361